VOLUME EIGHTY FIVE

# ADVANCES IN
# GENETICS

# ADVANCES IN GENETICS, VOLUME 85

Serial Editors

**Theodore Friedmann**
University of California at San Diego, School of Medicine, USA

**Jay C. Dunlap**
The Geisel School of Medicine at Dartmouth, Hanover, NH, USA

**Stephen F. Goodwin**
University of Oxford, Oxford, UK

VOLUME EIGHTY FIVE

# ADVANCES IN
# GENETICS

Edited by

## THEODORE FRIEDMANN
*Department of Pediatrics,*
*University of California at San Diego,*
*School of Medicine, CA, USA*

## JAY C. DUNLAP
*Department of Genetics,*
*The Geisel School of Medicine at Dartmouth,*
*Hanover, NH, USA*

## STEPHEN F. GOODWIN
*Department of Physiology, Anatomy and Genetics,*
*University of Oxford,*
*Oxford, UK*

ELSEVIER

AMSTERDAM • BOSTON • HEIDELBERG • LONDON
NEW YORK • OXFORD • PARIS • SAN DIEGO
SAN FRANCISCO • SINGAPORE • SYDNEY • TOKYO
Academic Press is an imprint of Elsevier

Academic Press is an imprint of Elsevier
225 Wyman Street, Waltham, MA 02451, USA
525 B Street, Suite 1800, San Diego, CA 92101-4495, USA
Radarweg 29, PO Box 211, 1000 AE Amsterdam, The Netherlands
The Boulevard, Langford Lane, Kidlington, Oxford, OX5 1GB, UK
32 Jamestown Road, London, NW1 7BY, UK

First edition 2014

**Notice**
No responsibility is assumed by the publisher for any injury and/or damage to persons
or property as a matter of products liability, negligence or otherwise, or from any use or
operation of any methods, products, instructions or ideas contained in the material herein.
Because of rapid advances in the medical sciences, in particular, independent verification of
diagnoses and drug dosages should be made.

ISBN: 978-0-12-800271-1
ISSN: 0065-2660

For information on all Academic Press publications
visit our website at store.elsevier.com

Printed and bound in USA
14 15 16 17   10 9 8 7 6 5 4 3 2 1

# CONTENTS

# CONTRIBUTORS

**Jared Andrews**
Department of Biology, Program on Disease Evolution, University of Louisville,
Louisville, KY, USA

**Jean-Christophe Billeter**
Behavioural Biology, Centre for Behaviour and Neurosciences, University of Groningen,
Groningen, The Netherlands

**Marianna Feretzaki**
Department of Molecular Genetics and Microbiology, Duke University Medical Center,
Durham, NC, USA

**Pushpendra K. Gupta**
Department of Genetics and Plant Breeding, Ch. Charan Singh University,
Meerut, UP, India

**Joseph Heitman**
Department of Molecular Genetics and Microbiology, Duke University Medical Center,
Durham, NC, USA

**Vandana Jaiswal**
Department of Genetics and Plant Breeding, Ch. Charan Singh University,
Meerut, UP, India

**Pawan L. Kulwal**
State Level Biotechnology Centre, Mahatma Phule Agricultural University, Rahuri,
MS, India

**Meghan Laturney**
Behavioural Biology, Centre for Behaviour and Neurosciences, University of Groningen,
Groningen, The Netherlands

**Ana C. Marques**
MRC Functional Genomics Unit, University of Oxford, Oxford, UK; Department of
Physiology, Anatomy and Genetics, University of Oxford, Oxford, UK

**Michael H. Perlin**
Department of Biology, Program on Disease Evolution, University of Louisville,
Louisville, KY, USA

**Kevin C. Roach**
Department of Molecular Genetics and Microbiology, Duke University Medical Center,
Durham, NC, USA

**Sheng Sun**
Department of Molecular Genetics and Microbiology, Duke University Medical Center,
Durham, NC, USA

**Jennifer Y. Tan**
MRC Functional Genomics Unit, University of Oxford, Oxford, UK; Department of
Physiology, Anatomy and Genetics, University of Oxford, Oxford, UK

**Su San Toh**
Department of Biology, Program on Disease Evolution, University of Louisville,
Louisville, KY, USA

CHAPTER ONE

# Neurogenetics of Female Reproductive Behaviors in *Drosophila melanogaster*

## Meghan Laturney and Jean-Christophe Billeter[1]

Behavioural Biology, Centre for Behaviour and Neurosciences, University of Groningen, Groningen, The Netherlands
[1]Corresponding author: email address: J.c.billeter@rug.nl

## Contents

*Advances in Genetics*, Volume 85
ISSN 0065-2660
http://dx.doi.org/10.1016/B978-0-12-800271-1.00001-9

1

## Abstract

We follow an adult *Drosophila melanogaster* female through the major reproductive decisions she makes during her lifetime, including habitat selection, precopulatory mate choice, postcopulatory physiological changes, polyandry, and egg-laying site selection. In the process, we review the molecular and neuronal mechanisms allowing females to integrate signals from both environmental and social sources to produce those behavioral outputs. We pay attention to how an understanding of *D. melanogaster* female reproductive behaviors contributes to a wider understanding of evolutionary processes such as pre- and postcopulatory sexual selection as well as sexual conflict. Within each section, we attempt to connect the theories that pertain to the evolution of female reproductive behaviors with the molecular and neurobiological data that support these theories. We draw attention to the fact that the evolutionary and mechanistic basis of female reproductive behaviors, even in a species as extensively studied as *D. melanogaster*, remains poorly understood.

## 1. INTRODUCTION

The study of the genetics of the vinegar fly *Drosophila melanogaster* has taught us a lot about the mechanisms that control reproductive behaviors. Male reproductive behavior in particular has been dissected using the fine scalpel of genetics for many decades. This approach has yielded tremendous insights in the cellular and physiological mechanisms underlying sensory transduction, as well as the neuronal connectivity and processing underlying male sexual behavior, providing general lessons for understanding the mechanistic and evolutionary basis of behavior (for recent reviews, see Billeter & Levine, 2013; Dickson, 2008; Pavlou & Goodwin, 2013; Siwicki & Kravitz, 2009; Villella & Hall, 2008; Yamamoto & Koganezawa, 2013). However, when it comes to understanding reproductive behaviors, the complement of this story is missing as we have inadequate knowledge of how females receive and interpret information communicated by males (Ferveur, 2010). This asymmetric level of understanding hampers the study of the interactive processes fundamental to sexual communication and reproduction.

The importance of studying female reproductive behaviors is obvious. From an ultimate perspective, understanding female mate choice is important for our understanding of the evolutionary process. Females are the gatekeeper of gene flow between species and show strong behavioral isolation when courted by nonconspecifics. Understanding the reproductive behaviors of *D. melanogaster* females thus not only sheds light on the genetic architecture of species-specific behavior, the molecular and neural basis of conspecific interactions, and the processes of selection within this species; but it also holds the key to evolutionary questions such as species formation and isolation by means of sexual selection. From a proximate perspective, one of the outstanding challenges of neuroscience is to understand how individuals identify others (e.g., member of the same species, sex, kin versus unrelated individuals, and familiar versus novel) and treat them differentially on the basis of their own identity (e.g., their genotype, sex, and experience) (Insel, 2010). *Drosophila* female reproductive behavior is a particularly ideal system in which to unravel such a quandary. The female brain must receive complex sensory signals about conspecifics, the physical environment, and her internal state; and process the information to control the differential treatment of individuals in a given social environment. What is perhaps less well appreciated is that females continue their choosy reproductive behavior beyond mate choice into postcopulatory feeding habits and choice of egg-laying sites. The study of female behaviors thus has the potential to tell us how complex cues from various sources are sensed, and how this information is integrated and processed, ultimately leading to a variety of behavioral outputs.

Here we attempt to review what we know about the genetic, cellular, and chemical basis of female reproductive behaviors: how virgin females locate their habitat, select their first love, and transition into maternal behaviors.

While writing this review, we noticed that genetic and neurobiological data on female behavior is often missing or anecdotal. As in-depth neurobiological studies of the female nervous system and the behaviors it supports are slowly beginning to emerge, we hope that this review, which contains more questions than hard facts, will isolate gaps in our knowledge that should be filled. We highlight evolutionary studies in female behavior that have postulated many interesting features of female reproductive neurobiology, whose mechanistic basis is still unknown. By intersecting work generated from a more evolutionary angle with work aimed at a stronger mechanistic understanding of the neurogenetic and neurobiological basis of female reproductive behaviors, we hope to highlight potential new areas we hope it will spur greater dialog between evolution and neurosciences. By trying to be Jack of all Trades, we are bound to miss several works of importance, and more worryingly, to have made some factual mistakes. We apologize in advance to those colleagues whose work we have overlooked.

We have structured this chapter around an imaginary adult female, whom we will follow from the moment of her choosing of a local habitat (Section 2). We next see how, once settled, she chooses her first mate (Section 3), transitions into producing offspring (Section 4) or chooses to mate with more males (Section 5), to finish on her choice of a site to have offspring (Section 6).

## 2. HABITAT SELECTION

Drosophila melanogaster females feed, mate, and produce progeny on the same site predictive of a strong interplay between habitat and behavior (Ferveur, 2010; Markow & O'Grady, 2005, 2008; Spieth, 1974; Zhu, Park, & Baker, 2003). The strong association of D. melanogaster with human activities, such as wine making and agriculture, probably explains how flies are introduced into new areas and why they are found throughout the world (Keller, 2007). Habitat selection for fruit flies might be as simple as locating a food source and then coping with the challenges that come with that specific locality. As we discuss the causes and consequences of habitat selection, we will see that fruit flies select their habitat on more factors than just food by using information about potential pathogens and the presence of other flies.

### 2.1 Drosophila melanogaster: the Yeast Fly

Drosophila melanogaster has been given many names owing to its attraction to various substances: fruit fly, pomace fly, and vinegar fly. If this species was to be named after its main attractant, it should really be called the *yeast fly*. Volatile chemicals issued from ripe fruits that demonstrably stimulate the fly olfactory system correspond to fermentation products of yeast growing on the fruit and not those of the fruit itself (Markow & O'Grady, 2005, 2008; Spieth, 1974; Zhu et al., 2003). Indeed, the odor of the yeast *Saccharomyces cerevisiae* growing on a synthetic minimal medium, including such volatiles as acetic acid, acetone, 2–methyl–1–butanol, 2–phenylethanol, and ethanol, is sufficient to attract flies (Baumberger, 1917; Becher, Flick, Rozpędowska, Schmidt, Hagman, & Lebreton, 2012; Bownes, Scott, & Shirras, 1988). The volatile components of vinegar (which is the result of fermentation by yeast and bacteria), namely acetic acid, 2–phenyl ethanol, and acetone, elicit pre-meditated flights and flies flow against a gradient of these compounds, showing that flies actively search for yeast source (Becher, Bengtsson, Hansson, & Witzgall, 2010). Although acetic acid can attract flies on its own, yeast volatiles function synergistically as the mixture of ethanol, acetic acid, and 2–phenylethanol in a ratio of 1:22:5 is six times more attractive (Becher et al., 2012; Zhu et al., 2003). These volatiles, which are cue of the presence of yeast, equally attract males as well as virgin and mated females (Becher et al., 2010). Females associate with a spectrum of yeast species but will exhibit preference for certain species, such as *S. cerevisiae* over *Pichia toletana* when given a choice (Anagnostou, Dorsch, & Rohlfs, 2010). This indicates that female flies perceive differences between yeast species.

Insight into the genetics and neurobiology of yeast attraction comes unexpectedly from the study of the deceptive pollination system of a lily flower. The Solomon's lily produce smells that mimic yeast volatile fermentation products (Becher et al., 2010; Stökl et al., 2010) which attract *Drosophila* to it and aids in the transfer of pollen from flower to flower. Fourteen volatiles emitted by the flower, corresponding to volatile compounds identified in ripe fruits and yeast, evoke a physiological response by the fly olfactory system (Stökl et al., 2010). The antennal lobes (Al) are a domain of the brain that receives direct input from the olfactory organ (Masse, Turner, & Jefferis, 2009). The Al has clear anatomical compartments, called *glomeruli*. Each glomerulus receives input from only one type of odorant receptor neurons (ORNs). Functional imaging of the physiological response of each of these glomeruli to the Lily's odors showed that 11 glomeruli are simultaneously activated (Stökl et al., 2010). Given that each glomerulus is innervated by neurons expressing the same odorant receptor (Or) gene (Couto, Alenius, & Dickson, 2005; Fishilevich & Vosshall, 2005), these results suggest that 11 Or genes must be involved in the attraction to yeast. Since the *Drosophila* genome contains 62 identified Or genes (Vosshall & Stocker, 2007), the response to yeast odor employs a large fraction of Or. One of these Or, Or42b, had previously been shown to be sufficient to mediate adult attraction to vinegar (Semmelhack & Wang, 2009) and another, Or85a, had been shown to be activated at higher concentrations of vinegar and sufficient to trigger aversion to toxic doses (Semmelhack & Wang, 2009). Therefore, it seems that the activation pattern of several different ORNs, rather than a single class of ORN, must mediate attraction to sites where yeast is present through a finely regulated mix of attraction and repulsion to different concentrations of odorants.

Attraction to yeast undergoing fermentation can be blocked if the site is infested with harmful microbes (Becher et al., 2010; Stensmyr et al., 2012). Many microbes are not only toxic but also may outcompete or even kill the yeasts that flies graze on. Being able to detect and avoid fruit colonized by harmful molds and bacteria is therefore essential for the localizing of an optimal habitat. This avoidance is mediated by the perception of a microbial compound called Geosmin by a single ORN: Or56a (Stensmyr et al., 2012). While attraction to yeast is regulated by several ORNs acting in concert, avoidance to microbes may be controlled by a single ORN.

Although the monitoring of the response to the Lily flower provided a useful tool used to uncover many unknowns about the neurobiology of yeast detection, it was only designed to detect changes in sensory neurons expressing Or (Stökl et al., 2010). An emerging function of a second class of odorant receptors also expressed in the fly olfactory system, called ionotropic receptors (Ir), will probably add more complexity to the response to the Lily flower (Benton, Vannice, Gomez-Diaz, & Vosshall, 2009). Exploration into the information integration of these two sensory networks will undoubtedly provide a more clear illustration of the neurobiology that supports habitat selection. The neurobiological basis of this attraction is under intense scrutiny and will continue to yield information about sensory integration. Given the importance of yeast for flies, such investigation will ultimately indicate the neuronal substrate for habitat selection. In the next section, we demonstrate that it is also the first step in the flies' ability to interact socially.

## 2.2 The Smell of Sociality

Beyond its importance as a food source, attraction to a common source, fermenting yeast, is a strategy that allows flies to find each other and interact socially. Females seek out yeast, whose nutrient-rich content is essential to their egg production and offspring development (Anagnostou et al., 2010; Baumberger, 1917; Becher et al., 2012; Bownes et al., 1988), and males probably because they represent good mating sites (Becher et al., 2012; Zhu et al., 2003). An aggregation system based on food alone would however be by itself suboptimal in an environment dense in fermenting yeast with low fly density, such as the beginning of the fruiting season in an orchard. Mathematical models indicate that too small a population of *Drosophila* would not manage to colonize and persist in a local environment (Etienne, Wertheim, Hemerik, Schneider, & Powell, 2002; Lof, de Gee, & Hemerik, 2009). A mechanism ensuring greater aggregation specificity was discovered by Bartelt, Schaner, Jackson, (1985). Chemical extracts from adult flies, be it males or females, do not attract other flies. However both males and females fly towards extracts from males combined with fresh yeast or even acetone and ethyl acetate (yeast-fermentation products), (Bartelt et al., 1985; Lebreton, Becher, Hansson, & Witzgall, 2012). This indicates that flies produce a food-dependent aggregation pheromone. Extracts from virgin females, however, are not attractive, indicating that the pheromone is male derived. The active male compound is *cis*-vaccenyl acetate (cVA), a compound produced in the male ejaculatory bulb (Brieger & Butterworth, 1970; Butterworth, 1969) and is equally attractive to males and females but only in the presence of food or yeast fermentation products (Bartelt et al., 1985; Schlief & Wilson, 2007; Wertheim, Dicke, & Vet, 2002a). Perception of this compound involves an odorant-binding protein (OBP) gene called *Lush*, as mutations in this gene blocks social aggregation (Xu, Atkinson, Jones, & Smith, 2005). This protein binds to cVA and is required for optimal association, with the Or67d OR (Gomez-Diaz, Reina, Cambillau, & Benton, 2013; Laughlin, Ha, Jones, & Smith, 2008). Or67d has been demonstrated to be an odorant receptor for cVA (der Goes van Naters & Carlson, 2007; Ha & Smith, 2006; Kurtovic, Widmer, & Dickson, 2007). It remains an open question as to how the presence of food is integrated with the perception of cVA to control aggregation; however, the interneurons connecting to the Ir84a ORN (detecting fruit volatiles) converge with those that receive input from the Or67d ORN, onto a part of the brain called the lateral horn (Grosjean et al., 2011). Whether these two receptors are necessary for aggregation, and whether the anatomical connection between these two olfactory pathways in the lateral horn reflects integration of social and food information for aggregation remain to be investigated. Furthermore, another odorant receptor Or65a also responds to cVA but its participation in aggregation remains to be tested (der Goes van Naters & Carlson, 2007; Ejima et al., 2007).

The use of the aggregation pheromone cVA offers a mechanism to render food sites on which flies are already present more attractive than food sites devoid of flies (Bartelt et al., 1985; Lebreton et al., 2012; Symonds & Wertheim, 2005; Wertheim et al., 2002a). The mechanism allowing deposition of cVA on particular sites remains unclear, but large amounts of cVA are deposited on food sites by males and even larger amounts by mated females (Bartelt et al., 1985). cVA is included in the ejaculate and transferred to females during mating (Butterworth, 1969), making it plausible that cVA is released in the environment during mating. Indeed, flies in the process of copulating have been demonstrated to attract other flies (Lebreton et al., 2012). cVA could also be released by females during egg-laying (Bartelt et al., 1985). The remaining question is what is the effective distance at which cVA works? Outdoor cage experiments indicate that food with fly odors consistently attract more flies than food without fly odors when flies were released 1 m from those food substrates (Wertheim et al., 2002a). However, wind tunnel experiments indicate that flies do not detect fly odors at a distance >30 cm (Lebreton et al., 2012).

## 2.3 Group Formation

Laboratory experiments indicate that cooperation in foraging begins with "primer" flies. The first wave of flies arriving in a new environment samples different food patches and settles on the favorable one. The second wave of flies arriving lands directly on the occupied food patch without sampling the other sites, indicating that the primer flies emit some signal trusted by the other flies (Tinette et al., 2004). Individuals show no predisposition for primer behavior (Tinette et al., 2004) suggesting primers are formed randomly, and most group members spend some of their time priming. The mechanism underlying food-searching behavior has yet to be identified but appears to involve the learning and memory system. *Dunce* flies, which have learning and memory disability caused by a mutation in a cAMP adenylate cyclase (Dudai et al., 1976), fail not only in locating a favourable food site as a primer fly but also aggregating on occupied food patch as a second wave fly (Tinette et al., 2004).

Once on a food patch, the density of flies becomes the next hurdle. Male behavior can regulate group size via aggregation and harassment: they charge, chase, and erect their wings at intruding males (Dow & Schilcher, 1975; Hoffmann, 1987) and intensely court females both of which lead to abandonment of the food patch by the receiver (Saltz & Foley, 2011; Wertheim et al., 2006). The balance of aggregation and aggression in males is determined mainly by the concentration of cVA. As group sizes increases, so too does the levels of cVA leading to an increase of male aggressive behavior. Interestingly, Or67d receptor is required for the social density-dependent aggression indicating a potential common neurological substrate for aggregation and aggression (Wang & Anderson, 2010; Wang et al., 2011). However, the release of cVA onto the food patch can potentially be modulated by both males and females. Males may regulate the amount of cVA transferred to females during mating, thus determining the amount females can release in the environment. Expression of cVA is indeed influenced by the social context (Kent et al., 2008), and male ejaculate size is modulated by the presence of other males (Garbaczewska et al., 2012; Sirot et al., 2011; Wigby et al., 2009). Mated females, on the other hand, who acquired cVA from males may also contribute to the control group size if they were able to control the release of cVA depending on fly density and substrate quality (Wertheim, 2005; Wertheim et al., 2002a, 2006).

Females may regulate group size through their own aggression. Qualitatively and quantitatively different from male aggression, females show aggressive behaviors toward one another when placed on a food source containing live yeast. A female may lunge at another female, erect her wings or push another by extending her legs, head-butting, and fencing (Ueda & Kidokoro, 2002). The genetic and neuronal basis of female aggression remains unclear. Although some aspects of *Drosophila* aggression are shared between male and female flies such as low-intensity "fencing," most are more frequent in one sex than the other. For example, "lunging" and "boxing" are mostly seen in male fights, whereas "shoving" and "head-butting" are characteristic of female fights (Nilsen, Chan, Huber, & Kravitz, 2004). These differences are due to sex-specific transcripts of the *fruitless* (*fru*) gene, who is transcribed in both sexes but only translated to produce the Fru[M] protein product in the brain of males but not in females (Vrontou et al., 2006). Female mutants, engineered to express Fru[M] proteins in their nervous system, show male-like patterns of aggression, and males that do not express Fru[M] show female patterns of aggression (Vrontou et al., 2006). This indicates that neurons controlling aggression may be shared between males and females but support different patterns of aggression due to the action of *fru* (Asahina et al., 2014; Chan & Kravitz, 2007; Fernández et al., 2010; Vrontou et al., 2006).

## 2.4 Conclusion

Flies are not little missiles randomly flying in the environment to find food and mate. They possess mechanisms that allow them to locate suitable food substrates, promote aggregation, and regulate population size. We will see in this chapter that aggregation of adults on a resource is the starting point for flies to find each other for both mating (Section 3) and remating (Section 5), and that aggregation improves the survival and growth of their offspring developing on these resources (Section 6) (Rohlfs & Hoffmeister, 2003; Wertheim et al., 2002a).

Habitat choice by *D. melanogaster* is influenced by a complex mix of social, nutritional, and genetic factors. Studies of habitat in *D. melanogaster* are an example of how integrating evolutionary and mechanistic methods can lead us to deeper understanding of biological phenomena. In the next section, we will examine how females, once they have chosen a location, chose their mate.

## 3. FIRST LOVE

Once landing on a fruit fly-colonized food patch, females are almost immediately courted by males (Wertheim et al., 2006). Despite the lack of conspicuous solicitation behavior by females, males are drawn to them and display intense courtship. Male courtship consists of approaching the female, mostly from the back, tapping her abdomen, extending one wing, and vibrating it to produce a song, circling the female while doing so, extending his proboscis to make contact with the female genitals and ultimately swinging his abdomen at the female to attempt copulation (reviewed by Yamamoto & Koganezawa, 2013). While this ritual is required by the female to accept mating, it can also become detrimental to her. At high population densities, intense male courtship harasses females. If she stays, she will have reduced fecundity and early mortality, probably due to disruption during oviposition and exhaustion resulting from constantly decamping (Markow & Manning, 1980; Partridge et al., 1987). If she leaves, she has reduced access to food and increased exposure to predators (Wertheim et al., 2006). Males are however not completely indiscriminate and exert a choice on which females they preferentially court (Long et al., 2009). In this section, we will review the mix of cues that signal female attractiveness. We will then see that attracting males allows females to sample the self-reporting cues that males display during courtship. Finally, we will review our knowledge of the genetics and neurobiology underlying the female decision to accept mating.

### 3.1 Female Sex Appeal

#### 3.1.1 Size

Cues signaling a female's attractiveness are most likely honest signals of her fecundity. Otherwise, over an evolutionary time-frame, males who mate with less-fecund females will be selected against and the alleles that support such decisions will decrease in frequency. Males given a choice between different females prefer the larger ones, which have higher life-time fecundity (Long et al., 2009). Size appears as an honest indicator of fecundity probably reflecting the size and thus capacity of the ovary, which occupy most of the female abdomen. Size is however constrained by the cost of exposure to heightened male harassment leading to reduced lifetime fecundity (Byrne & Rice, 2006; Long et al., 2009). Interestingly, the sex appeal of size is social context-dependent. The presence of other females matters to individual females because harassment is only focused on larger females when smaller females are present (Long et al., 2009), and between the male suitor and the target of his affection suggesting that flies are capable of social comparisons (Turiegano et al., 2012). Because larger females require longer courtship by males in order to accept mating (Turiegano et al., 2012), it is possible that female size might relate to receptivity.

What exactly makes size attractive is unknown; it could affect the amount of smell a female can produce, or affect her gait. Size is regulated by the insulin pathway, (Wu & Brown, 2006). Choosiness could be also experience-based; the frequency and intensity by which a female is courted could influence her choosiness. This renders it difficult to assign a precise genetic contribution to this sex-appeal factor. To our knowledge, mate preference assays with blind males experimentally exposed to a small and a large female have not been performed but this experiment would help sorting these issues.

### 3.1.2 Movement

The way a female moves, or does not move, contributes to her attractiveness. The *Shibire* temperature-sensitive mutation (*Shi^{ts}*) causes temporary paralysis at higher temperature but normal locomotion at room temperature (Grigliatti et al., 1973). When *Shi^{ts}* females are temporarily immobilized by application of restrictive temperature, they are courted half as much as at the permissive temperature (Tompkins, Gross, Hall, Gailey, & Siegel, 1982), showing that female movement stimulates male courtship. It is however difficult to separate the effect of female size, movement pattern, and gait as perception of all these cues are affected by male blindness. The recent identification of motion detection neurons in the visual system and the existence of tools to assay their specific function may help resolve this issue (Maisak et al., 2013). Moreover, part of the attractiveness of female movement appears to come from the noise it produces as female movement generates acoustic signals that arouse males (Ejima & Griffith, 2008). There is, however, nothing sex-specific in this signal as the noise made by males also enhances male arousal.

What exactly in the movement of females triggers male courtship? Dopamine is a regulator of overall locomotor activity levels. Reduction in amounts of dopamine by application of a pharmacological inhibitor of tyrosine hydroxylase, an enzyme controlling the final stage in the synthesis of dopamine, does slow down females but was not associated with a change in male courtship (Wicker-Thomas & Hamann, 2008). These data suggest that the overall level of locomotor activity is not the parameter of interest in female movement, but rather her gait or pattern of activity. Female activity patterns are indeed different from those of males: in isolation females constantly modulate their activity pattern, showing bouts of start and stop motion, whereas males have a steadier more constant walking pace (Belgacem & Martin, 2002, 2005; Gatti, 2000; Martin et al., 1999). This difference in locomotion is controlled by a small cluster of neuroendocrine insulin-producing neurons in the pars intercerebralis (PI), a neurosecretory center sitting atop the brain midline (Belgacem & Martin, 2002). Feminization of these neurons in an otherwise male results in female-like pattern of locomotor activity (Gatti et al., 2000). Ablation of the insulin-producing neurons in the adult stage (ablating too early results in abnormal development) gives males a female pattern of activity, but has no effect in females indicating that female movement is the default state (Belgacem & Martin, 2002, 2005). The PI regulates locomotion via insulin circulating throughout the body because transplanting these neurons in the abdomen of a male is sufficient to feminize locomotor activity (Belgacem & Martin, 2002). The insulin signal targets the corpus allatum (CA), a gland located in the first thoracic segment, as demonstrated by the necessity of the insulin receptor in the CA for male mode of locomotor activity (Belgacem & Martin, 2007), and ablation of the CA results in female-like pattern of activity in males (Belgacem & Martin, 2005). CA is likely to control locomotor activity via differential release of juvenile hormone (JH) in males because feeding males with a JH inhibitor gives them a feminized pattern of locomotor activity (Belgacem & Martin, 2002). Insulin-producing cells in the PI control JH expression by the CA. Sexually dimorphic activity is thus controlled by a neuroendocrine axis composed of the PI and the CA. The last-known output of this pathway is JH but because this hormone has multiple physiological effects, it remains unknown what the final output controlling locomotor activity is.

That sexually dimorphic locomotor activity of females is important for their attractiveness has, however, not yet been demonstrated. If masculinization of PI neurons confers male-typical locomotor patterns, females with a male gait should be less attractive than control females. However, this experiment would not tease apart the role of acoustic cues produced during walking and the visual cue of movement.

### 3.1.3 Pheromones

Looks are not everything, and are not necessary for female attractiveness, since flies can mate in the dark (Gailey, Lacaillade, & Hall, 1986). Bastock and Manning (1955) noted that although males orient toward females, they do not perform any courtship behavior at a distance >5 mm. Computer-mediated observation shows that males and females interact at a distance of 2.5–3.5 mm (Branson, Robie, Bender, Perona, & Dickinson, 2009). It appears that close proximity cues "release" further courtship steps (Bastock & Manning, 1955),which are probably chemicals perceived via the olfactory or gustatory system (Gailey et al., 1986; Krstic, Boll, & Noll, 2009; Trott, Donelson, Griffith, & Ejima, 2012). Furthermore, transfer of female chemical extracts either by blowing air through a chamber containing females into a chamber (Averhoff & Richardson, 1974; Shorey & Bartell, 1970) or placing extracts on a piece of filter paper (Tompkins, Hall, & Hall, 1980) can trigger males to initiate courtship toward nearby flies. This indicates that female volatiles can confer momentary attractiveness to flies of any sex.

What are the attractive chemicals produced by females? Flies display hydrocarbon molecules at the surface of their body, called cuticular hydrocarbons (CHC) (Everaerts, Farine, Cobb, & Ferveur, 2010; Farine, Ferveur, & Everaerts, 2012; Ferveur, 2005; Jallon, 1984; Levine, Billeter, Krull, & Sodhi, 2010; Yew, Cody, & Kravitz, 2008; Yew et al., 2009). There are between 30 and 50 species of CHC and most are sexually dimorphic in either presence or amounts (Ferveur, 2005; Jackson, Arnold, & Blomquist, 1981; Jallon, 1984; Yew et al., 2009). To determine what specific CHC compound renders female attractive, dead male flies washed in an organic solvent (removing their natural CHC profile) were perfumed with CHC fraction from females or synthetic HC (Antony & Jallon, 1982). The strongest CHC to induce courtship is 7,11-heptacosadiene (7,11-HD), a 27-carbon-long diene made only by females (Antony, Davis, Carlson, Pechine, & Jallon, 1985). The 29-carbon-diene 7,11-nonacosadiene (7,11-ND) and the 25-carbon-monoene 7-pentacosene (7-P) also induce courtship by males, but at lower efficiency (Antony et al., 1985). There are thus single compounds in the CHC profile of females that can stimulate male courtship.

The diene 7,11-HD is produced through the action of a female-specific enzymatic pathway located in cells just under the surface of the abdomen called the oenocytes including the activity of *desaturase1 (desat1)* (Marcillac, Bousquet, Alabouvette, Savarit, & Ferveur, 2005a), *desaturaseF (desatF)* (Chertemps, Duportets, Labeur, Ueyama, & Wicker-Thomas, 2006), and *elongaseF (eloF)* (Chertemps, Duportets, Labeur, Ueda, Takahashi, & Saigo, 2007). Knock-out of *desat1* or knock-down of *desatF* and of *eloF* by RNA interference (RNAi) result in reduction in 7,11-HD and a concomitant reduction in courtship latency by males (Chertemps et al., 2006, 2007; Marcillac, Grosjean, & Ferveur, 2005b). The female-specific expression of *desatF* and *eloF* is under the control of a developmental regulator gene called *doublesex (dsx)*, which encodes a transcription factor that differentially regulates gene expression between males and females controlling most aspects of sexually dimorphic development. The female form of Dsx, Dsx$^F$, is sufficient and necessary to activate the expression of 7,11-HD (Waterbury, Jackson, & Schedl, 1999) and directly binds the promoter of *desatF* and likely of *eloF* as well (Chertemps et al., 2006, 2007; Legendre, Miao, Da Lage, & Wicker-Thomas, 2008; Shirangi, Dufour, Williams, & Carroll, 2009). Forcing the expression of Dsx$^F$ in the male oenocytes feminizes his CHC profile and renders him attractive to other males (Fernández et al., 2010; Ferveur et al., 1997), showing that CHC can trump all other sexually dimorphic female signals. These experiments suggest that CHC expressed at the surface of females make the most of their appeal.

That the diene 7,11-HD is the aphrodisiac signal of females is at first look a closed case. In a surprising turn of events, ablation of the oenocytes or elimination of CHC by genetic means results in female flies that are not only still attractive (Billeter, Atallah, Krupp, Millar, & Levine, 2009; Savarit, Sureau, Cobb, & Ferveur, 1999), but actually are more attractive than wild-type females (Billeter et al., 2009). Female 7,11-HD is thus not necessary for attractiveness. These observations lead to a model where 7,11-HD and other CHC are superimposed on another unknown attractive pheromone, not made by the oenocytes. The ability to detect chemicals is limited by technology: a new analytical chemistry method will doubtless reveal the presence of additional pheromones (Levine et al., 2010; Yew et al., 2009).

What is the role of female-specific CHC if they are not necessary for attractiveness? *Drosophila melanogaster* coexist with different species of *Drosophila*, especially *Drosophila simulans*, which are similar in appearance. Because mating can occur between female *D. melanogaster* and male *D. simulans*, yet only yield sterile females, male *D. simulans* should be able to recognize females from his species. A key to the function of 7,11-HD is that *D. melanogaster* females but not *D. simulans* make this compound (Jallon & David, 1987). In absence of 7,11-HD, females become attractive to males from other species (Billeter et al., 2009; Savarit et al., 1999) and in the presence, blocks courtship from males from other species who avoid contact with 7,11-HD (Billeter et al., 2009; Coyne, Crittenden, & Mah, 1994; Savarit et al., 1999).

Another possible function of 7,11-HD is in intraspecific sexual conflict. As we have seen in Section 2, females receive cVA from males during mating. Acquisition of this pheromone via semen, together with other male CHC rubbed on the female, renders females unattractive to other males, probably in an attempt by males to prevent females from mating with other males (Jallon, 1984; Scott & Richmond, 1987; Yew et al., 2009). Females devoid of CHC but perfumed with cVA remain unattractive to males for a much longer period than females with intact CHC profile indicating that CHC mitigate the antiaphrodisiac effects of cVA. Coperfuming females devoid of CHC with 7,11-HD and cVA is sufficient to restore wild-type level of female attractiveness. One role of 7,11-HD is thus to compensate the antiaphrodisiac effect of cVA imposed by males (Billeter et al., 2009).

The 7,11-HD pheromone may be understood as a way to repel males from other species, and as a way for a female to remain attractive after mating, rather than merely as an aphrodisiac. However, the relevance of 7,11-HD for female attractiveness is underscored by its rapid evolution both between species and within *D. melanogaster*. The enzymatic pathway that controls its expression is variably expressed between species. The promoter of *desatF* shows repeated loss and gain of the sex-specific *dsx*-binding element, making it expressed in both males and females in some species (Shirangi et al., 2009). There is also intraspecific variation in the amount of two heptacosadiene (HD) isomers: 7,11-HD is predominant in most laboratory strains, whereas 5,9-HD levels are relatively high in Caribbean and sub-Saharan strains (e.g., Tai; Ferveur, Cobb, Boukella, & Jallon, 1996). The 7,11-HD:5,9-HD ratio is entirely controlled by *desat1* and *desat2*, two closely linked genes on chromosome 3 (Coyne et al., 1994; Dallerac, Labeur, Jallon, Knipple, Roelofs, & Wicker-Thomas, 2000). A deletion in the promoter of *desat2* has been correlated with high 7,11-HD and low 5,9-HD, whereby an active *desat2* promoter results in high 5,9-HD (Takahashi, Tsaur, Coyne, & Wu, 2001). The functional consequence of this difference is still unclear. Females with 7,11-HD mate in general faster than females with 5,9-HD, perhaps because they stimulate more courtship by males (Ferveur et al., 1996).

The CHC profile of females plays other roles in controlling the attractiveness of a female beyond conveying information about sex and species identity. CHC indicate the age and sexual maturity of a fly. Mature female-like sex appeal is present in unreceptive immature females and males (Jallon & Hotta, 1979). Young males and females have similar CHC profile containing longer chain CHC with a more complex mix of dienes. At day 4, flies become fully sexually dimorphic and the total amount of CHC reaches its maxima (Arienti, Antony, Wicker-Thomas, Delbecque, Jallon, 2010). A brain signal controls the CHC profile because decapitation, which does not kill the female, within an hour of eclosion prevents the maturation of the CHC profile (Wicker & Jallon, 1995). This factor may be dopamine because mutation in the *dopadecarboxylase* (*Ddc*) gene, which controls the biosynthesis of the monoamines dopamine and serotonin, prevents full maturation of the female CHC profile, an effect that can be rescued by feeding the fly dopamine or directly injecting it into the nerve cord of a decapitated female (Marican, Duportets, Birman, & Jallon, 2004). The dopamine signal is likely to be relayed from the nervous system to the oenocytes through ecdysone because sexual dimorphism of CHCs is reduced in *ecdysone-less* mutants (Wicker & Jallon, 1995). The physical environment also plays a role in attractiveness, as CHCs are also an indicator of the microorganisms borne by the female. The CHC profile can be affected by the commensal microbes present in the organism, and this correlates with a change in attractiveness and mate choice (Sharon et al., 2010).

Control of female attractiveness is very finely regulated by the CHC system, probably using a mix of hydrocarbon pheromones acting in combination that still need to be elucidated. As females continuously display a CHC profile, it makes it difficult for them to control their attractiveness in the very short term. There is however a report that females release an abdominal droplet just before mating that rapidly arouses the male fly (Lasbleiz, Ferveur, & Everaerts, 2006). Other pheromonal systems beside CHC probably act to control female attractiveness and much remains to be discovered. One area of interest is whether females may be able to affect the pheromones they receive from males during mating by degrading them or changing their structure. An early report on the role of cVA as an antiaphrodisiac reported its modification by *esterase-6*, a carboxylesterase found in the male ejaculate, into *cis*-vaccenyl alcohol (cVOH) (Mane, Tompkins, & Richmond, 1983). This cVOH compound is a longer lasting antiaphrodisiac than cVA when applied directly on a female (Costa, 1989). However, later studies questioned whether cVA was ever transformed into cVOH in vivo; (Vander Meer, Obin, Zawistowski, Sheehan, & Richmond, 1986). Since females produce *esterase-6* (Chertemps et al., 2012), it remains possible that this enzyme can affect the chemistry of pheromones that regulate the display or turn-over of pheromones.

## 3.2 Receptivity

Females initially react to male courtship by running or flying away, or by exhibiting a series of rejection behaviors whose occurrence depends on her developmental stage and reproductive status (Connolly & Cook, 1973; Ewing & Ewing, 1984). Younger females flick their wing at the male which produces an acoustic signal (Paillette, Ikeda, & Jallon, 1991) and kick their legs toward the courting male, while mated females may extrude their ovipositor (Bastock & Manning, 1955; Chen, Stumm-Zollinger, Aigaki, Balmer, Bienz, & Böhlen, 1988; Connolly & Cook, 1973; Manning, 1967; Markow & Hanson, 1981; Nakano et al., 2001). Successful courtship progressively decreases female movement until she comes to a standstill, apparently signaling mating acceptance (Spieth, 1974), followed by the male's attempt to copulate. Females show choosiness in the males they mate with, selecting males with certain characteristics over others. In this section, we will review our knowledge of what makes a female receptive to one male but reject another.

### 3.2.1 Evolutionary Principles Underlying Female Choosiness

Before we examine the mechanisms underlying female receptivity, it is helpful to explore the factors contributing to why females are the choosier sex and view female receptivity in an evolutionary context. Compared to males, *D. melanogaster* females bear a higher cost to reproduction: exposure to constant male courtship (Partridge & Fowler, 1990), tissue damage from contact with male genitals (Kamimura, 2007), and seminal fluids transferred in the ejaculate (Chapman, Liddle, Kalb, Wolfner, & Partridge, 1995; Wigby & Chapman, 2005) all reduce female longevity. Although females can resist copulation, males can force copulation upon newly emerged females causing these females early mortality and reduced fecundity (Dukas & Jongsma, 2012; Seeley & Dukas, 2011). Thus, females must exert selectivity when it comes to choosing whom to mate with. We will introduce a few evolutionary concepts that highlight the challenges of understanding female receptivity.

Females exercise their mating choice by biasing mating with males of certain phenotypic trait. What is considered a suitable mate for a *Drosophila* female? This question is hard to answer because the evolution of what we call "preference," the type of males that females bias mating with can in theory be the result of several different modes of selection. For instance, the theory of sexual conflict (Chapman, 2006) predicts that adaptation in one sex can lead to counteradaptation in the other. We have seen above that mating and male harassment incurs a cost to female lifespan and fecundity. Females should coevolve to resist this cost of mating, implying that males that are successful at gaining mating are those better able to overcome female resistance, and not only those that can simply attract females. This is exemplified in the case of large males who court females more intensely and successfully increase the mating frequency of these females. However, females produce less eggs and live less long when housed with such large males; therefore, preferred males actually reduce female fitness (Friberg & Arnqvist, 2003; Pitnick, 1991). The trait that is preferred in males, i.e., size, may be associated with harmful traits such as higher harassment, wounding during copulation, and transfer of more toxic seminal fluids (Friberg & Arnqvist, 2003).

Female mate preference can also be linked to ecological conditions. It may be beneficial for females to prefer males with traits linked to genes that confer adaptation to local microclimates. This would prevent genetic exchange with nearby populations adapted to different climates and promote assortative mating (mating with genetically similar males). Such scenario is observed in the "evolution canyon" in Israel where two nearby slopes of the canyon possess different microclimates: one slope receives intense exposure to sun making it drier and hotter than the other slope. If female mating preferences function to increase the likelihood of her offspring to be well suited for the environment, then females should mate assortatively with males from their slope. Indeed flies from different slopes of the canyon show preference for sexual partners originating from the same slope (Korol et al., 2000). Assortative mating creates a risk of inbreeding, which often causes a reduction in the fitness of offspring. However, *D. melanogaster* adult females prefer mating with their brothers over unrelated males, which supports an inbreeding preference (Loyau, Cornuau, Clobert, & Danchin, 2012b) or lack of avoidance (Tan, Løvlie, Pizzari, & Wigby, 2012; Tan et al., 2013) in this species.

A final complication to understanding the "rules" by which females choose their partners is phenotypic plasticity demonstrated in the context of a changing social environment. For instance, females avoid mating with males that are semen-limited, which incur a cost of mating without the full benefit of acquiring a great amount of sperm by avoiding males whom they have witnessed recently mating (Loyau, Blanchet, Van Laere, Clobert, Danchin, 2012a). However, females may also use social information to copy the mate choices of other females by preferentially mating with males of the successful phenotype even though they did not have a prior preference for this trait (Mery et al., 2009). Furthermore, females may initially avoid mating with familiar males (Ödeen & Moray, 2007) but remate with familiar males (Tan et al., 2012), or mate with a phenotype they have already encountered (Dukas, 2005). Whatever the adaptive significance of these observations, it is obvious that female mate choice is both complex and flexible.

The aim of this subsection was to illustrate that several evolutionary processes may be at play in the shaping of female mate preference, which may complicate the genetic analysis of female behavior. The following section delves into mechanisms allowing females to detect male traits that influence her mate preference and ultimately choice.

### 3.2.2 What Women Want: Sensory Basis to Female Receptivity

Females assess a variety of cues displayed by males during courtship, including visual (During following and circling of the female), tactile (tapping the female with the forelegs), acoustic (performance of a species–specific courtship song), and chemosensory. Females likely summate multiple sensory cues in order to discern that the courting male is quantitatively robust and qualitatively of her species (Bastock & Manning, 1955). Despite the complexity of female mate choice, the genetic, cellular, biochemical, and biophysical basis of some male traits selected by females have been uncovered. Here we will review the male traits that females base their mate choice on and the sensory inputs they use to sample these traits. We will focus on identified sensory inputs detecting the male courtship song and pheromones as, to our knowledge, not much is known about the nature of visual, taste, and tactile cues beside the fact that they matter. We hope that the focus on acoustic and chemosensory signals will give general principles that can be applied to other modalities.

#### 3.2.2.1 Acoustic Cues

It is generally thought that the male courtship song is the primary cue for female choice (Crossley, Bennet-Clark, & Evert, 1995; Villella & Hall, 2008). This is for instance supported by the observation that males with abnormal courtship song but normal pheromonal profile achieve mating less rapidly than those that have an abnormal pheromonal profile but normal song (Rybak, Sureau, & Aubin, 2002). Males produce the courtship song by unilaterally extending a wing and vibrating it (Koganezawa, Haba, Matsuo, & Yamamoto, 2010). This produces a song composed of two sub-types: *pulse song* and *sine song*. The pulse song is characterized as trains of "pulses". The amount of time between these trains of pulses is termed the interpulse interval (IPI). This interval is species–specific and lasts approximately 34 ms in *D. melanogaster*. Pulse song is occasionally interrupted by a humming called sine song, which has a frequency of 160 Hz. The function of these various song parameters for female receptivity has been studied using playback of artificially manipulated courtship song. The pulse song positively influences the female receptivity to males whose wings have been removed (Kyriacou & Hall, 1982). The sine song function is less clear as its playback in the absence of male primes females to accept copulation faster when subsequently presented with males, but only after a very long expo-sure time of 15 min (Kyriacou & Hall, 1984; Rybak et al., 2002). A function for the sine song in increasing male attractiveness is still implied by the fact that males modulate which song is prominent depending on female move-ment; the more the female moves, the more pulse song the male performs (Trott et al., 2012). While the courtship song is a major cue for female mate choice, it is relatively invariable in terms of its parameter between *D. mela-nogaster* populations (Gleason, 2005; Ritchie, Yate, & Kyriacou, 1994; Turner, Miller, & Cochrane, 2013). What does differ between males is the intensity of pulse song production, so called "duty cycles" (Talyn & Dowse, 2004). Larger males produce longer pulse songs and are preferred by females. Male size preference may thus be linked to their production of a more emphatic song and not to an intrinsic difference in song parameters.

A key to the function of song in female receptivity is species recogni-tion. Sibling *Drosophila* species have different male song structures that are the main contributors to female recognition of males from her own spe-cies (Kyriacou & Hall, 1986; Wheeler et al., 1991). Songs differ between *D. melanogaster* and *D. simulans* on multiple components: different IPI (34 ms in *D. melanogaster* and 50 ms in *D. simulans*; Cowling & Burnet, 1981); male pulse song fluctuating rhythm in IPI (55 s in *D. melanogaster* and 25 s in *D. simulans*; Kyriacou & Hall, 1980); and species–specific rhythmic fluctua-tion (55 s in *D. melanogaster* and 35 s in *D. simulans*; Kyriacou & Hall, 1982). Species differences in IPI rhythms are controlled by the circadian system (Kyriacou & Hall, 1980, 1986; Ritchie, Halsey, & Gleason, 1999; Wheeler et al., 1991). The clock system regulates the behavioral and physiological rhythms of the fly, such as when it sleeps and when it mates. *Period* (*per*) is at the core of the timing mechanism as *per* mutations disrupt the period of the circadian rhythm (Hardin, 2011). Small differences in Per amino acid (aa) sequence between *D. simulans* and *D. melanogaster* are responsible for differences in IPI (Wheeler et al., 1991).

A critical question in the context of this review is how do females recognize songs from their species. With evolving male courtship songs, female preferences need to adjust accordingly or vice versa (Greenacre, Ritchie, Byrne, & Kyriacou, 1993). Courtship song IPI rhythms and species-specific female preference could be genetically coupled via *per*. Hybrid *D. melanogaster/D. simulans* females show selective preferences for artificially generated songs carrying intermediate characteristics of two males (Kyriacou & Hall, 1986). This indicated that Per might couple production and perception of species-specific aspect of the courtship song, but that relationship remains ambiguous (Butlin & Ritchie, 1989). This coupling does not hold for within species song perception. *per* mutant females with a longer period than wild type were presented with artificial songs matching those found in *per* mutant male flies, in presence of wingless males (Greenacre et al., 1993). In this case, mutant females always preferred the song produced by wild-type males indicating that song perception is under different genetic control than song perception and thus that the two are not genetically coupled (Greenacre et al., 1993).

It is unclear how the female processes the male song to affect her decision to mate, but progress has recently been made on understanding song detection by the nervous system (Kamikouchi, 2013). Sensory modalities are encoded by receptor neurons housed in specialized sensory organs at the periphery of the fly. These so-called first-order neurons send their projections to separate areas of the brain. The song is detected by the fly ear, called the Johnston's organ (JO), located in the second antennal segment (see Kamikouchi, 2013 for a recent review of the working of this organ). This organ contains 480 chordotonal neurons (Kamikouchi, Shimada, & Ito, 2006). These first-order neurons convert mechanical forces resulting from the rotation of the antennae caused by gravity and air pressure on a feathery antennal structure called the *aristae* into electrical signals (Göpfert & Robert, 2001; Kamikouchi et al., 2009). The structure of the antennae and the JO are mechanically tuned to respond best to the pulse-frequency of males from their species (Riabinina, Dai, Duke, & Albert, 2011). While this allows females to amplify the song from their own males, it does not explain how females can discriminate information in the IPI. The neuroanatomy of song perception has been delineated using a genetic approach. The *nanchung* and *inactive* (*iav*) genes encode ionic channels that are required for sound and gravity transduction (Eberl, Hardy, & Kernan, 2000; Kamikouchi et al., 2009). Mutant *iav* females show reduced receptivity, which is consistent with a role in audition (O'Dell, 1993). The *nompC* gene encodes an ionic channel that appears to amplify mechanical input to the transduction channel (Lehnert, Baker, Gaudry, Chiang, & Wilson, 2013). Mutations in *nompC* only affect sound and not gravity perception (Eberl et al., 2000; Kamikouchi et al., 2009; Lehnert et al., 2013). Identification of this gene allowed entry into the system for song perception. Tracing of the expression of *nompC* shows expression in two of five subsets of Johnston's organs. These two subsets respond preferentially to sounds rather than gravity. Subset A responds to a wide range of frequencies, while subset B responds to low frequencies (Kamikouchi et al., 2009). The projection of these first-order JO neurons was traced to the brain, by fusing the promoter of *nompC* to the green fluorescent protein. JO subsets AB innervate a part of the frontal brain called the antennal mechanosensory and motor center, specifically in two adjacent zones called AB (Kamikouchi et al., 2006). These two zones make synaptic contacts with second-order neurons (interneurons) originating from the inferior ventrolateral protocerebrum (VLP) (Kamikouchi et al., 2006, 2009; Lai, Lo, Dickson, & Chiang, 2012). Simultaneous physiological recordings of the response of JO and their second-order neurons to pulse song are beginning to show how females discriminate songs from their own species (Tootoonian, Coen, Kawai, & Murthy, 2012). The tuning of JO and second-order neurons to pulse song is matched; the more pulses in the song, the more firing of both JO and second-order neurons. This graded response is observed in IPI ranging from 20 to 90 ms. However, when the IPI is below 50 ms, the second-order neurons graded response obtains a new characteristic; the overall membrane depolarization is sustainably enhanced during the pulse train (called a DC shift). This DC shift in second-order neurons matches the IPI of *D. melanogaster* (34 ms), and is not reliably induced by the IPI of *D. simulans* (50 ms) (Tootoonian et al., 2012). It is therefore possible that some species-specific information coding takes place in those second-order neurons. This neural coding has however only been established for one type of second-order neuron connecting to JO responsive to song. There are several more groups of neurons connecting to these JO, who respond to different sound frequencies and whose neural coding await analysis (Kamikouchi et al., 2009; Lai et al., 2012; Tootoonian et al., 2012).

Unfortunately, there is no direct demonstration that JO neurons AB or that *nompC* are required for female receptivity. However, the behavioral response of males to the courtship song was studied. Males start courting each other when they hear a courtship song, probably as a reflex induced by the perception that if someone is emitting a courtship song, there must be a female around (Eberl, Duyk, & Perrimon, 1997). Silencing of subset B, but not other subsets of JO organs, suppressed the response of males to the courtship song, supporting the idea that this population is responsible for courtship song perception (Kamikouchi et al., 2009). Unfortunately, no specific test of subset A was possible due to lack of tools.

The field of auditory signaling is rapidly evolving and no doubt that the neuronal coding of courtship song for female receptivity will be soon unraveled through a combination of electrophysiology and genetics. With regards to genetics, a new approach demonstrating changes in gene expression after hearing the courtship song isolated several new genes with roles in response to song (Immonen & Ritchie, 2011). Study of the function and expression of these genes will probably help dissect how song is processed by the female central nervous system.

Finally, a new conserved acoustic signal was recently discovered that involves quivering of the male abdomen that generates vibrations born by the substrate at a frequency of 6/s with great regularity (Fabre et al., 2012). While males extend their wing as the female is moving away, quivering happens mostly when the female is immobile and might be the cause of her stopping. It is thus interpreted as a receptivity-enhancing male signal. This new acoustic system may work via the same neurons of the antennae that detect courtship song, or may use other vibration-sensitive organs in the body.

### 3.2.2.2 Pheromones

In concert with acoustic signals, chemosensory signals from males are critical for female receptivity. The first indication that female receptivity is induced by the scent of males came from females bearing the *OlfactoryD* (*OlfD*) and *smellblind* (*sbl*) mutation, which cause reduced olfactory function. These females do not slow down during the time they are courted, and have a lengthened latency to mate, indicating that olfaction is important for receptivity (Gailey et al., 1986; Tompkins et al., 1982). *OlfD* and *sbl* are the alleles of the sodium channel *paralytic* (*para*) (Lilly, Kreber, Ganetzky, & Carlson, 1994), important for sensory transduction, among other things in olfactory neurons. The olfactory system of females is located in the third antennal segment, just below the fly's ear. In keeping with the notion that olfaction is important for female receptivity, surgical removal of the third antennal segment reduces female receptivity (Grillet, Dartevelle, & Ferveur, 2006).

Given the requirement of an intact olfactory system for receptivity, males must emit a chemical substance that causes females to slow down during courtship. Males genetically manipulated to lack oenocytes, abdominal glands that produce CHC, are devoid of CHC and have lengthened latency to mate and are disfavored by females over wild-type males (Billeter et al., 2009). This indicates that a substance, probably a CHC, made by the male oenocytes is attractive to females. Males display over 30 CHCs on their body surface (Everaerts et al., 2010; Ferveur, 2005; Yew et al., 2009). Analysis of different population of *D. melanogaster* shows that CHC are under sexual selection and correlate with female mate choice (Grillet et al., 2012). The predominant CHC displayed by males is 7-tricosene (7-T), which females only make in small amount (Jallon, 1984). This conspicuous sexual dimorphism led some to propose that it acts as a female attractant. Evidences for this are ambiguous. Females of most populations have a shorter copulation latency with males naturally abundant in 7-T than with males with low amounts; moreover, copulation latencies become indistinguishable if the low 7-T male in the presence of synthetic 7-T added on blotting paper suggesting that 7-T increases male mating success (Scott, 1994). The discovery of a mutation that affects the CHC profile of males, *desaturase-1* (Marcillac et al., 2005a), reducing the amount of 7-T further supports the role of 7-T as a female attractant, as such males are disfavoured over the wild-type controls by females (Grillet et al., 2006). However, *desat1* does not only affect 7-T level, but levels of most of the 30–50 CHC made by males (Marcillac et al., 2005a). So the causal link between 7-T and female preference remains to be demonstrated. Furthermore, a more recent report indicated no correlation between female mate choice and levels of 7-T (Scott et al., 2011) and not all *D. melanogaster* populations have high 7-T males and females with corresponding preferences (Grillet et al., 2012; Haerty, Jallon, Rouault, Bazin, & Capy, 2002; Scott, 1994). So the function of 7-T remains an open question.

A way to link 7-T to female receptivity would be to identify the receptor for this compound. One candidate is the putative taste receptor gene *Gr32a*, which is expressed in the taste system of the legs and mouthparts; however, females mutant for this receptor show no defect in receptivity (Miyamoto & Amrein, 2008). Females do not make much contact with males prior to copulation, lessening the likelihood that 7-T is detected via the Gr32a receptor for precopulatory mate choice. However, 7-T has recently been shown to have some degree of volatility and to be deposited onto the substrate on which males walk (Farine et al., 2012). It is thus possible that 7-T is detected by females during courtship either via an olfactory receptor in its volatilized form or by a taste receptor when on the ground (Farine et al., 2012; Sakurai, Koganezawa, Yasunaga, Emoto, & Yamamoto, 2013). This is further supported by the observation that females whose antennae were surgically removed mate faster with *desat1* males than control males, indicating that an intact olfactory system is important for CHCs and perhaps 7-T discrimination (Grillet et al., 2006).

### 3.2.2.2.1 First-Order Olfactory Neurons Influencing Female Receptivity The third antennae segment harbors hair-like structures called *sensillae* that house first-order olfactory receptor neurons. There are three morphological types of sensillae: basiconic, coelonic, and trichoid (Vosshall & Stocker, 2007). To generalize, trichoid sensillae respond to fly odors (Clyne, Grant, O'Connell, & Carlson, 1997; der Goes van Naters & Carlson, 2007), basiconic to fruit odors (de Bruyne, Foster, & Carlson, 2001), and coelonic to amines and acids (Silbering et al., 2011). The olfactory system is made of thousands of ORNs, most of which expressing one of 62 odorant receptors (Vosshall & Stocker, 2007) or one of 66 ionotropic receptors (Rytz, Croset, & Benton, 2013). ORs are expressed in trichoid and basiconic sensillae and Irs are expressed in coelonic sensillae. Sensory neurons expressing OR or Irs send projections to different zones, called *glomeruli*, of a structure called the *antennal lobe*. Each glomerulus is an area of dense synaptic connection between first-order olfactory neurons and second-order neurons called projection neurons. Each glomerulus receives innervation from only one type of receptor neurons resulting in the anatomical segregation of signals from different receptors (Avila, Bloch Qazi, Rubinstein, & Wolfner, 2012; Couto et al., 2005; Fishilevich & Vosshall, 2005; Rytz et al., 2013; Silbering et al., 2011). Physiological and anatomical evidences indicate that a fraction of the olfactory system responds to odors emitted by flies. The physiological response of trichoid sensillae to fly odors (der Goes van Naters & Carlson, 2007), as well as the sexual dimorphism in numbers between males and females (Shanbhag, Müller, & Steinbrecht, 1999; Swanson, Wong, Wolfner & Aquadro, 2004), make them good candidate for housing receptors for male substances that increase female receptivity. More specifically, three glomeruli—DA1, Va1v, and VL2a—have sexually dimorphic size; they are bigger in males than females (McGraw, Gibson, Clark, & Wolfner, 2004; Stockinger, Kvitsiani, Rotkopf, Tirián, & Dickson, 2005). These glomeruli are innervated by the trichoid ORN Or67d, Or47b, and the coelonic ORN Ir84a (Couto et al., 2005; Fishilevich & Vosshall, 2005; Grosjean et al., 2011).

The functional meaning of this sexual dimorphism remains unknown, but it attracted investigation in the possible function of the olfactory receptor expressed in these neurons as sex-pheromone receptors.

The trichoid sensilla expressing the Or67d gene had been identified as the receptor for cVA (Ha & Smith, 2006), which we saw previously is an aggregation pheromone produced in the reproductive tract of males and transferred to females during mating (Section 3). A genetic knock-out of this receptor had the unexpected effect of reducing female receptivity, while enhancing male courtship toward other males (Kurtovic et al., 2007). A role for Or67d in female receptivity is further supported by the observation that physiological hyperactivation of the Or67d neuron by genetic means increases female receptivity (Ronderos & Smith, 2010). Wild-type Or67d thus normally promotes female receptivity, while blocking male courtship, in response to cVA. The physiological response to cVA is abrogated in Or67d mutants, indicating that Or67d is a receptor for cVA and that cVA is a male attractant for females. That cVA is a male compound that stimulates female receptivity was slightly unexpected because cVA is made by males in their ejaculatory bulb and it remains unclear whether males display this compound on the surface of their cuticle prior to mating (Everaerts et al., 2010; Yew et al., 2008). Nevertheless, cVA is the only identified male chemical attractant so far with a cognate receptor.

The successful identification of an olfactory receptor affecting female receptivity based on anatomical data spurred research into the function of Or47b, which innervates sexually dimorphic glomerulus Va1v. Knock-down of Or47b reduces female receptivity, identifying this receptor as a sensory input for this behavior (Sakurai et al., 2013). Unlike Or67d, the ligand for Or47b remains unknown. One complication is that Or47b responds to both male and female odors (der Goes van Naters & Carlson, 2007) giving no clue as to the nature of the ligand.

The last of the olfactory receptors innervating a sexually dimorphic glomerulus is Ir84a, which connects to glomerulus VL2a. The ligand of this receptor opens up a new concept in the control of sexual behavior. Ir84a responds to phenylacetic acid and phenylacetaldehyde, which are aromatic substances produced by fruits (Grosjean et al., 2011). Males court females at higher intensity in the presence of food, an effect that can be recreated by exposing flies to phenylacetic acetic. Genetic knock-out of Ir84a in males results in lowering of courtship intensity in the presence of phenylacetic acetic. This specific fruit odor is thus an aphrodisiac for males, which fits well with the idea that the life cycle of *Drosophila* is closely connected to its food source. Unfortunately, no receptivity phenotype was uncovered in Ir84a mutant females leaving it open whether Ir84a participates in female receptivity (Grosjean et al., 2011).

**3.2.2.2.2 Effect of Support Sensillae** Once an olfactory receptor is activated by an odorant, it needs to be quickly inactivated in order to gain temporal resolution. Inactivation of odorant signals can be mediated by odor-degrading enzymes secreted by support cells into the lymph that fills the olfactory sensillum. The carboxylesterase *esterase-6* (*est-6*) is expressed by support cells into sensilla and is known to hydrolyze cVA in vitro. This suggests that est-6 may be an odor-degrading enzyme that regulates how long cVA stays on the sensillae. The response duration of sensilllae that express Or67d to cVA is lengthened in *est-6* mutants indicating a problem with signal termination (Chertemps et al., 2012). The expression of *est-6* is much lower in the female antennae than in the male antennae (Chertemps et al., 2012). It is therefore possible that the sexually dimorphic expression of Est-6 affects the sensitivity to cVA and participates in the different effect of cVA on the sexes. This hypothesis however remains to be tested.

A role for an odor-degrading enzyme has been evoked for female mate choice. Besides its role in hydrocarbon synthesis in the oenocyes, *desat1* is also expressed in supporting cells of the trichoid sensillae (Bousquet, Nojima, Houot, Chauvel, Chaudy, & Dupas, 2012; Marcillac et al., 2005b). Expression of *desat1* in the antennae is necessary for perception of sexually dimorphic CHC by males (Bousquet et al., 2012; Grillet et al., 2012; Marcillac et al., 2005b). The mechanism through which *desat1* operates is unknown, but it may be acting in a dynamic interaction between 7-T or other CHCs with their cognate olfactory receptors in a manner similar to that of Est-6 (Bousquet et al., 2012). Perhaps it is affecting pheromone perception in females, as *desat1* mutant females appears less able to discriminate between *desat1* mutant males and control males (Grillet et al., 2006). The exact function of odor-degrading enzymes in female receptivity remains an open question, but answering it may hold the key to understanding variation in female preference and even incipient speciation (Grillet et al., 2012). Females from different populations may, for instance, degrade attractive male CHC at different rate affecting the length of time this compound remains associated with its cognate receptor.

**3.2.2.2.3 Second-Order Neurons Influencing Female Receptivity**
We have seen that second-order neurons of the auditory system may affect the female processing of song from different species (Tootoonian et al.,

2012). Perhaps a similar change in neuronal coding happens in the olfactory system to determine sexually dimorphic response to sex pheromones? The Or67d receptor neurons responsive to cVA sends projections to the DA1 glomerulus of the antennal lobe, where it makes synaptic contact with DA1 projection neurons (second-order neurons) (Datta et al., 2008; Ruta et al., 2010). Despite sexual dimorphism in size and the differential effect of cVA on male and female receptivity (inhibitory in the former and stimulatory in the latter), no difference in the electrophysiological response to cVA was recorded between males and females second-order neurons (Datta et al., 2008). This indicates the absence of sex-specific processing of cVA information at the level of first- or second-order neurons. We will see in the next section that differences in this circuit emerge at the level of third-order neurons.

The lack of sexually dimorphic difference between second-order neurons responding to cVA in males and females might not extend to all the olfactory system as revealed by analysis of the *spinster* (*spin*) gene. Mutation in *spin* causes virgin females to reject courting males and rarely mate despite being actively courted (Nakano et al., 2001; Sakurai et al., 2013; Suzuki, Juni, & Yamamoto, 1997). The pattern of spin female rejection resembles that of a unreceptive virgin rather than that of a female being in a postmated state, indicating that the mutant phenotype is due to a problem with the perception of the male quality and not due to the female being in a postmated unreceptive state (Nakano et al., 2001). *spin* is expressed in different female tissues and affects different processes such as oogenesis and neuronal survival (Nakano et al., 2001; Sakurai et al., 2013). Receptivity of *spin* mutants can be rescued by expressing wild-type *spin* in neurons only, indicating that receptivity issues are caused by nervous system defects (Sakurai et al., 2013). *spin* is expressed in six clusters of neurons in the brain, two of which, spin-A and spin-D, are necessary for female receptivity (Sakurai et al., 2013). The spin-D cluster contains interneurons that project to six olfactory glomeruli, including Or47b and Or98a, which are both necessary for female receptivity (Sakurai et al., 2013). No gross anatomical defects were observed in these second-order neurons in *spin* mutants. *spin* is a component of the signaling pathway associated with the mTOR pathway that composes the core system for the homoeostasis of nutritional states (Nakayama, Kaiser, & Aigaki, 1997; Ribeiro & Dickson, 2010; Vargas, Luo, Yamaguchi, & Kapahi, 2010). One role of *spin* in controlling receptivity could be to affect signal transduction in those second-order neurons in response to nutritional state in a female-specific fashion (Sakurai et al., 2013).

### 3.2.3 Neuronal Network Underlying Female Receptivity

We have seen that a variety of signals are used by females to gauge the identity of their Courters, and that some of the sensory structures detecting these signals have been identified and their neuronal projections traced back to second-order neurons in the brain. However, these sensory structures do not explain, by themselves, how a female chooses a male. This is illustrated by the observation that males that produce song but abnormal CHC, or abnormal CHC but normal song, may achieve mating, but males which produce neither will always fail to mate (Rybak et al., 2002). This indicates that females integrate both these male sensory cues and that not one of them is absolutely required for receptivity. Therefore, female receptivity must be controlled by an area of the brain that not only receives input from several sensory modalities but also is capable of integrating this information and projecting to motor neurons that control copulation acceptance.

The brain center controlling female receptivity has been broadly mapped using sex mosaic animals called *gynandromorphs*. In *D. melanogaster*, individuals with two X chromosomes with a normal complement of autosomes are female, whereas individuals with one X chromosome are male. Genetic tricks can be employed so that female embryos randomly lose one of their X chromosome in some cells during development, becoming a mosaic of male and female tissues (Hall, 1979). Such gynandromorphs can be tested for their receptivity to copulation with males, and then sacrificed to determine which portion of their nervous system has to be female for the fly to be receptive. Such an approach revealed that to be receptive, a gynandromoph needs to be genetically female in a pair of bilaterally symmetrical clusters in the dorsal anterior brain (Szabad & Fajszi, 1982; Tompkins & Hall, 1983). This part of the brain is not a sensory center, but rather contains neurons that receive input from them. The first lesson from this finding is that first-order sensory neurons do not have to be female for a fly to be receptive to copulation with males. This matches well with the observation that mutations affecting sensory function do not block female receptivity, but rather reduce the probability of mating. Multiple male cues detected by different sensory systems must be integrated by a higher order receptivity center. This receptivity center was surmised 30 years ago to represent parts of the female brain that integrate input from receptors stimulated by the courting male's sex pheromones and song. Do we know more now? The answer is not really, but recent reports indicate progress in that direction.

Fine-grain analysis of sexual dimorphism in the brain has been made possible by analysis of the *fruitless (fru)* and *doublesex* genes, which are pivotal regulators of sex determination. Mutations affecting *fru* result in severe defects in all aspects of male courtship leading to male behavioral sterility, but have no effect on female behavior (Billeter, Rideout, Dornan, & Goodwin, 2006a; Villella & Hall, 2008; Yamamoto & Koganezawa, 2013). A class of *fru* sex-specific transcripts are only translated in males giving rise to the male-specific Fru$^M$ proteins (Lee, Foss, Goodwin, & Carlo, 2000; Usui-Aoki et al., 2000). Lack of translation of these transcripts in females explains the lack of female-specific *fru* mutant phenotypes (Demir & Dickson, 2005). This lack of female phenotype should make neurogenetic analysis of *fru* useless for understanding female behavior. However, *fru* sex-specific transcripts are expressed in about 2000 neurons in both males and females (Lee et al., 2000; Ryner et al., 1996; Stockinger et al., 2005; Usui-Aoki et al., 2000) and blocking synaptic transmission in these *fru*-expressing (*fru$^+$*) neurons affects sexual behavior in both males and females (Kvitsiani & Dickson, 2006). Silencing *fru* neurons in virgin females makes them unreceptive and triggers egg-laying, while it blocks male sexual behavior (Kvitsiani & Dickson, 2006). Therefore, although *fru* is not translated in females and is thus unlikely to have a function, its expression acts as a marker for neurons that influence sexual behavior in both sexes.

What explains the difference in behavior between the sexes if the same populations of neurons are involved? An extensive survey of the expression pattern of *fru*-expressing neurons in males and females reveals that *fru* is expressed in sensory neurons, interneurons, and motor neurons suggesting that the *fru*-expressing neurons form a connected circuit controlling sexual behavior from sensory input to motor output (Billeter & Goodwin, 2004; Cachero, Ostrovsky, Yu, Dickson, & Jefferis, 2010; Kimura, Ote, Tazawa, & Yamamoto, 2005; Manoli et al., 2005; Stockinger et al., 2005; Yu, Kanai, Demir, Jefferis, & Dickson, 2010). Finer grain analysis by warping high-resolution tracings of individual neurons onto a common reference brain indicated that one-third of neurons expressing *fru* are sexually dimorphic (Cachero et al., 2010). If *fru* is indeed expressed in neurons forming an interconnected circuit, differences in neuronal morphology between males and females imply differences in interconnectivity that would change pathways of neural processing, thus causing sexual dimorphic behaviors (Cachero et al., 2010). The mechanisms that underlie these differences involves, among other things, differential cell death of neurons during development, whereby expression of the *fru$^M$* isoform in males protect some neurons from dying, while those perish in females (Kimura et al., 2005; Kimura, 2011). Most of this dimorphism is in the projection patterns of third-order neurons innervating the protocerebrum (Cachero et al., 2010; Yu et al., 2010), in the same broad area where Tompkins & Hall (1983) mapped the center for female receptivity. In keeping with the prediction that a center for receptivity should integrate many sensory signals, this area of the brain receives input from most sensory systems (Yu et al., 2010). Unfortunately, the function of these neurons in female behavior has not yet been investigated.

How does a change in morphology of third order neurons affect differences between males and females? Activation of *fru* neurons that control wing extension in males also activates wing extension in females (Clyne & Miesenböck, 2008). Sexual dimorphism in the sensory system, such as in cVA sensing neurons, does not show differences in physiological response between the sexes and thus does not explain differences in sex-specific behavior (Datta et al., 2008; Kurtovic et al., 2007). The neural pathways for sensory processing and motor output may not be fundamentally different between males and females. Recent experiments indicate that males and females may fundamentally differ in behavior because of sexual dimorphism in third-order neurons. Third-order neurons in an area of the protocerebrum called the lateral horn are placed in the right way to receive input from the olfactory, gustatory, auditory, and visual neurons (Cachero et al., 2010; Yu et al., 2010). cVA-responsive PNs send projection to those sexually dimorphic third-order neurons. This includes two sex-specific populations of third-order neurons which are both connected to cVA-responsive PNs in male and female brains but come from different developmental origins and have different dendritic placements (Cachero et al., 2010; Kohl, Ostrovsky, Frechter, & Jefferis, 2013). These male- and female-specific populations of lateral horn neurons develop under the control of *fru*, whereby expression of male $Fru^M$ proteins results in the presence of the male group and absence of $Fru^M$ results in the presence of the female group (Kohl et al., 2013). The multiple dimorphic targets of cVA olfactory input could explain how a pheromone that acts through the same sensory inputs may elicit different behaviors in the two sexes (Kohl et al., 2013). The sex-specific projection differences generated during development may alter a circuit, which has shared elements between males and females. It has, so far, not been possible to test the behavioral consequences of the sexual dimorphism of these two groups of male and female neurons. One might argue that changing the sex of a few neurons could never suffice to change the sexual behavior of a female. However, masculinization of a small group of neurons located in a region intrinsic to the lateral protocerebral complex—called P1 neuron—can generate low levels of courtship behavior in an otherwise female fly (Kimura, Hachiya, Koganezawa, Tazawa, & Yamamoto, 2008). There are thus reasons to believe that manipulation of a small group of higher order neurons might be sufficient to observe behavioral differences.

The VLP, located just below the lateral horn, may also act as another multimodality processing center because it receives input from the sound, taste, and visual systems (Kamikouchi et al., 2009; Lai et al., 2012; Miyamoto & Amrein, 2008). Because males and females respond differentially to the courtship song, males move faster, and females slow down (Crossley et al., 1995), it is likely that the sexual dimorphic pattern of connectivity observed in the lateral horn also extend to this region.

Differences in nervous system development are not solely under the control of *fru*. Another gene called *dsx* is also differentially expressed between the sexes. *dsx* is expressed in a variety of tissues, including the pheromones producing cells, gonads, genitals, and also the brain (Rideout, Dornan, Neville, Eadie, & Goodwin, 2010). Blocking neuronal communication in *dsx* neurons in the brain resulted in reduced receptivity in females, and attempt to dislodge males during mating (Rideout et al., 2010), indicating a fundamental function of *dsx* neurons in female receptivity. *dsx* expression overlaps that of *fru* in several but not all neurons (Billeter, Villella, Allendorfer, Dornan, Richardson, & Gailey, 2006b; Lee, Hall, & Park, 2002; Rideout, Billeter, & Goodwin, 2007; Rideout et al., 2010; Robinett, Vaughan, Knapp, & Baker, 2010). *dsx* is again heavily expressed in the protocerebrum, where it participates in sexually dimorphic neuronal development together with *fru* (Kimura et al., 2008; Rideout et al., 2010). Because the expression of *dsx* in the brain is relatively sparse (Lee et al., 2002; Rideout et al., 2007, 2010), dissection of the function of *dsx*-expressing neurons controlling female receptivity offers great prospect for identifying the center for female receptivity.

The functional significance of differences in third-order neuronal connectivity in the anterior dorsal protocerebrum for female receptivity has not yet been tested. If this area of the brain is indeed the center for receptivity, one could expect that several of the genes involved in female receptivity are expressed in neurons connected to this area. Projection neurons from *spin*, which innervate olfactory neurons involved in female receptivity, do project to this region (Sakurai et al., 2013) lending further support to the hypothesis that the anterior dorsal protocerebrum is one of the sites for female receptivity. So far all of this is conjecture but as the tools to refine analysis of neurons expressing *fru*, *dsx*, and other genes affecting female receptivity become available, we expect that it will not be long before this hypothesis can be tested. If anatomical differences in third-order neurons do indeed have a functional effect on receptivity, it will support the idea that the difference between male and female behavior is largely connected to the way their brains select and suppress responses to the same stimuli (Cachero et al., 2010; Kohl et al., 2013).

### 3.2.4 Do Genes Affecting Female Receptivity Indicate the Neuronal Substrate for Female Receptivity?

We have seen above that the study of the *fru* and *dsx* genes opened a window to the brain centers that control sexually dimorphic behaviors. However, these genes have so far mostly been studied in the context of male behavior. In anticipation of more studies on the function of *fru* and *dsx* neurons in female behavior, can we learn something from other genes known to affect female receptivity?

The *icebox* (*ibx*) genetic variants causes lowered mating receptivity, without apparently causing other behavioral defect (Kerr, Ringo, Dowse, & Johnson, 1997). This mutation was later found to be an allele of *neuroglian*, a gene encoding cell adhesion molecules critical for nervous system development (Carhan, Allen, Armstrong, Goodwin, & O'Dell, 2005). Examination of the brain of *ibx* mutants shows severe defects in neuronal development affecting several brain structures. Those defects are observed in both males and females. The mutation thus seems to have a larger involvement in nervous system function and may not say much about the specific control of female receptivity.

*Chaste*-mutant females resist male courtship, decamping and flying away from them more than wild-type females, and have reduced probability of mating with males (Juni & Yamamoto, 2009). *Chaste* is an allele of *muscleblind* (*mbl*), a gene involved in the development of neurons and muscles. Since the Mbl protein is known to regulate splicing and cell death, perhaps a reduction in *mbl* expression leads to developmental deficits in the neural circuitry, underlying female sexual receptivity. So far this does not say much about the specific control of female receptivity.

The *retained/dead ringer* (*retn*) mutation causes a reduction in females receptivity, while *retn* males are behaviorally normal (Ditch et al., 2004). *retn* is expressed in the mushroom bodies (MBs), suboesophageal ganglion, ventral ganglion, and developing photoreceptors. *retn* affects the development of sex-specific neurons and their path-finding ability (Ditch et al., 2004). There is however no information yet on the function of neurons expressing this gene and their connection with female receptivity, again not giving much information about how female receptivity is controlled.

*dissatisfaction (dsf)* female mutants are also unreceptive to males (Finley, Taylor, Milstein, & McKeown, 1997). If they mate, they try to dislodge copulating males by flicking their wings, bucking, and kicking at the male, which closely resembles the effects of silencing *dsx* neurons (Rideout et al., 2010). Although *dsf* females ovulate and mature eggs reach the uterus, they remain there unfertilized and degenerate. The 2–4 motor neurons that innervate the uterus in wild-type females do not reach the uterus in *dsf* females, indicative that *dsf* controls neuronal development. *dsf* encodes a tailless-like nuclear receptor (Finley et al., 1998). The *dsf* expression pattern in the protocerebrum of females is within the anterior dorsal protocerebrum focus for female receptivity; *dsf* might be expressed in at least some of the neurons involved in female receptivity and control their sex-specific development as it does for neurons in the abdominal ganglion (Finley et al., 1998). Unfortunately as for *retn*, no further experiments have been performed on this gene.

Mutations in the *painless (pain)* gene increase sexual receptivity (Sakai, Kasuya, Kitamoto, & Aigaki, 2009). *pain* encodes an ionic channel involved in thermal, mechanical, and chemical nociception (Tracey, Wilson, Laurent, & Benzer, 2003). Expression of *pain* in sensory organs suggests that this mutation may affect perception of male cues that reduce female receptivity, such as species-specific cues.

The few paragraphs above are a list of genes that affect female receptivity. They however do not tell us much about how female receptivity functions. Most of the genes have been discovered over a decade ago, yet have very few publications linked to them. This paints a rather bleak picture of progress made in understanding the genetics of female receptivity. However, recent work on the *spin* gene offers a new methodology to delve deeper in this question, which we hope will resurrect the study of those other female receptivity genes. We have seen in the previous section that the *spin* mutation leads to enhanced mate refusal (Suzuki et al., 1997). Like many of the genes mentioned above, *spin* affects female receptivity. Its wide expression in glia, neurons, and reproductive organs does not give a clear insight in the tissues controlling receptivity. To identify the central nervous system neurons that control sexual receptivity, Sakurai et al. (2013) produced mosaic animals in which a small number of *spin*-mutant homozygous cells were created at random in an otherwise *spin* heterozygous brain. They screened over 900 such mosaic virgin females and selected those that reject courting males. Using histological methods, they identified neurons that were *spin* mutant vs those that were *spin* wild-type and correlated the genotype of those cells with mutant vs wild-type female behavior. Correlating spin mosaic female behavior and histology allowed the authors to identify clusters of cells that were important for female receptivity. Out of the broad expression of *spin* in the brain, expression in two clusters of neurons located in the suboesophageal ganglion, and in the dorsal anterior protocerebrum is required for receptivity. We hope that this approach, which is a refinement on the gynandromorph studies that initially identified a putative center for female receptivity, will be applied to other female receptivity mutants to further map the clusters for female receptivity.

### 3.2.5 Modulation of Female Receptivity

The sections above described the use of genetics to identify the neuronal substrates for female receptivity, and to identify the neuronal circuitry coding and integrating sensory information. At the beginning of this section, we highlighted the phenotypic plasticity of female mate preference across different social contexts. The circuitry for female preference may be plastic too, and is likely modulated by internal states and environmental conditions. These differences are often mediated by the effect of hormones, neurotransmitters, and neuromodulators that modify neuronal dynamics, excitability, and synaptic efficiency (Bargmann, 2012). In this section, we will discuss the effect of hormones and neurotransmitters on female receptivity, and the modulation of receptivity by the circadian system.

#### 3.2.5.1 Hormones and Neurotransmitters that Modulate Female Receptivity

An important global regulator of female behavior is JH. JH is made in the CA and implantation of active CA into female abdomens shortly before eclosion accelerates the maturation of female receptivity by 1 day (Manning, 1967) suggesting that increasing the levels of JH enables early female receptivity. To further support this, genetic ablation of the CA was found to delay maturation of sexual receptivity (Riddiford, 2012). JH appears to regulate the timing of female sexual behavior and, as we will see in Section 4, postcopulatory changes in female physiology.

The peptide with the most striking effect on female receptivity is probably SIFamide. Females lacking SIFamide are hyperreceptive, mating with males nearly instantly. Both knock-down of this peptide by RNAi and ablation of the four cells that make this peptide in the brain produce these effects (Terhzaz, Rosay, Goodwin, & Veenstra, 2007). What is striking is that those four neurons send projections throughout the female brain, making connections to most neuropils (exception being MB stalks, lobes, and calyces). The neurites entering the antennal lobe arborize in most of the glomeruli, which suggests that the abnormal behavioral phenotypes are associated with olfactory processing defects (Carlsson, Diesner, Schachtner, & Nässel, 2010).

Female receptivity is controlled not only by sensory pathways that detect male traits but also by pathways that detect internal states such as hunger. These correlated responses have to be regulated in concert. One such mechanism that may modify signaling of different neuronal networks is dopamine. Pharmacological depletion of the neurotransmitter dopamine has far-reaching effects such as reduced fecundity (Neckameyer, 1996), receptivity (Neckameyer, 1998), pheromone production, and locomotion (Wicker-Thomas & Hamann, 2008), showing that this molecule can simultaneously control several aspects of female reproductive behavior. Similarly, the monoamine serotonin (5HT) regulates a host of biological processes including sleep and aggression: application of a 5HT-7 receptor antagonist caused decreased female receptivity (Becnel, Johnson, Luo, Nässel, & Nichols, 2011). In the adult female, ecdysteroids are produced in the ovary (Handler, 1982). Ecdysteroid signaling has been implicated in adult processes such as oogenesis (Carney & Bender, 2000) and may coordinate oogenesis and female receptivity (Ganter et al., 2012).

### 3.2.5.2 Looking for a Good Time? Temporal Aspect of Female Receptivity

Female receptivity is dependent on a host of male cues, but internal factors also affect the female's willingness to mate. One such internal cue is time of the day. Reports of the time of highest female receptivity are somewhat contradictory. In one report, female receptivity is highest in the morning, drops around dawn, and increases again during the night (Sakai & Ishida, 2001). In another report, mating activity is lowest at midday increasing to a maximum by late afternoon and remaining high through the night (Tauber, Roe, Costa, Hennessy, & Kyriacou, 2003). Irrespective of the exact preferred time for mating, reports agree that drop in receptivity at certain time of the day is controlled by clock genes expressed in females. Females mutant for the clock genes *period* (*per*) or *timeless* (Hardin, 2011) show no change in mating receptivity throughout the day in constant darkness (Sakai & Ishida, 2001). That *per* carries information about rhythmicity in mating was further supported by studies of species difference in mating rhythms. A transgenic study switching *per* between *D. melanogaster* and *Drosophila pseudoobscura* shows mating rhythms between species are different and controlled by this gene (Tauber et al., 2003).

Observation of groups of flies over several days revealed that male courtship is more intense at night (Hardeland, 1971; Fujii, Krishnan, Hardin, & Amrein, 2007), although isolated males and females typically have higher locomotor activity during the day (Fujii et al., 2007). Males seem to adapt their courtship rhythms to females through olfactory cues. Males with an impaired olfactory system spent less time near females during the night (Fujii et al., 2007; Lone & Sharma, 2012). This indicates that general locomotor activity differences in male–female coupling compared to the activity of male–male or female–female pairs are due to female pheromones perceived by males.

The circadian system modulates emission and perception of pheromones. Male sex pheromones vary in intensity during the day under control of the circadian clock (Krupp et al., 2008, 2013). The experiment has not been done yet, but it is likely that females also express pheromones rhythmically, which might explain differences in male courtship intensity at different times. Another hypothesis that could explain rhythms in female receptivity is linked with oscillation in activity of the olfactory system. The olfactory system has its own clock mechanism, which regulates its sensitivity to various odorants (Krishnan et al., 2001; Krishnan, Chatterjee, Tanoue, & Hardin, 2008). Olfactory responses have highest sensitivity at night, which is also when flies court at the highest intensity.

## 3.3 Conclusion

This section makes it clear that we have little more than circumstantial knowledge about the mechanisms underlying female receptivity. The difficulty in studying female receptivity stems in part from the variety of cues a female uses to choose which male to mate with. Receptivity is a huge complex system involving integration of several sensory inputs with internal states. Adding to this complexity is the prediction that females from different populations and in different environments are bound to have preferences for different male traits, making it very hard to generalize what a given female finds attractive. We are however optimistic about the future of this field of study. Dissection of the individual mechanisms enabling perception of male cues by females is underway, such as how females perceive male courtship song and pheromones. Identification of the first- and second-order neurons involved is allowing tracing the projection of these neurons to deeper part of the brain where sensory integration takes place.

This will eventually allow designing testable hypotheses about the location and nature of the neuronal substrate underlying female receptivity. Once identified, this neuronal substrate can then be tested in different populations of females and in different environments to understand the origin of different female preferences.

In this section, we have examined what leads a virgin female to accept her first mate. After finding the right habitat, mating is but the second step in female reproductive behavior. Next we will review the changes that a virgin female undergoes once she has mated.

## 4. THE POSTMATING RESPONSE

A mated female is fundamentally different from a virgin. Once mated, females exhibit a host of changes at almost all systemic levels. Compared to a virgin, a mated female has a different transcriptome, proteome, pattern of neuronal activity, and exhibits different behaviors. The postmating response (PMR) is a collection of behaviors that are consistently produced by mated females. In the literature, the classic PMR highlights two main behavioral changes: virgin females are sexually receptive and lay few eggs; and mated females exhibit low receptivity and lay many eggs. The PMR also includes changes in other faculties such as feeding and food preferences, ovulation, general activity level, and of course, sperm storage/utilization. How a female behaves after mating can strongly influence offspring quantity and/or quality and, consequently, has strong fitness implications. Despite its clear importance, the mechanisms of female postcopulatory changes in physiology are incompletely understood. However, advances in genetics have enabled manipulation of gene sequence and expression, visualization of neuronal projections, and most recently exploitation of neuronal activity. All these tools have allowed researchers to dive into the mysterious world of female postcopulatory behaviors.

### 4.1 How is It Studied?

The majority of studies investigating the mechanisms of the PMR have centered on postmating receptivity and egg-related behaviors (ovulation and ovipositioning). These studies determined that the ejaculate (the combination of sperm and seminal fluids) produced by males and transferred to female during mating initiates the PMR. Within the seminal fluid are seminal fluid proteins (SFPs), a term which includes all the proteins produced in the male reproductive tract and transferred during copulation. A subclass of

SFP is accessory gland proteins (Acps), which includes only those molecules made within the accessory gland (AG).

To determine which of these proteins are involved with this response, researchers have employed one of two experimental designs. One, a female is singly paired with a male that has been genetically manipulated causing an incomplete male ejaculate. These deficiencies have been both general creating males without sperm or seminal fluid (Chapman, Choffat, Lucas, Kubli, & Partridge, 1996) and specific creating males lacking a particular gene product (Chapman, Bangham, Vinti, Seifried, Lung, & Wolfner, 2003b; Liu & Kubli, 2003; Wigby & Chapman, 2005; Wigby et al., 2011). The other methodology has included injecting fractions of the male ejaculate or synthetic peptides representing those in the male ejaculate directly into the abdomen of females. To test postmating receptivity, females are isolated from the male after mating and are exposed to a novel wild-type male 4–48 h later (Chapman et al., 2003b; Liu & Kubli, 2003; Yang et al., 2009) or exposed to novel virgin males 3–6 h after injection (Chen et al., 1988; Fleischmann, Cotton, Choffat, Spengler, & Kubli, 2001; Saudan et al., 2002; Schmidt, Choffat, Klauser, & Kubli, 1993; Soller et al., 2006). In ovulation tests, females are mated to either wild-type males or males mutant for specific Acps (Chapman et al., 2003b) or injected with a specific ejaculate extract and ovulation is determined either by dissecting the female at various time points (Chen et al., 1988), by squeezing the abdomen 2–3 h later to force ovulated eggs out (Prokupek, Kachman, & Ladunga, 2009; Saudan et al., 2002; Schmidt et al., 1993), or by cooling the female 2 h later, which has the effect of spontaneously triggering egg-laying (Bono, Matzkin, Kelleher, & Markow, 2011; Fleischmann et al., 2001). In oviposition tests, again the females are mated in pairs or injected, placed alone in a food vial, and provided fresh food every 24 h. The number of eggs laid in the food are recorded and compared among groups (Chapman et al., 2003b; Chen et al., 1988; Cole, Carney, McClung, Willard, Taylor, & Hirsh, 2005; Saudan et al., 2002; Schmidt et al., 1993; Smith, Sirot, Wolfner, Hosken, & Wedell, 2012).

## 4.2 The Male Ejaculate and the Postmating Response

Multiple reviews have been written outlining the production, transfer, and effect of the components within the male ejaculate (Isaac, Li, Leedale, & Shirras, 2010; Kubli, 2003; Ravi Ram & Wolfner, 2007; Rodríguez-Valentín et al., 2006; Wolfner, 2009). A recent and comprehensive review outlines the various SFPs that have been identified in many insect species including *D. melanogaster* (Avila, Sirot, Laflamme, Rubinstein, & Wolfner, 2011).

Although over 100 SFPs have been identified, the relationship between most of them and the female response has not been investigated. Here, we review the literature examining the components of the ejaculate and the consequence on postcopulatory female behavior. We highlight the correlated female behavioral changes to single SFPs and focus on their effects. This will provide background for a comprehensive review of the neuronal circuitry supporting the female PMR, reviewed in Section 1.3.4: The Neurobiology of Sperm and Seminal Fluid.

## 4.3 Receptivity, Egg Production, and Egg-Laying: All or Nothing?

The collection of behaviors that are part of the PMR are triggered in concert. Very few examples exist where a single behavioral aspect of the PMR was induced in absence of the others. Since the beginning of the research in this field in the 1960s, sexual receptivity and egg-related behaviors associated with the PMR (oocyte development, ovulation, and oviposition) have almost always been studied together. And over the years, evidence continues to show that components of the ejaculate that decrease receptivity also increase egg production and oviposition. However, with new technologies and novel experimental designs, researchers have uncovered an important distinction. Although females become less sexually receptive directly after mating and may remain so for many days (regaining receptivity when sperm stores are exhausted due to egg-laying), the biological basis of the short- and long-term changes in receptivity are independent processes and may be governed by different SFPs and therefore possibly different physiological processes and neuronal substrates within the female. Females mated to males that did not transfer sperm within their ejaculate were found to exhibit a short-term response in sexual receptivity but failed to remain unreceptive and were likely to remate ~24 h after copulation (Manning, 1967). Based on this finding, the two different biological events were labeled the "copulation effect" for the short-term reduction of receptivity and the "sperm effect" for the long-term reduction in receptivity as it requires the presence of sperm (Manning, 1967). Investigation into the molecular basis of these two responses uncovered a vast array of molecules transferred to females with varying outcomes.

Sex peptide (SP) was first isolated from seminal fluid and identified as the regulator of the PMR in 1988. Chen et al. injected different protein fractions purified from the male AGs into virgin female abdomens and showed that one fraction caused females to behave as if she has mated. That fraction contained a single peptide of 36 aa, which was sufficient to induce rejection behavior typical of mated females such as ovipositor extrusion and reduced receptivity for 24–72 h (Chen et al., 1988). Females mated to genetically modified males that produce and transfer sperm and SFPs but are SP deficient via either knock-out of the gene encoding SP (Liu & Kubli, 2003) or knock-down by RNAi (Chapman et al., 2003b; Fricke, Wigby, Hobbs, & Chapman, 2009; Isaac et al., 2010; Lee, Rohila, & Han, 2009; Wigby & Chapman, 2005) failed to produce the "sperm effect" and only a weak short-term "copulation effect". These experiments showed that SP is the molecular basis to prolonged reduced receptivity. It is important to note here that the time course of the two methods of SP exposure and correlated female receptivity are conflicting. After injection of SP, the pattern of female receptivity corresponded to the "copulation effect" rather than the "sperm effect". This relationship was clarified with the discovery that SP is bound to sperm from which is slowly released over days causing the prolonged reduction of receptivity and thus is the molecular mechanism underlying the "sperm effect" (Peng et al., 2005a). This finding also highlights the important difference between manual injection of specific components and transfer from the male ejaculate. From these experiments, it was concluded that the sperm and copulation effects are induced by different components of the ejaculate and should be regarded as different physiological postmating events.

Both manual injection of SP (Chen et al., 1988) as well as natural reception of SP via the male ejaculate (Avila et al., 2011; Chapman et al., 2003b; Fricke et al., 2009; Liu & Kubli, 2003; Rubinstein & Wolfner, 2013; Wigby & Chapman, 2005) cause an increase in ovulation and oviposition. SP thus not only influences long-term sexual receptivity but also increases egg-related behaviors. Further evidence of the physiological connection between these two behaviors comes from inspection of the different functional domain of SP. Injection of fragments of this 36-aa-long peptide reveals that the decrease in receptivity and increase in egg-laying are both caused by the same region of the peptide. A peptide without the 7 aa at the N-terminus evoked a similar PMR as the entire peptide. Therefore, the 29-aa-peptide containing the C-terminal aa and the disulfide bridge is responsible for both behaviors (Schmidt et al., 1993). It is unlikely that there are different receptors for the two behaviors because injecting females with different fragments of SP evoked either both behaviors or no changes at all (Schmidt et al., 1993). It is also unlikely that SP initiates reduction in receptivity and increase in egg-laying by interacting with different areas of the nervous system. This is because ectopic expression in several different tissues of a membrane-bound form of SP, which can only act on the cells in which it is expressed, always either evoked both behaviors or no changes at all (Nakayama et al., 1997). Furthermore, the PMR is also influenced by DUP99B. Functional assays of both molecules showed that the same aa sequence that is responsible for the effect of SP is responsible for the effect of DUP99B (Schmidt et al., 1993). Therefore, it is likely that SP and DUP99B induce a decrease in receptivity and increase in egg-laying through the same signaling sequence (Saudan et al., 2002), implying again that the triggering of these two aspects of the PMR are not separable.

Other SFPs are thought to act on long-term receptivity and egg-production behaviors through their influence on the amount of SP released from the pool of SP bound to sperm. Although *Acp36DE*-null mutants males transfer normal amounts of sperm to females, females mated to mutants store significantly less sperm than females mated to wild type (Block Qazi & Wolfner, 2003; Neubaum & Wolfner, 1999) and have increased receptivity already 24 h after mating (Neubaum & Wolfner, 1999). Although it has been shown that Acp36DE enters the sperm storage organs (SSOs) before sperm does and is suggested to promote sperm accumulation within these structures, the mechanism of this process remains unknown (Bloch Qazi & Wolfner, 2003). The most parsimonious explanation for the decreased sperm storage and the positively correlated influence on the sperm effect is that the reduction of sperm storage and consequently SP released over time results in a reduced sperm effect PMR. Furthermore, Acps that are predicted to influence the association of SP to the sperm tails and its gradual release have also been shown to influence the sperm effect. An RNAi knock-down screen of candidate Acps-coding genes provided evidence that candidate genes *CG1652*, *CG1656*, *CG17575*, and *CG9997* were involved in the PMR. Females mated to modified males show a significant reduction in long-term PMR as they have increased remating and decrease egg-laying at time points great than 24 hours. (Ram & Wolfner, 2007). As these gene products have been shown to localize SP to the SSOs (Ram & Wolfner, 2009) it is most likely that these molecules influence the PMR through their action on SP release from sperm and not via a direct interaction with a dedicated female receptor for these peptides.

Acps do not always have an effect on the female PMR. Females mated to *Acp29AB*-mutant males have significantly less sperm within their SSOs 4 days after mating compared to females that mated with wild-type males; however, females of both groups showed similar long-term PMR in (Wong et al., 2008). As Acp29AB is not found bound to sperm (Wong et al., 2008) and mutant males sperm is easily dislodged from female SSOs after females remate with a different male (Wong et al., 2008), the peptide may participate in efficient release of sperm from the storage during ovulation. The unaffected female PMR is at first puzzling as other mutants that influence sperm storage do influence such behaviors (as Acp36DE mentioned above). The reported mean number of stored sperm at day 4 was ~300 sperms stored in females mated to controls and 200 in females mated to mutants (Wong et al., 2008). In another study that visualized SP through antibody staining of sperm found in the female storage organs, tails of sperm became "spotty" at day 2 and by day 5, tails look extremely sparse (Figure 1C and 1D in Peng et al., 2005a). Taken together, although there are significantly less sperm in storage on day 4 after mating to *Acp29AB* mutants, the consequential difference in level of SP in the female may be too small to have a physiological effect on the PMR.

From these studies, it is clear that SP is the biological agent within the ejaculate that directly causes the long-term response associated with the female PMR. However, SP is not the end of the story. As we have already seen, molecules that influence the ability of SP to localize to and be cleaved from sperm, proper storage of sperm within the female reproductive tract, and can also affect the female PMR.

### 4.3.1 Short-Term Decreased Receptivity

Although it is clear that the long-term decrease in receptivity is established though SP, the relationship between SP and the copulation effect, the decrease in receptivity seen within the first 24 h after mating, is not clear. Females mated to mutant males lacking SP showed receptivity similar to mated females after 4 h were intermediate between mate controls and virgins after 12 and 24 h, and were indistinguishable from virgins after 48 h (Liu & Kubli, 2003). Confirming these results, females mated to SP knock-out males show a decrease in receptivity at 4 (Peng et al., 2005a) and 6 h after copulation (Fricke et al., 2009). These results suggest that immediately following mating and lasting <48 h, factors associated with mating other than SP must be causing the short-term decrease in remating behavior. However, if SP had no influence on short-term remating, the presence of SP within the ejaculate should not have any bearing on the response of females: females mated to SP+ or SP-null males should all have the same remating rate at any time point within the first 24 h, but this is not the case. The propensity of females mated to SP knock-out males to remate is significantly higher than females that mated with SP+ males suggesting that this peptide does have an effect within the first 6 h of mating (Avila et al., 2012; Chapman et al., 2003b; Fricke et al., 2009; Hasemeyer, Yapici, Heberlein, & Dickson, 2009). It is clear that a portion of SP transferred to females during copulation is not bound to sperm and may be quicker to reach its target in the female thus impacting the short-term response (Peng et al., 2005a). Furthermore, ubiquitously expressing (Aigaki, Fleischmann, Chen, & Kubli, 1991) or injecting SP (Chen et al., 1988) directly into the female abdomen causes females to decrease receptivity within a few hours providing evidence that if SP can reach its cellular target quickly within the female after a natural mating, it could contribute to the short-term effect on mating as well.

Regardless of the relationship between short-term response and SP, other components of the mating experience must contribute to the decrease in receptivity. Compared to the factors that influence the long-term PMR, much less is known about those that contribute within the first 24 h. One Acp has been shown to have influence over the propensity to remate within this window. Wild-type females that were mated to *PEBII* knock-down males were quicker to remate compared to those females mated to wild-type controls (Bretman, Lawniczak, Boone, & Chapman, 2010). Interestingly, this gene also contributes to the formation of the mating plug, a gelatinous secretion deposited in the female reproduction tract during mating. Thus, it is possible that problems in sperm storage (and ultimately SP storage) may be driving the increase in mating as the mating plug could facilitate such a process frequency. However, females mated to these male showed no reduction in progeny indicating that sperm was stored normally. Therefore, an another hypothesis is that the product could be this gene product could be exerting its effects on short-term female receptivity through the mating plug acting on stretch receptors in the posterior end of the uterus, making her unwilling to remate. Alternatively, PEBII could also act as a pheromone acting upon the central nervous system of the female to reduce remating via a plug-independent pathway.

The lack of identified Acps influencing female short-term reduction in receptivity is not due to lack of effort. A number of studies have attempted to locate the specific molecule but continue to come up short. Due to sequence similarities between SP and DUP99B, DUP99B became an excellent candidate for the short-term reduction in receptivity (Saudan et al., 2002). Moreover, investigation into the association of this molecule and sperm showed that DUP99B associates with the head of the sperm and only for first few hours after mating suggesting that this molecule is quickly released into the female (Peng et al., 2005a). Specific genetic manipulations involving the *paired* mutation produces males that lack AGs but develop morphologically normal testes, ejaculatory ducts, and bulb and therefore produce and transfer sperm and molecules made in the ejaculatory duct and bulb (including DUP99B) but no Acps (referred to hereafter as Acp-null mutants) (Xue & Noll, 2000). These males have been used to determine the role of DUP99B in short-term PMR. Females were mated to either mutants or wild-type males and then were presented with a novel wild-type male immediately after initial copulation event (Rexhepaj, Liu, Peng, Choffat, & Kubli, 2003) or 12 h later (Xue & Noll, 2000). Females previously mated to mutants males all mated within 1 h compared to none that had been mated to wild-type males (Xue & Noll, 2000; Rexhepaj et al., 2003). Based on these results and confirmation that these mutants produce normal amounts of DUP99B, the copulation effect must be an Acp-mediated process and DUP99B does not reduce short-term receptivity. This result is surprising as injection of the DUP99B directly into the female abdomen evoked a PMR (Saudan et al., 2002).

A few possible explanations exist for this. First, although Acp-null males produce DUP99B, it could not be confirmed that these mutants transferred the molecule to females during copulation (Rexhepaj et al., 2003). Next, it may take more than 1 h but <12 h for DUP99B to exert its influence. Therefore, specific DUP99B-null males (generated in Rexhepaj et al., 2003 and tested for receptivity 24 h after initial copulation) could be tested at different time points within the 12 h window. To be sure SP did not also influence receptivity, these males should be generated in an SP-mutant background. However, it should not be overlooked that this molecule may indeed not be involved in reduction of receptivity and could highlight the importance of methodology and the innate difference of an organic copulation event versus injection of the ejaculate directly into the female abdomen.

Whatever the relationship between DUP99B and receptivity, DUP99B may still be involved in the female PMR. Rejection behavior (number of ovipositor extrusions within a 10 min period) was also recorded by Rexhepaj et al. (2003). Although females mated to mutant or control males showed equal number of ovipositor extrusions, only those females that were mated to the Acp-null males remated (Rexhepaj et al., 2003). Therefore, counter intuitively, rejection behavior and receptivity to remating may be influenced by different male ejaculate components. Rejection being influenced by DUP99B (or other seminal products produced in the ejaculatory tract or bulb; Takemori & Yamamoto, 2009) but not remating. It is also possible that DUP99B weakly influences all behaviors of the PMR and that the rejection and remating receptivity behavioral tests do not have the same sensitivity for detecting significant effects of this peptide.

The relationship between DUP99B and the copulation effect remains unclear and the molecular components of the ejaculate that exert such a response remain unidentified despite intense investigation (Bellen & Kiger, 1987; Mueller, Page, & Wolfner, 2007; Ram & Wolfner, 2007; Yang et al., 2009). One reason for the lack of identification of SFPs associated with short-term decrease in receptivity might be linked with the way it is studied. For example Short-term receptivity may be more susceptible to social context. In nature, *Drosophila* mate in groups when aggregated on food patches (Markow & O'Grady, 2008; Spieth, 1974, 1993) causing females to be exposed to multiple males and females after their virginal mating. Experiments done in groups of females continuously exposed to males show females remating multiple times within a mating arena containing food (Billeter, Jagadeesh, Stepek, Azanchi, & Levine, 2012; Krupp et al., 2008; Kuijper & Morrow, 2009), which differs from the traditional remating paradigm which isolates the female directly after mating. However, the social experimental design is also not completely natural: females and males are restricted to the enclosed arenas, which may artificially increase the remating rate (see convenience polyandry in next section). However if short-term reduction in receptivity is more sensitive to social context than the long-term receptivity modulated by SP, then the behavioral assays that assess this behavior should reflect that. The switch from virgin-like to maternal state could also be under female control and forcing the female into isolation by removing her from all forms of social context (placing her alone in a vial) could induce the female to transition more quickly than a female in a social setting. Although Acps within the ejaculate may affect this process, the effect is lost due to isolation as male- and female-derived components would be working in the same direction. Future experiments should explore mating females to Acp-mutant males within different social contexts or different assays to explore this possibility.

### 4.3.2 Short-Term Ovulation and Egg-Laying

Short-term ovulation and ovipositioning (<24 h) are governed by male-derived molecules that may not require the presence of SP. Ovulation can be induced by the SFP ovulin (Acp26Aa) and *CG33943*, a protein-coding gene with no predictive domains, which stimulates ovulation 1 day after mating (Herndon & Wolfner, 1995). Ectopically expressing ovulin or either one of its C-terminal cleavage products causes eggs to be released by the ovary and enter the reproductive tract (Heifetz, Vandenberg, Cohn, & Wolfner, 2005). Although the mechanism is unknown, ovulin and/or its cleavage products may elicit egg release either by interaction with neuromuscular target in the lateral oviducts (Heifetz, Lung, Frongillo, & Wolfner, 2000) or via the neurodendocrine system (Heifetz & Wolfner, 2004). Similarly, females mated to knocked-down *CG33943* males fail to induce the short-term mating response as they lay significantly less eggs within the first 24 h compared to female mated to control males (Ram & Wolfner, 2007).

### 4.3.3 Change in Feeding and Food Preference

Another female PMR is a change in eating behavior. SP causes increased food intake by mated females (Carvalho, Kapahi, Anderson, & Benzer, 2006), and specifically yeast (Avila et al., 2012; Fleischmann et al., 2001; Ribeiro & Dickson, 2010; Vargas et al., 2010). The possible functions of this increased consumption of yeast may be related to the high postmating rates of oogenesis and therefore required aa-rich diet (Drummond-Barbosa & Spradling, 2001). However, the depletion of protein resources is not the cause of the PMR preference shift because SP-induced yeast preference is still present in females that cannot produce eggs (Ribeiro & Dickson, 2010). Furthermore, SP has been shown to alter the expression levels of nutrient-sensing genes of the mTOR pathway in females 3 and 6 h after mating (Ribeiro & Dickson, 2010; Vargas et al., 2010). This pathway is thought to link ATP and aa levels within the individual with food consumption behaviors (Itskov & Ribeiro, 2013). The SP signal within the ejaculate most likely modifies the internal nutrient-sensing pathways because experimental modification in expression of genes within this pathway caused an increase in yeast consumption (Ribeiro & Dickson, 2010; Vargas et al., 2010).

### 4.3.4 Change in Activity

A reduction in general activity level has also been noted in mated females (Tompkins et al., 1982). This reduction in movements may have evolved as a mechanism in females to reduce courtship by surrounding males as movement is attractive to males (Section 3) (Tompkins et al., 1982). A virgin female moving away from a courting male may actually encourage continued courtship rather than merely signal rejection. Furthermore, the architecture of the activity of females also changes depending on their mating status. Although both virgin and mated females show a spike in activity during light-transition periods (at dawn and at dusk) as well as low activity levels during dark phase, mated females show an increase in activity during light phase as compared to virgin females (Isaac et al., 2010). When females were mated to SP-deficient males, the activity of the female more closely resembled that of a virgin female with low amount of activity during the light phase compared to females that mated with control males (Isaac et al., 2010). These results were interpreted as virgins benefiting from low activity during the day to minimize risks to environmental conditions such as dehydration, whereas mated females benefit from high locomotor activity to satiate increased nutritional demands generated by increased egg production and the pursuit of appropriate egg-laying site.

The SFPs that females receive during copulation initiate and maintain the postcopulatory changes seen in female behavior. Although SP has received most attention and is associated with most aspects of behavioral change, multiple SFPs have been identified which influence the female PMR. With the vast array of molecular techniques available in the *Drosophila* tool kit and new experimental designs, identification of new SFPs that influence female PMR is still very likely. However, in order to achieve a full understanding of the mechanisms that support the transition from virginal to maternal behavioral patterns, we must determine the targets of the male-derived chemicals and the consequence of their interaction. In the next section, we dive deeper into the physiological postcopulatory changes in the female as we explore how the female central nervous system responds.

## 4.4 Neurobiology of Sperm/Seminal Fluid Effects

The majority of research attempting to understand the PMR has centered on the components of the male ejaculate and much less on the contribution of females to their own response to mating (Section 4.1–4.4). This bias may have been born from the ease of manipulating male ejaculate composition. Each seminal fluid component is a product of a single male gene and can be studied individually (Avila et al., 2011), but the female behavioral response probably includes an array of sensory neurons equipped with different receptors for different male seminal fluids (Hasemeyer et al., 2009; Rezaval et al., 2012; Yang et al., 2009), neuronal circuitry possibly involving neuromodulators (Avila et al., 2012; Heifetz & Wolfner, 2004), and eventually activation of specific behavioral patterns involving motor neurons, all of which is likely to involve the expression of multiple genes with varying effects. The male-biased perspective has created an image of the female as an arena in which male-derived molecules and sperm combat for the grand prize of fertilization privileges. However, data is starting to accumulate indicating that females are not the passive victims of male-generated peptides as once thought. Instead, they are more likely active participants who collect and integrate information from their environment, their mate, and their internal state to transform from virgin to preform maternal behavioral outputs. Therefore, the PMR does not need to be seen as a result of molecule oppression but can also be seen as a decision-making process by the female, where cost and benefits are weighted to optimize female fecundity in different conditions.

Much research has been done to understand sexual conflict; the tug of war that exists between the sexes fueled by different interests and ways in which they increase their fitness (Chapman, 2006). Although the details of the who-is-in-control-of-what debate remain unclear, one thing is certain: the female PMR requires the involvement of both males and females. Therefore, to understand the postcopulatory physiological changes in the female, we must determine not only the genes and corresponding gene products within the male ejaculate (reviewed in Section 4.3), but also the genes that produce the cellular components and neuronal circuitry within the female that the male-derived molecules interact with. Within the last decade, the molecular basis of the female PMR has slowly been unveiled with identification of genomic and proteomic changes that happen during and after copulation (Mack, Kapelnikov, Heifetz, & Bender, 2006; McGraw et al., 2004; Prokupek et al., 2009; Swanson et al., 2004), the female genome and its influence on PMR such as female postcopulatory receptivity (Giardina, Beavis, Clark, & Fiumera, 2011) and offspring production (Chow et al., 2010). This change has fueled new discussions about the female PMR as an interaction between the sexes (Prokupek et al., 2009; Schnakenberg, Siegal, & Bloch Qazi, 2012; Wolfner, 2009). Here we review the relevant literature to identify the female cellular/neuronal substrate that interacts with the male ejaculate.

### 4.4.1 Gynandromorphs, the Central Nervous System, and the Postmating Response

One of the first tools used to investigate the neurobiology of the PMR were gynandromorphs. As we have seen in Section 3.2.3, gynandromorphs can be of great use to determine the areas of the brain involved in female mating behavior. In the context of female PMR, this can be done by identifying gynandromorphs that perform normal PMR and identify which areas of the nervous system are genetically female. If a given area is always genetically female in gynandromorphs that performed a normal PMR and male in the rest of the gynandromorphs, then it is highly likely that this region is associated with this normal female behavior. All gynandromophs that were receptive produced a normal PMR, which may indicate that the neuronal circuitries that support virginal receptivity and remating are the same. For instance, the dorsal anterior brain needs to be genetically female in order to support normal sexual receptivity in a virgin and femaleness of this center also correlated with the control for postmating receptivity because gynandromorphs with feminized dorsal anterior showed a normal PMR (Tompkins & Hall, 1983). However, using gynandromorphs to investigate the neurobiological basis of the PMR has limitations. To study postcopulatory behaviors using flies that are gender mosaics, the focal fly must have a number of prerequisite feminized tissues. First, the focal fly must elicit courtship from a male which demands a feminized abdomen to emit female pheromone (Jallon & Hotta, 1979), be capable of receiving a male which requires female genitalia, and be receptive to male which requires the feminization of some specific brain regions. If the neural circuitry supporting the PMR is located within any of these regions, it would be impossible to unambiguously identify them with the use of gynandromorphs which require a comparison of male and female tissues. The fact that all gynandromorphs that showed a normal PMR had to be receptive as virgins either means the circuitry for pre- and postmating sexual receptivity are the same, or that PMR-specific circuitry is within these required-feminine regions. Although unable to disentangle these issues, gynandromorphy studies showed that virginal receptivity, insemination detection, and egg-related behaviors all require a female central nervous system (Szabad & Fajszi, 1982; Tompkins & Hall, 1983).

### 4.4.2  Sex Peptide Receptor

It is clear that SP and the closely related DUP99B (two members of the SP pheromone family; Ding, Haussmann, Ottiger, & Kubli, 2003) evoke at least a partial PMR in females; however, the question of how the signal produces behavioral changes in females still remains. Radiolabeled or alkaline phosphatase-labeled SP and synthetic DUP99B were injected into females to visualize their association with target female tissue. Both peptides interact with the same regions in the head: strong labeling of the antennal nerve, pharyngeal nerve, antennal mechanosensory center, the periphery of the suboesophageal ganglion, cervical connective, and weak labeling of the antennal lobes; thoracic ganglion; and the genital tract: the oviduct and the uterus (Ding et al., 2003; Giardina et al., 2011; McGraw et al., 2004; Ottiger, Soller, Stocker, & Kubli, 2000; Saudan et al., 2002). However, this type of technique has some potential caveats. SP and DUP99B are normally transferred to the female during copulation within the ejaculate. This accumulation of sperm and SFPs in the female reproductive tract is a controlled process. For example, structural changes of the female reproductive tract and stages of sperm storage within the SSOs follow stereotypic systematic changes (Adams & Wolfner, 2007). Furthermore, both these molecules associate with the sperm and therefore require cleaving in order to be activated. The manual injection of molecules such as SP and DUP99B is not necessarily a similar delivery method and therefore the binding pattern of injected labeled molecules may not be comparable and thus will not reflect that of the postcopulated female. Interested in the binding pattern of SP within a mated female, Ding et al. (2003) sectioned females after mating at different time points and applied the alkaline phosphatase-labeled SP probe to compete with the natural SP acquired by females through mating. If no probe interacted with the tissue sections, then it was assumed that SP from the male was present and blocked the probe. In the uterus, a signal was found only after 7 h, where the pattern and signal intensity of the oviducts and central nervous system remained unaffected suggesting that SP did not interact with these tissues in vivo. Overall, these results show that SP and DUP99B have the potential to interact with many different tissues types throughout the body in order to evoke the PMR in females. However, more experiments investigating the timing of the physiological response of the female and the cellular components of the other male–derived molecules that are known to influence female postcopulatory mating behavior are required. For instance, during the first 24 h after the virginal mating, females housed in groups composed of males of different genetic backgrounds remate more often than females in groups containing males from the same inbred strain (Billeter et al., 2012; Krupp et al., 2008). As SP seems to be the trigger for the short-term PMR, it is possible that females could modify their behavior through modification of SP interactions. Therefore, it would be interesting to determine any changes in SP-binding patterns of a mated female in isolation and within a social context. Alternatively, one other SFP has been shown to influence the short-term decrease in postcopulatory receptivity (Bretman et al., 2010). To investigate if PEBII influences female postcopulatory receptivity through a plug-independent or –dependent pathway, this male–derived molecule could be visualized within the female reproductive tract to determine if it only is associated with the mating plug or if it also interacts with the female central nervous system.

A major turning point in the quest for the cellular target of SP was the identification of its receptor (Yapici, Kim, Ribeiro, & Dickson, 2008). It was identified through an RNAi knockdown screen for neuronal genes required to produce a normal PMR. Females with central nervous system-specific knock-down of *CG16752* laid significantly less eggs after mating and remated significantly more 48 h later compared to controls (Yapici et al., 2008). Knock-down females directly injected with SP were significantly more receptive 5 h later compared to controls (Yapici et al., 2008). The gene, hereafter referred to as *sex peptide receptor (SPR)*, is a G-protein-coupled receptor that is activated by both SP and DUP99B (Yapici et al., 2008). *SPR* is expressed in both the female reproductive tract (the common oviduct and the spermathecae) as well as in the central nervous system (predominately in the suboesophageal ganglion and the ventral nerve cord). Although *SPR* was not found to be expressed in the male reproductive tract, it was found to be expressed in the male central nervous system suggesting that SPR may interact with other ligands and may function in other processes than the female PMR (Yapici et al., 2008). This idea is in keeping with the observation that the PMR can be mapped to a subset of *SPR*+ neurons that coexpress *fru*, a gene which is expressed in neurons that control sexual behavior in both males and females and *pickpocket (ppk)* an ionic channel expressed in sensory neurons involved in taste and mechanosensation (Hasemeyer et al., 2009; Yang et al., 2009; Yapici et al., 2008). With use of neuronal activity manipulation tools, 6–8 peripheral nervous system neurons in the uterus that project to the brain were shown to convey the signal of SP for the PMR (Hasemeyer et al., 2009; Yang et al., 2009). When SP interacts with these neurons, neuronal activity is reduced and this change in activity is most likely the signal to higher order brain centers that the female has mated (Hasemeyer et al., 2009; Yang et al., 2009).

We have seen in Section 3.2.3 that *dsx* is a gene that controls morphological sexual differentiation as well as precopulatory sexual behavior. When *dsx*+ neurons are silenced, females also fail to produce the normal PMR (Rideout et al., 2010). These mated females laid no eggs over a 5 day postcopulation period and remated significantly more than controls. Furthermore, as discussed in Section 3, these females have longer copulation latency, a shorter copulation duration, and move significantly more than controls during this virginal mating (Rideout et al., 2010). Together, the results show that *dsx* is involved in both pre- and postcopulatory female reproductive behavior. The expression of *dsx* overlaps with some *fru* neurons in the CNS, but only a few of the sensory neurons of the reproductive tract, where *dsx* is expressed in 11 neurons and *fru* in 6 (Rezaval et al., 2012). To determine if the neurons previously identified overlapped with *dsx* circuitry females were engineered to express a transgenic form of SP that is tethered to its membrane allow to target the effect to SP-specific cells (Nakayama et al., 1997). When virgin females expressing tethered SP in *dsx+* neurons were exposed to males, females were significantly less receptive, extruded their ovipositor more often, moved less during courtship, and moved less in general compared to controls—a behavior pattern typical of a mated female (Rezaval et al., 2012). Of the roughly 27 neurons that express *dsx* in the Abg, it was determined using an intersectional method that about nine of these neurons coexpress a *Tdc2*, marker for neurons that produce octopamine and project to the female reproductive tract (lateral and common oviduct, uterus, and the SSOs; Rezaval, Nojima, Neville, Lin, Goodwin, 2014). Artificial activation of these nine neurons significantly reduced receptivity and increased ovipositor extrusion and egg-laying—showing induction of the PMR in a virgin female (Rezával et al., 2014). When these neurons are silenced in mated females, reduction in receptivity and increase in egg-laying are not observed, showing that these neurons are both necessary and sufficient for these aspects of the PMR (Rezával et al., 2014).

When SPR was knocked down in *dsx+* neurons, mated females had increased remating, extruded their ovipositor less, moved more during courtship and in general compared to mated controls—a behavior pattern typical of a virgin female (Rezaval et al., 2012). Thus, SPR expression in *dsx+* neurons is necessary and sufficient to produce the PMR. *dsx* circuitry also supports downstream neural pathways projecting from the central nervous system to the reproductive tract. When these neurons are silenced, mated females have a higher propensity to remate and lay significantly fewer eggs compared to mated controls (Rezaval et al., 2012). Unlike the SP-responsive neuronal cluster in the uterus that once associated with SP decreases activity, which induces the PMR, these neurons increase activity after mating to produce the PMR. These neurons originate in the abdominal ganglion (Abg) and project to the SOG and the reproductive tract most likely conveying information about the presence of SP or another SFP (Rezaval et al., 2012).

SPR has also been linked to the feeding change in mated females. Mated females have increased yeast preferences compared to virgin controls. However, mated SPR-null females act as virgins and do not show yeast preference supplying further evidence that food preference is part of the normal PMR. Restoring SPR specifically in *ppk+* rescued the yeast preference in mated females (Ribeiro & Dickson, 2010). Although this preference switch results in consumption of more yeast which is used during egg production, the preference itself is not elicited by ovulation as females unable to make fully developed eggs still show the preference switch (Ribeiro & Dickson, 2010). Moreover, the preference switch also involves the TOR/S6K signaling pathway suggesting that feeding decisions involves internal sensing. As SP has been shown to alter the expression levels of genes of the mTOR pathway (Gioti et al., 2012), and functioning SPR is required for this behavioral change, neurons that express SPR may either modify expression levels of these genes or participate in a signaling pathway to modify the expression in other cells, which modifies the internal nutrient-sensing pathways and results in greater yeast consumption behavior. Changes in feeding behavior are paralleled by changes in digestive tract physiology (Cognigni, Bailey, & Miguel-Aliaga, 2011). The excrement of mated females is more concentrated and acidic, indicating that more proteinaceous nutrients are extracted from the food by mated than virgin females. This is most likely due to the increased nutritional need as a mated female increases their activity (less sleep during the day; Isaac et al., 2010) and the metabolic requirement of egg production. However, this change in physiology is not a direct effect of diverting resources from the gut to the ovaries, as mutant females unable to produce eggs still demonstrate the postcopulatory change. With the use of SP-null males, it was determined that these changes are due to the SP that the females received within the ejaculate (Cognigni et al., 2011).

A recent report suggests that the SPR story may be more complicated than first thought. SPR-mutant females have been shown to be able respond to SP when SP is directly expressed in neurons or when the blood–brain barrier is disrupted suggesting that although SPR is required for the SP response, it may not function as its receptor but as a mechanism to deliver the peptide to the nervous system (Haussmann, Hemani, Wijesekera, Dauwalder, & Soller, 2013). A confirmation of these results with visualization of the binding pattern of SP within SPR females with either a working or disrupted blood–brain barrier would help to clarify these results.

The identification of the SPR and the neurons in the reproductive tract that express different combination of *fru*, *dsx*, and *ppk* responsible for eliciting the female PMR represents the first molecular entity of the neuronal circuitry underlying female postmating behavior. The next step in the signaling pathway of the PMR will be uncovered with the confirmation of their neuronal targets. From this starting point and armed with the genetic tools available in this species, mapping the rest of the circuitry that supports the PMR may not be too far away. Advancement in this field should be guided by experiments that make use of female genetic mutants (reviewed in Section 4.4.4) as well as the gynandromorph results. For example, Szabad et al. (1982) suggested that the area of the central nervous system that detected insemination was most likely in the head. Together with the findings that the neurons in the uterus responsible for the PMR project into the Abg and may target the suboesophageal ganglion (SOG) (Hasemeyer et al., 2009; Rezaval et al., 2012) suggests that the SP signal is sensed in the uterus and relayed to neurons in the SOG that have their cell bodies in the head or relayed again to third-order neurons that do. One main goal of research in this area should be to uncover the uncoupling of the SP signal to identify the unique circuitry that supports the different behaviors of the PMR.

### 4.4.2.1 Unique circuitry

Although the circuitry of the PMR is far from resolution, some results hint that unique circuitry supports the different aspects of the PMR. Within the reproductive tract, *ppk*, *dsx*, and *fru* are coexpressed in different combinations (Rezaval et al., 2012), which provides a tool to access different neuronal populations. For example, silencing *ppk* neurons inhibit egg deposition, presumably by impeding egg transport along the oviducts (Yang et al., 2009), yet females can still lay eggs when activity of *fru*$^+$ neurons is blocked (Yang et al., 2009) suggesting that neurons with different genetic expression patterns direct distinct PMRs.

Recently, *SPR*-mutant females have been found to increase ovipositioning in *dsx*$^+$ neurons via SP (Haussmann et al., 2013). Expressing membrane-bound SP (mSP) in *dsx*$^+$ neurons within a mutant SPR female elevates the level of oviposition but receptivity is not affected. Compared to wild-type virgin baseline of about 5 within 48 h, these females laid 20 eggs. This shows that the presence of mSP influences female reproductive behavior, but the mechanisms such as another receptor for this molecule has not yet been identified. A nice complement to these results would be temporal resolution to this effect. For example, placing mSP expression under the control of heat-shock in *dsx*+ neurons in SPR-mutant females and measuring oviposition rate (Rezaval et al., 2012).

Furthermore, there is some evidence that the circuitry underlying post-copulatory receptivity and egg-laying behavior are at least somewhat independent downstream of Acp-responsive neurons. Females mated to Acp-null males showed no decrease in postcopulatory receptivity at any time points (0.5, 1, 2, 3, 4, and 5 days) suggesting that this behavior is an Acp-mediated process. This may indicate that the decrease in both short- and long-term postcopulatory receptivities is mediated by Acps. However, egg-laying behavior was consistently higher than virgin controls but lower than controls mated to wild-type males at all time points suggesting that egg-laying is influenced by both Acp- and non-Acp-mediated processes. However, it has not been confirmed that the lack of developed AGs actually reduced the production of Acps (Rexhepaj et al., 2003; Xue and Noll, 2000).

### 4.4.3 Neurobiology of Egg Production

Gynandromorph studies were the first to identify gross regions of the central nervous system involved in egg production. Findings from these studies suggest that egg-laying behavior may be controlled by multiple regions of the nervous system that exist both inside and outside of the head. The region(s) of the central nervous system that controls the transfer of eggs from ovaries to uterus, the release of sperm from SSOs, and ovipositioning is not within the head because decapitated females accomplish all these behaviors. However, the rate at which a virgin female lays eggs must exist within the brain as flies with egg-laying capabilities but male brains laid similar number of eggs as a mated female (Szabad & Fajszi, 1982). The regions responsible for this increase may be the MBs as MB-ablated (but otherwise normal) females have a higher virgin ovipositioning but not ovulation rate (Fleischmann et al., 2001).

Another contribution from the gynandromorphs was the identification of JH and the areas of the brain that control its production important in postmating behavior. Szabad & Fajszi (1982) identified a subset of gynandromorphs that produced eggs (ovulate) but did not lay them (oviposit). These specific gynandromorphs were implanted with hormone-producing glands (CA–corpus cardiacum complexes and ring glands) of wild-type females or performed a topical treatment of JH analog and almost all the experimental flies began to lay eggs (Szabad & Fajszi, 1982). It is of interest that neither treatment changed virginal receptivity. Further investigation showed that the sex of the CA–corpus cardiacum complexes were not important for the regulation of ovipositioning.

After mating, SP enters the female reproductive tract and eventually the hemolymph. As the $N$-terminal of SP interacts with the CA–corpora cardiaca complexes to stimulate JH in vitro, it is very likely that this process also occurs in vivo (Moshitzky et al., 1996). It is important to note here that DUP99B, which is most similar to SP at the C-terminus but differs at the N-terminus, does not elicit JH production in vivo. Egg production proceeds based on a balance of JH and 20-hydroxy-ecdysone (20E) by eliciting vitellogenic oocyte progression through stage 9, the putative control point of oogenesis in virgin females. The interaction of SP with the CA–corpora cardiaca complexes may represent the first step in the SP-dependent ovulation-specific pathway. Although SP has been shown to be more or less the general PMR regulator for all associated behaviors, the physiological pathways for each must split at some point. As application of JH analog does not increase ovipositioning or reduce receptivity for ovulation, this may be that point (Kubli, 2003).

As JH is also important for male mating behavior (Wilson, DeMoor, & Lei, 2003), it is likely that this region and its ability to produce JH is not sexually dimorphic, but that the neuronal circuitry supporting either the amount or timing of JH is. One candidate region is the PI as it is known to have sexually dimorphic function (Martin et al., 1999), it has been shown to communicate with CA–corpus cardiacum complexes (Belgacem & Martin, 2002), and cauterizing this region stops mated females from ovipositing (Boulétreau-Merle, 1976). Alternatively, SPR could be more richly expressed in the CA–corpus cardiacum complexes contributing to the increased JH synthesis in females. Although males also express SPR in a very similar pattern to females, males show less prominent binding in the entire thoracic ganglion including where the CA–corpus cardiacum form a complex (Ottiger et al., 2000). This hypothesis can be addressed by confirming the sexual dimorphism expression pattern of SPR in this region.

Ovulation is the process through which eggs are produced by the ovaries and released into the oviducts. Several studies have indicated that the neuromodulator tyramine and its derivative octopamine are involved in the regulation of this process (Cole et al., 2005; Lee et al., 2009; Middleton et al., 2006; Rubinstein & Wolfner, 2013; Tompkins & Hall, 1983). Octopamine is synthesized from tyrosine by actions of tyrosine decarboxylase (dTdc1 and *dTdc2*) and tyramine beta-hydroxylase (Tbh). Mated females, mutant in *dTdc2* and consequently lacking both of these, were found to be sterile because they can release eggs from the ovaries but not deposit them (Cole et al., 2005; Monastirioti, 2003). Octopaminergic and tyraminergic neurons that project from the thoracic-abdominal ganglion innervate the reproductive tissues (Middleton et al., 2006) including the SSOs (Avila et al., 2012) as well as the common and lateral oviducts, adult brain, and the nerve cord (Cole et al., 2005). Octopamine is released through vesicles at buttons and most likely encourages ovulation by contracting the ovarian muscles and relaxing oviductal muscles (Rodríguez-Valentín et al., 2006). The receptor for octopamine that is critical for the increased ovulation after copulation is *oamb* which has two isoforms (K3 and AS) produced by alternative splicing of the last exon. Both isoforms produce female sterility due to egg retention when mutated (Lee, Seong, Kim, Davis, & Han, 2003), both activate an increase in intracellular $Ca^{2+}$ (Balfanz, Strünker, Frings, & Baumann, 2005; Han, Millar, & Davis, 1998; Jallon & Hotta, 1979; McGraw, Clark, & Wolfner, 2008), and K3 stimulates a cAMP increase (Han et al., 1998). Although *oamb* is expressed in the brain, thoracic–abdominal ganglion and the reproductive tract, the critical site for ovulation is the oviduct epithelium (Lee et al., 2009). Although SP has been shown to increase and sustain long-term ovulation rates, the mechanism by which this is accomplished is unknown. However, many of the molecules involved have been identified. Ovulin has been shown to increase short-term egg-laying likely through interaction with these neurons. Ovulin has recently shown to relax oviduct muscles and indirectly cause an increase in octopaminergic synaptic sites most likely through directly or indirectly increasing neuronal activity (Rubinstein & Wolfner, 2013).

After ovulation sperm is released from the SSOs, fertilization can take place. This process also involves the use of neurons expressing tyramine and octopamine. Inhibiting both affected release from both organs, but selectively inhibiting octopamine (by mutation of the gene responsible for the enzyme of chemical synthesis from tyramine to octopamine) inhibited release from the seminal receptacle (SR) (Avila et al., 2012). Future research should determine which neuronal populations they receive their information from or if these neurons can respond to JH, SP, or other SFPs. Therefore, it is possible that neurons that express *SPR* and *dTdc2* represent the beginning and end of the neuronal circuitry supporting the postcopulatory increase in egg production.

Much like the neuronal circuitry underlying egg production, ovulation, and fertilization, the circuitry for the rest of the PMR is slowly revealed bit by bit. Usually, the first clues to the neurons involved in a particular behavior are through the use of single gene mutants that have very obvious mutant phenotypes. Although alone these studies cannot tell us much about the wiring of the PMR, they will help guide the future research.

### 4.4.4 Genetic Mutants and the Neurobiology of the PMR

As the cellular substrate of the PMR is only partially elucidated and the process in which information is processed remains elusive, genetic mutants that show abnormal PMR either after a natural mating or injection of SP can be used as tools to identify further subsections of the neuronal circuitry underlying the PMR.

Females mutant for *dunce*, a gene that encodes a cAMP phosphodiesterase, have a higher remating propensity, do not lay eggs, and do not respond to SP (Chapman et al., 1996). Furthermore, expressing the wild-type *dunce* allele prior to SP injection partially rescues the PMR suggesting that this gene is involved in the physiological response to SP but not the development of the neural circuitry that supports its response (Fleischmann et al., 2001). Because the SP signal must be integrated in *dnc+* neurons, and the wild-type *dnc* allele is expressed in (but not exclusive to) the MB (Nighorn, Healy, & Davis, 1991), researchers set out to determine if this brain structure was a vital hub in the neuronal circuitry supporting the SP signaling cascade. MB-ablated females injected with SP, however, showed the normal PMR of increase ovulation/oviposition, and decreased receptivity (Fleischmann et al., 2001). Therefore, the SP signal from the male ejaculate that eventually increases egg-related behaviors and decreases sexual receptivity does not require this brain structure. Combining these results suggests that (1) SP alleviates this suppression without the use of the MBs and (2) *dnc* is involved in the SP-dependent PMR in another part of the nervous system.

Mutations in other genes that produced females with a phenotype similar to that of *dnc* (decrease in oviposition but not ovulation or egg-retainers and increase in sexual receptivity) such as *egghead* (*egh*) have also been fruitful in providing insight into PMR circuitry. Like *dnc*, females mutant for *egh* do not respond to SP (Soller et al., 2006). When *egh* is selectively rescued in a subset of *egh*-mutant neurons in the ventral nerve cord that coexpress *Apterous (Ap)*, mutant females were able to respond to SP as witnessed by decreased receptivity and increased oviposition after SP injection. These interneurons form a common fascicle and project from the ventral nerve chord to the central brain. The behavioral data was supported by anatomical findings, which showed that *egh* mutants exhibit innervation defects in $Ap^+$ neurons connecting to the central brain. Perhaps they are *dsx* neurons that have recently been shown to be critical for mediating PMRs. These *dsx* neurons lie in the abdominal ganglion (*Abg*), some of which target posterior regions of the brain (Rezával et al., 2012). As these neurons are afferent interneurons, they most likely receive information from sensory neurons (possibly the *dsx+*, *ppk+*, and *fru+* neurons that express SPR in the reproductive tract) and bring information toward the central nervous system (Rezával et al., 2012).

Finer grain mapping could be achieved by analysis of other genes necessary for the PMR. Virgin females that overexpressed *sarah* (*sra*; previously called *nebula*), *norpA*, *broad*, *grapes*, or a targeted candidate gene (*CG11700*, *CG4612*, *CG30169*, and *CG3961*) within the nervous system with use of the Gal4–UAS system had increased ovulation rates compared to controls (Ejima et al., 2004). Again, overexpression of *sra* within the central nervous system of virgin females caused mated-like behavior: low receptivity and rejection behavior including ovipositor extrusions. Using genetic mutant studies, researchers found that females homozygous for a hypomorphic allele had abnormal egg-production at both virgin and mated states. Compared to wild-type virgins, mutant females had higher ovulation rates; however, compared to mated wild-type females, mated mutant females had lower ovulation rates and could not produce viable eggs (Ejima et al., 2004). *sra* is expressed in the central nervous system of larvae and in the brain and nurse cells of the oocytes of adults (Ejima et al., 2004). The protein product of *sra* interacts with CanB2 and Pp2B-14D (the calcineurin regulator Sra plays an essential role in female meiosis in *Drosophila*, Takeo et al., 2006) and involved in the egg development.

Finally, a long list of genes have been identified by microarray analyses that are either up- or downregulated in mated/unmated flies or females that have been exposed to a specific Acp (Domanitskaya, Liu, Chen, Kubli, 2007; Fedorka, Linder, Winterhalter, & Promislow, 2007; Gioti et al., 2012; Innocenti & Morrow, 2009; Kapelnikov, Rivlin, Hoy, & Heifetz, 2008; Mack et al., 2006; McGraw et al., 2004, 2008; Peng, Zipperlen, & Kubli, 2005b; Prokupek et al., 2009; Short & Lazzaro, 2013). However, the generation of lists like these does not provide much insight on how the postcopulatory changes are generated. Genes important for the construction of the neuronal circuitry, for example, are most likely expressed during development long before the female has any interactions with conspecific males. Similarly, other approaches such as association studies have provided evidence that there are female genotypic effects for female PMR (Clark, Aguadé, Prout, Harshman, & Langley, 1995; Clark, Begun, & Prout, 1999; Fiumera, Dumont, & Clark, 2005, 2007; Civetta, Rosing, & Fisher, 2008). Although these experiments do not show causational relationship between a candidate gene and the female PMR, they do suggest that genes within a specified region may be involved. Once these genes are identified and their role in PMR has been confirmed, they will be most helpful in the mapping PMR neuronal circuitry. For all the genes listed above, investigation into the expression pattern within the reproductive tract and CNS may help to determine how the gene is involved in the PMR and the neuronal circuitry underlying these behaviors in females.

## 4.5 SFPs and Natural Genetic Variation

Now that we understand which components of the ejaculate and the neuronal network in the females that support the PMR, we can use these to determine how much genetic variation exists within populations, how this variation produces variation at the cellular and/or behavior level, and eventually if variation in female PMR contributes to the process of evolution, species formation, and species maintenance.

Once the genes that give rise to these SFPs were identified, the next step toward understanding the female PMR is to determine if natural genetic variation exists in the gene sequence, gene expression, or the amount of the gene product transferred to females during copulation and the correlated behavioral response in females. If variation at any level influenced the PMR of the female, then it may hint to the underlying function of the peptide and interacting female components. Furthermore, if we can understand the variation, we can start to understand how evolution has shaped these molecules. Unfortunately, variation at these three levels has only been investigated for SP, future studies should be done to determine if the genes and/or expression vary for others and if this variation contributes to variation we see in the female response.

A few studies have investigated the genetic variation for SFPs. Fiumera et al. (2007) performed association test with the third chromosome substitution lines to determine if genetic variation in 13 previously identified male reproductive genes influenced female postcopulatory phenotypes. Of the genes investigated, a polymorphism in CG6168 was associated with female willingness to remate. Although this gene was already identified as a gene expressed in the male reproductive tract and its product transferred to females during copulation, its function to influence female PMR was unknown. On the other hand, two other genes previously identified as PMR genes also showed associations. Polymorphisms in esterase-6 were associated with changes in female egg-laying behavior. This relationship was previously found by Saad, Game, Healy, and Oakeshott (1994) that looked at variation in esterase-6 activity levels in 18-field–derived lines and found that they were negatively correlated with the number of eggs laid by female (Saad et al., 1994). Fiumera et al. (2007) also determined that variation in SP influences the variation that was seen in female willingness to remate. Although other labs have also found a relationship between SP gene expression and refractory period in female remating (Smith, Hosken, ffrench-Constant, & Wedell, 2009), others have not. With the use of chromosome extraction lines, Chow et al. (2010) identified multiple polymorphisms of SP, some of which affect the aa sequence and others that may alter the expression level of the peptide. Although they determined that levels of SP mRNA differed significantly between lines, they found no association between transcript abundance and PMR (egg-laying rate and remating rate) and no association between transcript abundance and SP protein levels. Similarly, Smith et al. (2012) determined that there is no relationship between the variation in transcript abundance and number of eggs laid (Smith et al., 2012). Although it is clear that both the sequence of the gene and upstream regulatory elements vary (Chow et al., 2010; Cirera & Aguadé, 1997; Fiumera et al., 2007), and different lines produce different levels of protein product (Avila et al., 2011; Chow et al., 2010; Ram & Wolfner, 2007; Smith et al., 2009, 2012), it is unclear if variation at this level can give rise to the variation we see in female postcopulatory behavior. It is possible that the lack of consistency that it found in previous research could be due to the indirect link of male gene expression and female behavioral response. Perhaps, instead of variation at the genetic level, we should focus on the variation in the amount of gene product that is actually transferred to the female.

Although the amount of SP transferred to females during copulation may not be determined by the level of genetic expression of SP (Ram & Wolfner, 2007; Smith et al., 2012; Wigby et al., 2009; Wolfner, 2002) or in other words, more transcription does not mean more peptide donated to the female, the volume of SFPs found within the female reproductive tract after copulation may still have a male genetic basis. Lines selected for large and small AG size produced relatively more and less SP and transferred relatively more and less volume of SP to females during copulation, respectively. Armed with this tool, Wigby et al. (2009) investigated whether females that mate with males that donate significantly more SP produce a different PMR than those that mate with males that donate a wild-type amount. The authors performed a "multiple-mating competition assay" over a 10-day period. In this assay, 6 flies (2 females, 2 males, and 2 males from either the large AG selection line, small AG selection line, or control line) are housed together without interruption and number and paternity of progeny were counted. They determined that the males from the large AG sired significantly more offspring than males from the control or small

AG lines. However, because the frequency of copulation events was not recorded, it is possible that the increase in offspring production was due to increased copulation events. The authors noted that there were no differences in premating competitive ability because each day for a 3-h period, assays were observed and mating pairs were noted. However, mating outside these limited windows was not included in the calculation and mating occurs at different frequencies during the day (Sakai & Ishida, 2001). As the raw data was not presented, it is difficult to know how many copulations were observed in total. Although the selection line for large AG did show a response, it is impossible to determine if the selection regime influenced other sexually selected traits like pheromone profile or courtship song production as these traits were not measured. Furthermore, although virgin large AG males contain greater amounts of SP within their AGs compared to control and small AG males, and they transfer more SP during their virginal mating, it was not determined if over a 10 day period females mated with large AG males consistently receive more SP than those mated to control and small AG males. Therefore, from this experiment, it cannot be said with certainty if variation in volume of SP transferred to females during copulation influences female PMR (Fiumera et al., 2007; Wigby et al., 2009; Wolfner, 2002).

The contents of the ejaculate transferred to females during copulation is not only influenced by the genotype of the male but also his social environmental. Males can alter the amount of SFPs they transfer depending on his precopulatory social environment. Males that are housed with potential "revile" males donate more SFPs during copulation than males housed alone (Fiumera et al., 2007; Smith et al., 2012; Wigby et al., 2009), and the degree of female's PMR (specifically reduced receptivity and egg production) depends heavily on the male's prior experience to rival males (Bretman, Fricke, & Chapman, 2009). Males can also alter the amount of sperm within the ejaculate based on rival males (Garbaczewska et al., 2012) or based on mating status of female (Sirot et al., 2011).

As the molecular and neurobiological basis of the postcopulatory physiological changes are determined, so too is the circuitry underlying decision-making behavior. After mating, what a female does and the decisions she makes influences not only her fitness but also the resulting fitness of her mate. In the next section, we will review the types of choices a female faces after her first mating and the mechanisms that she may use to make such choices possible.

## 5. FEMALE POSTCOPULATORY MATE CHOICE: CRYPTIC VS NONCRYPTIC

It is not always appreciated that sexual selection continues after copulation. What the female does after the virginal mating, and each subsequent mating, can impact the fertilization success of the first male she mated with. Postcopulatory mate choice (PCMC) is any decision regarding reproductive behavior that is made after mating and includes not only decisions that can be clearly observed such as the degree to which a female is polyandrous (noncryptic PCMC) but also decisions that are made about the fate of the male sperm within the female reproductive tract that determines offspring production and paternity (cryptic PCMC). There is evidence in the literature that both noncryptic and cryptic PCMC have a genetic basis. The genes involved and the neural network supporting these processes can be identified. Exploration into the molecular basis of PCMC could indicate how it evolves and influences the genetic make-up of *Drosophila* populations.

### 5.1 Noncryptic

It is generally accepted that *Drosophila* females mate several times during their lives (Markow, 2011). However, the notion that females mate multiple times within one reproductive episode (before the sperm from the previous male has been fully used) is often questioned or ignored by researchers. This is an oversight since female polyandry not only has been observed in both natural (Imhof, Harr, Brem, & Schlötterer, 1998; Markow, 2011) and laboratory settings (Billeter et al., 2012; Krupp et al., 2008; Kuijper & Morrow, 2009) but also alters offspring production and genetic make-up (Billeter et al., 2012; Lefevre & Jonsson, 1962) and therefore has strong evolutionary consequences. Several nonmutually exclusive hypotheses have been suggested to explain female polyandry within an evolutionary framework of cost versus benefits (reviewed by Arnqvist & Nilsson, 2000; Birkhead & Pizzari, 2002; Brown, Bjork, Schneider, & Pitnick, 2004; Byrne & Rice, 2005; Chapman, Arnqvist, Bangham, & Rowe, 2003a; Gowaty, Kim, Rawlings, & Anderson, 2010; Jennions & Petrie, 2000; Markow, 2011; Partridge et al., 1987; Salmon, Marx, & Harshman, 2001; Slatyer, Mautz, Backwell, & Jennions, 2011). One concept which contributes to the confusion surrounding the actual mating system (monogamous vs polygamous) of *D. melanogaster* females is the calculation of the cost–benefit ratio. With the identification of the numerous potential costs in terms of reproductive output incurred by females that multiply mate (reviewed below), it seemed unlikely that the behavior could be profitable. However, genetic techniques have improved making more hypotheses available to be empirically tested. The cost–benefits analyses discussion is now in the laboratory in order to identify possible mechanisms with molecular biology techniques.

The cost of sex to females has been linked to the diversion of resources to increased egg-production and postcopulatory immune response. Copulation can be costly to the female due to the change in energy allocation from somatic maintenance to reproduction. The lifespan cost of producing offspring is supported by the finding that both females with inactive or absent ovaries (Flatt et al., 2008) and those unable to lay eggs due to removal of an appropriate substrate (Partridge et al., 1987) live longer than their fecund female controls. One possible mechanism is oxidative stress as it increases as a result of egg production following mating (Salmon et al., 2001). However, females that are mated but are unable to produce mature eggs due to the *ovo* mutation also suffer a cost of mating (decreased lifespan) (Barnes, Wigby, Boone, Partridge, & Chapman, 2008) indicating there are egg-production-independent costs as well. The act of copulation itself has the potential to incur a cost to the female. Females can be physically (Kamimura, 2007) but also chemically damaged by the SFPs received during copulation. The ejaculate that males transfer to females along with sperm contains about 100 different molecules that vary in their detrimental effects to the female: some are protective and some toxic (Chapman et al., 1995; Mueller et al., 2007). SP is an important component of the ejaculate with regard to the female PMR (Section 4), but with use of genetic mutants, researchers were able to disentangle the Acp-related costs of copulation from the general cost of reproduction (such as increased egg-production). Females that were housed with SP-null males mated more frequently, but did not lay less eggs or die faster compared to control females (Wigby & Chapman, 2005); and twice mated *dunce*-mutant females (who do not show the normal female PMR to SP including increased egg production) have a shorter life span than once mated dunce females (Chapman et al., 1996). This supports the notion that SP is the molecule within the male ejaculate that incurs the cost of mating to females.

Another potential cost of mating in *D. melanogaster* is the increase in the female immune response: if females divert resources to the immune system after mating, they become unavailable for normal cell maintenance which could contribute to their decreased longevity. Although females upregulate immune-related genes after mating (Innocenti & Morrow, 2009), the mechanism is still unknown. The presence of sperm, Acps, and nonsperm/Acp factors has all been linked to this effect suggesting that the ejaculate, micro-organisms in the female reproductive tract, or the Acps that signal to the female to upregulate could all contribute to the change in the transcriptome (McGraw et al., 2004). Although the first two hypotheses still require testing, the last proposition was confirmed. Females mated to SP-null males failed to show the typical change in expression of these genes (Gioti et al., 2012) and transgenic females that constitutively express SP (thus not needing to mate with males to acquire this peptide) also show high levels of antimicrobial peptide genes (Peng et al., 2005b). The identification of immune response genes that are differentially induced after infection in virgins and mated females may lead to the identification of the mechanisms of the tradeoff between mating and immune defense (Short & Lazzaro, 2013); however, a direct link has not been made (Morrow & Innocenti, 2012). A carefully designed experiment that mimics the genetic expression signature of mating specifically on immune-related genes to determine the impact on female longevity is still required to resolve unambiguously if the immune response is costly to the female.

Whatever the source of the cost of copulation and reproduction, one thing is for sure: sex is costly. The costs of mating accelerates quickly with each subsequent mating (Kuijper, Stewart, & Rice, 2006) and females are most fecund when given the minimal male exposure required to fertilize her eggs (Rice et al., 2006). Therefore, selection may act on the standing genetic variation within a population on remating rate and favor females that tend to mate less often. Evidence that variation in remating behavior is at least partially explained by genetic variation comes from selection studies. Both fast and slow remating rate can be selected for suggesting that the behavior has a genetic basis (Gromko & Newport, 1988). Experimentally skewing the sex ratio of the population over several generations alters the selection pressures because females in high male:female ratio populations are more frequently courted and remate more often compared to equal and low male:female ratio populations (Wigby & Chapman, 2004). Although females showed no differences in survival when housed in the absence of males, females that were selected in high male:female ratio populations lived longer than females from the other populations when housed with wild-type males (Wigby & Chapman, 2004) suggesting that there is a genetic basis of female resistance to the cost of mating. Although the propensity of female remating behavior has been shown to be a major factor contributing to the genetic variation in female resistance (Linder & Rice, 2005), the authors of this study did not identify remating rate as a causal variable for longevity. This could be due to the method of observing number of copulations: 10 observations of at least 20 min apart were made twice a week. This severely underestimates the actual frequency of mating for all flies within the experiment. Furthermore, as daily rhythms of mating behavior is under genetic control (Billeter et al., 2012; Krupp et al., 2008; Sakai & Ishida, 2001; Tauber et al., 2003), this selection regime may have also influenced timing of mating, possibly causing failure to detect a difference between the two groups. Nevertheless, an important finding by the researchers was the ability to select for a more resistant female phenotype through higher intensities of sexual conflict. Taken together, these results indicate that there is genetic variation supporting female's willingness to remate and resistance to longevity-related costs of mating for females that multiply mate.

The cost of mating reported in most studies in a laboratory setting is a decrease of 4–8 days out of a roughly 30 day lifespan (Markow, 2011). Thus, mated females have a measurable decrease in lifespan. However, does this matter outside of the laboratory? Wild-caught females taken into the lab and allowed to lay eggs were found to produce offspring from 4 to 6 different males (Imhof et al., 1998) demonstrating that females remate multiply within one reproductive episode. Female polyandry is thus not a lab artifact but a natural life history trait. This is not surprising as females from most species engage in copulation with several males; polyandry rather than monandry is the observed norm (Holman & Kokko, 2013). So what is the missing piece? Why do the theories predict monogamy but we continue to observe polyandry? One possibility is that females from natural populations fail to pay the cost of mating. Although we could not find data on the life expectancy or fecundity of *D. melanogaster* females in natural conditions, in order to pay the cost of remating, females from a natural population must live more than 24 days and continue a consistent rate of offspring production. It is possible, but in our opinion highly unlikely. Therefore, polyandry may be selected for because females simply do not survive long enough to suffer the consequences. That is not to say that the findings from the lab are irrelevant; only the importance of this cost in wild situation may not contribute enormously to the evolution of female polyandry. This questions is further complicated by the observation that the lifespan of mated female caught directly in the wild is longer that that of virgin females caught directly from the wild indicating a cost of virginity rather than a cost of mating in a nonlab setting (Markow, 2011).

### 5.1.1 Female Polyandry is Advantageous

The cost of sex incurred by females and findings that female polyandry is under genetic control begs the question: why do females remate? Multiple reviews have been written (Arnqvist & Nilsson, 2000; Birkhead & Pizzari, 2002; Chapman et al., 2003a; Gowaty, 2012; Jennions & Petrie, 2000; Partridge et al., 1987; Singh, Singh, & Hoenigsberg, 2002; Slatyer et al., 2011; Yapici et al., 2008) and experiments conducted (Arnqvist & Kirkpatrick, 2005; Brown et al., 2004; Byrne & Rice, 2005; Fowler & Partridge, 1989; Gowaty et al., 2010; Markow, 2011; Salmon et al., 2001) in the attempt to understand just this. For years, female polyandry has puzzled biologists because the advantages gained by females are not obvious. Based on the asymmetrical investment into gamete formation, optimization of female reproductive output is achieved via maximizing resource accumulation, not copulation events. Without any clear benefits, why do females remate?

Even without immediate material benefits, remating can still be advantageous. Due to the lower cost of sperm production than egg production, males benefit from a higher frequency of matings compared to females (Bateman, 1948; but see Gowaty, Kim, & Anderson, 2012). Therefore, females are courted more often than that of their optimal remating frequency resulting in inevitable rejection behavior toward the male. When rejection behavior becomes more costly than mating, there would be an advantage for polyandrous females. Support for the hypothesis comes from research in other species such as the water strider. Rowe (1992) reasoned that the increase in female polyandry in a skewed sex ratio condition of 3 males to 1 female was due to the cost of rejection being outweighed by the cost of mating. The same phenomenon of increased remating occurs when populations of D. melanogaster are manipulated in a male-biased fashion (Wigby & Chapman, 2004). There is however no direct demonstration that the females do not benefit from remating in other ways than avoiding male harassment.

Polyandry may be advantageous through beneficial increases in offspring production and quality. Although females have a finite number of eggs they can produce and one insemination event usually supplies a sufficient amount of sperm, female remating increases offspring production as twice mated female produce slightly, but significantly, more offspring compared to once mated females (Lefevre & Jonsson, 1962). Subsequent matings may provide additional male-derived Acps required to sustain ovulation and oviposition machinery thus optimizing sperm utilization (Cameron, Day, & Rowe, 2007). The relationship between Acps, progeny production, and polyandry has not been tested directly. However, the relationship between Acp and progeny production is clear: Acps turn on machinery to increase offspring production (Xue & Noll, 2000). Polyandrous females that either repeatedly mated with the same male or allowed to mate with multiple males showed no differences in fitness-related measurement which suggests that females multiply mate in order to retain more SFPs and not to diversify her progeny (Brown et al., 2004; Jennions & Petrie, 2000; Mueller et al., 2007). A metaanalysis that was performed on 122 experimental studies and assessed the fitness effects of multiple matings on female insects found that lifetime offspring production increases as a consequence of mating multiple times, and these benefits far outweighed any costs to the female such as shorter lifespan (Arnqvist & Nilsson, 2000; Slatyer et al., 2011). Furthermore, female polyandry increases fecundity in closely related species *D. pseudoobscura* (Chapman et al., 1995; Gowaty et al., 2010; Slatyer et al., 2011) and does not cause female to live less long in nature (Gowaty et al., 2010). If nothing else, multiple mating may simply provide certainty of fertilization (Fisher, Doff, & Price, 2013; Jennions & Petrie, 2000). For example, when females were mated with recently twice mated males, sperm transfer was reduced and the resulting postcopulatory receptivity remained high (Lefevre & Jonsson, 1962). This suggests that inadequate ejaculate size can contribute to remating.

Alternatively, additional mating may provide female not just more offspring but better offspring as well. With each new mate, the genetic diversity of the sperm increases within the female reproductive tract. Although these males may be comparable in terms of the male traits used during prezygotic mate choice as they all successfully obtained copulation, the sperm within and between ejaculates may vary for traits used during postzygotic mate choice. Evidence that males favored during precopulatory mate choice were not favored during fertilization suggests a multistage mate choice procedure by females (Pischedda & Rice, 2012). One possible function of the multistage mate choice system is to produce high-quality offspring: prezygotic selection allows for the most attractive males to mate, and postcopulatory selection mechanisms may provide a more reliable way to selecting the best or most compatible sperm (Jennions & Petrie, 2000). Although some studies fail to find differences between offspring fitness of polyandrous females and serial monogamous (for example, Brown et al., 2004) suggesting that mating with multiple males does not increase fitness of the mother, others are able to link increased female mating frequency with high lifetime reproductive success of their daughters (Priest, Roach, & Galloway, 2008). Therefore, it is possible that in order to fully identify the benefits of female polyandry, next generation fitness levels need to be assessed.

### 5.1.2 Genetic Basis of Variation in Female Polyandry

To determine if female polyandry has a genetic basis in *D. melanogaster*, researchers applied artificial selection techniques and biased offspring production to either fast- or slow-to-remate females. Behavioral response to this selection pressure suggests that timing and propensity to remate is at least partially under genetic control (Pyle & Gromko, 1981). Furthermore, studies that made use of outbred strains with relatively high genetic variation showed the degree to which females remate had a genetic basis (Linder & Rice, 2005), and comparisons between inbred lab strains showed that there are line typical remating frequency and latency (Billeter et al., 2012; Lawniczak & Begun, 2005; Lüpold et al., 2013). A QTL analysis was completed and identified three regions of the genome, 57B, 87B-E, and 8D-9A, containing hundreds of confirmed and predicted genes, were associated with refractoriness to remating (Lawniczak & Begun, 2005). Future studies will be required to fine map these regions down to identify the genetic information that was driving the relationships in this study. Giardina et al. (2011) identified nonsynonymous polymorphisms in two genes (*CG9897* and *CG11797*) through association study and chromosome substitution lines: the variation within these genes contributes significantly to the variation in remating behavior seen within the remating paradigm. *CG9897*, a predicted serine endopeptidase, is expressed in the SSOs (Prokupek et al., 2009) and *CG11797* (aka *Obp56a*) is an OBP, a protein that influences the processing of odors in olfatory sensillae. OBPs have previously been associated with mating behavior such as the case with lush that binds to cVA and together interacts with OR67d (Ronderos & Smith, 2010). Neither of these studies directly show a causal relationship between identified genes and remating behavior. Future studies that directly manipulate the sequence of these genes are required to determine the role they play in female polyandry.

Comparisons between inbred lines, QTL analysis, and mutagenesis studies indicate that female polyandry has a genetic component. Future research for this field should include identifying the genes within the regions of the QTLs and the regions significantly associated with remating and focus efforts onto how the gene functions within the female to contribute to her remating phenotype. Further investigation into the natural variation that exists at a population level within these genes at either the sequence or expression level and how that variation leads to changes in mating behavior is also required.

### 5.1.3 Environmental Factor that Influences Female Polyandry

Female remating rate is influenced not only by the expression pattern of her genes or the organization of her nervous system but also by the environment that she is in. The social environment and the number and genetic variability of the flies around her influences female mating behavior. To determine the effect of the social context, this classic paradigm including a single male and female had to be replaced with a new more social setup. Although high social densities were rarely shown to increase remating (Harshman, Hoffmann, & Prout, 1988), manipulating the sex ratio (Wigby & Chapman, 2004) or genetic diversity of the males (Krupp et al., 2008; Billeter et al., 2012) can increase mating behavior. This context-dependent increase in female polyandry may be caused by females perceiving social genetic diversity through olfaction as females carrying the *Orco* allele (impaired olfactory system) failed to show the effect (Billeter et al., 2012). It is possible that either the male CHC profile was modified as this male trait has been shown to be influenced by his social context (Kent et al., 2008; Krupp et al., 2008), or males are simply perceived to be more attractive in direct comparison to other males in the social context.

Other factors in the environment other than potential mates can also influence remating rate. Part of the PMR is a change in feeding behavior: mated females consume significantly larger meals than virgin controls and this change is regulated through SP (Carvalho et al., 2006). Although the absence of food does not seem to deter virgin or sperm-depleted females to mate, recently mated females will not remate if food is absent (Harshman et al., 1988), and remate more often with more food (Trevitt, Fowler, & Partridge, 1988) or higher quality food (Chapman & Partridge, 1996). Therefore, the nutritional status of the female (Fricke, Bretman, & Chapman, 2010) and food availability in the environment determines how a female responds to the male ejaculate in terms of the normal PMR (Fricke et al., 2010). One proposed mechanism of nutritional-dependent remating is through the insulin/insulin-like growth factor-like signaling (IIS) pathway as it has been shown to be responsive to nutritional status (Ikeya, Galic, Belawat, Nairz, & Hafen, 2002). Ablation of the *Drosophila* insulin-like peptide (DILP)-producing median neurosecretory cells (MNCs), and knock-out of DILP genes (*dilp2, dilp3*, and *dilp5*) significantly reduced remating (Wigby et al., 2011). The MNCs that coexpress all three DILP genes are located in PI (Söderberg, Carlsson, & Nässel, 2012), a brain area that has been predicted to influence egg-production in mated females by communicating with

CA–corpus cardiacum complexes causing the female to produce more JH and ultimately producing more mature eggs (Kubli, 2003). As mated females have higher nutrient needs in order for optimal egg production, it is possible that the food-related receptivity is governed by the egg-laying process and there is no direct link between receptivity and nutritional state. However, females with ovaries that produce underdeveloped eggs also show the same decrease in receptivity in response to SP (Barnes, Boone, Partridge, & Chapman, 2007) suggesting that egg production is not the direct cause for the normal postmating decrease in receptivity. Since the reduction in remating for MNC-ablated females was shown to be dependent on the receipt of SP, it is possible that SP interacts directly with or upstream of the PI which could signal the CA–corpus cardiacum complexes and also to a central controlling mating behavior. Research confirming that SP changes the neuronal activity of MNCs within the PI, and where these neurons project to is required.

### 5.1.4 Sexual Conflict in Female Polyandry

Although polyandry provides females with numerous benefits, it reduces the number of offspring the first male she mates with will sire. As a result of polyandry, male–male conflict arises between the different ejaculates in the female reproductive tract giving rise to the phenomenon of sperm competition. There are multiple examples of how males manipulate the components of their ejaculate in order to maximize their fitness by influencing some aspects of the female PMR. Some SFPs within the male ejaculate contribute to the formation of the mating plug—a gelatinous secretion by the male temporarily closing the female reproductive tract. Although the plug is completely contained within the female, it could function to encourage sperm storage, impair copulation/sperm transfer by rival males, or decrease receptivity. PEB-me, PEBII, and the PEBIII proteins made in the ejaculatory bulb make up the posterior region of the mating plug, and proteins from the AG make up the anterior (including Acp36DE) region (Bretman et al., 2010; Lung & Wolfner, 2001). Although Acp36DE has been associated with sperm storage (Avila & Wolfner, 2009), females mated with PEBII knock-down males did not show any deficiencies in offspring production suggesting that the anterior and posterior regions are functionally different (Bretman et al., 2010). Females that mate with PEBII knock-down males formed abnormal mating plugs and remated significantly faster than females that mated with controls suggesting that the mating plug itself reduces postmating receptivity or this gene has pleiotropic effects (also decreases remating through an unidentified plug-independent mechanism) (Bretman et al., 2010).

If Acps aid sperm fertilization success within the female reproductive tract in the presence of other males' sperm, then increasing perceived competition should also increase Acp gene expression. However, males reared in groups of four produced significantly less Acp26Aa (ovulin) and Acp62f (Acp for sperm defensive behavior), and no different levels of SP compared to those reared in isolation (Fedorka, Winterhalter, & Ware, 2011). This result indicates that ovulin, Acp62f, and SP are not differently expressed around rival males in a manner that would indicate they are used in sperm competition. Alternatively, the increased number of males in the social context may have decreased expression of ovulin in an attempt to reserve stores for future mating. If the production of ovulin is costly, males may want to limit the production and transfer to females. As this Acp is associated with egg-production, one hypothesis is that the presence of rival males indicates to the focal male that females he encounters will most likely have previously mated or will mate again in the near future. Accordingly, males transfer 20% less ovulin to females that previously mated 24 h prior than males that are mated to virgins (Sirot et al., 2011) suggesting that males can potentially high-jack ovulin received by previous/future rivals. Furthermore, some studies show that transfer of Acp36DE by the first male facilitated storage and use of sperm from Acp36DE-null males (Chapman, Neubaum, Wolfner, & Partridge, 2000; and for a review, see Hodgson & Hosken, 2006). However, there are a few problems with this explanation. One, males also decreased expression of the gene encoding Acp62f which is associated with defensive behavior in sperm competition. It is difficult to see how downregulation of Acp62f could be beneficial in this context unless it has another unknown function. Next, males who interacted with rival males 24 h prior to mating increased the amount of ovulin and SP transferred to females (Wigby et al., 2009), which goes in the opposite direction of the gene expression data on Acp62f. The differences between the experiments designs such as amount of time males interacted with rival males, mating status of the focal female, and genetic background of the males used probably contributed to this discrepancy indicating the high plasticity of male seminal peptide expression.

## 5.2 Cryptic Choice

We established above that *D. melanogaster* females should be considered polyandrous and therefore mate with more than one male within one reproductive episode. A conflict therefore arises between males and females for control over polyandry, which is favored by females but strongly discouraged by males. Polyandry results in the mixing of the males' ejaculates in the female's reproductive tract allowing ejaculate–ejaculate and ejaculate–female interactions. Studies have shown that the fertilization success of each successive male is not random. The fertilization set, the sperm stored within the SSO and competing for fertilization, does not contain equal number of sperm from each partner. In a twice mated female, the storage organs contain more of the second male's sperm, a phenomenon known as last-male sperm precedence (Manier et al., 2010). To study this, researchers mate a female to the first male, isolate the female, and allow her to lay eggs from 24 h to 4 days, introduce new male, and if she mates, count and compare offspring sired by each male (Lüpold et al., 2013). However, as the number of copulations increase and the timing between copulations decrease, last male sperm precedence breaks down (Billeter et al., 2012), which suggests that the standardized assay provides a unique context rather than a generalized condition. Regardless, this paradigm has been useful to understand male and female contributions to competitive fertilization. Male reproductive success can be measured with either defensive ability (the proportion of offspring sired by the first male after the female has remated, aka P1) or offensive ability (the proportion of offspring sired by the second male, aka P2). The battle of the ejaculates, known as *sperm competition*, is well studied, including its the genetic architecture (Fiumera et al., 2007), and the components of the male ejaculate has been extensively studied (for reviews, see Avila et al., 2011; Ravi Ram & Wolfner, 2007; Schnakenberg et al., 2012). However, females are not simply passive arenas in which males fight out sperm wars. Evidence is mounting that suggests females exert at least some control over the outcome of sperm competition making the result of fertility success an interaction of male and female contribution. Using the classic sperm competition paradigm, researchers have shown that offspring paternity is influenced by the mother's genotype (Clark et al., 1999; Chow et al., 2010; Giardina et al., 2011) as well as age (Mack, Priest, & Promislow, 2003) and reproductive tract morphology (Amitin & Pitnick, 2007; Miller & Pitnick, 2002). By mating two standard males to genetically different females, it was determined that the genotype of the female influences the outcome of sperm competition (Clark & Begun, 1998; Clark et al., 1999). QTL analysis identified regions of the genome that contribute to propensity to use first or second male sperm and 33F/34A and 67F–69A corresponding to nearly 300 genes (Lawniczak & Begun, 2005).

Another form of cryptic PCMC is the social context-dependent female fecundity found in the *Oregon-R* strain. When females mate with males of their own strain, in the presence of *Canton-S* males, females forgo their fecundity and produce little or no offspring (Billeter et al., 2012). Although these studies suggest that females contribute to the outcome of sperm competition and skew the paternity of her offspring, no candidate genes for this process or mechanism has been determined.

Many mechanisms by which females bias the paternity of her offspring have been proposed. Possible targets of cryptic female choice include changes in reproductive tract biochemistry, neurophysiology, and morphology that interact with ejaculate to bias the paternity (Lüpold et al., 2013) and interactions between male SFPs and sperm storage morphology (Schnakenberg et al., 2012) or female-derived components (Wolfner, 2009), (Wolfner, 2011). Based on the evidence from previous studies, we review the possible mechanisms by which females exert preference or bias offspring production.

### 5.2.1 Modification of Acps

A substantial portion of Acps are predicted proteases, protease inhibitors, and lipases (Swanson, Clark, Waldrip-Dail, Wolfner, & Aquadro, 2001), which are hypothesized to cleave other SFPs to modify their biological activity (Wolfner, 2002). However, as SFPs are found in their modified state within the female reproductive tract, it has been suggested that female-derived molecules may also participate in their activity (Ravi Ram & Wolfner, 2007). Swanson et al., (2004) identified genes that are richly expressed in the female reproductive tract which are the most likely candidates to interact at the molecular level with the components of the male ejaculate, including Acps and sperm surface proteins. Although this screen produced 526 genes that could potentially influence the female PMR, the authors argue that the list of most likely candidates are those that show evidence of positive selection and based on characteristics of the gene product such as having a transmembrane domain or signal sequences, including *CG4928, CG10200, CG16707, CG7415*, and *CG3066* (Swanson et al., 2004). Mack et al. (2006) and McGraw et al. (2004) identified 1783 genes that were differentially expressed in mated females and virgins and 539 genes in the lower reproductive tract, respectively. The functions of these genes include transcriptional factors, signal transducers, enzymes, proteases, and protease inhibitors (Mack et al., 2006; McGraw et al., 2004). Therefore, female-derived molecules may act to protect sperm, destroy it, or a combination of the two (Wolfner, 2009). Prokupek et al. (2009) found that a quarter of genes expressed in the SSOs have a predicted serine protease function. This study also found that after mating, both ST and SR express JH hydrolases (*JHeh2* and *JHeh3*) genes that encode for enzymes that inactive JH with a possible function to counter the effects of biosynthesis of JH stimulated by SP.

Although these studies provide indications that females–derived molecules interact with the ejaculate, no gene product has been experimentally shown to modify a male-derived product but the candidate gene lists produced previously is an excellent place to start. In the attempt to identify the functions of these molecules and their role in the female PMR, this list of peptides and female proteins has been prioritized with use of a new tool in bioinformatics (Findlay et al., 2014). Peptides and proteins that must interact to affect the PMR would evolve in unison across species. The Evolutionary Rate Covariation tool was used to identify interacting seminal fluid peptides and female proteins. This process lead to the identification of three new males proteins influencing the association of SP with sperm and three female proteins that may affect SP dissociation from stored sperm (Findlay et al., 2014).

### 5.2.2 Sperm Acquisition and Usage

Females may be able to bias the paternity of their offspring by controlling how sperm is stored during and after copulation and utilized during ovulation and fertilization. For instance, if females are capable of biasing the relative amount of sperm from each male that is stored in her SSOs, she would then be able to impact the resulting paternity of her offspring. A review by Schnakenberg et al. (2012) describes the female reproductive tract including the two SSOs: the paired mushroom-shaped sperm spermathecae (St) and the SR with the latter as the primary source of sperm used for fertilization. It outlines the process of storage and presents the genes in both the male and female genome that have been demonstrated to influence sperm storage and sperm precedence. Although there is no clear indication that sperm storage is actively manipulatable by females, evidence is starting to mount that it is at least partially controlled by females. One, an intact feminine central nervous system is required for proper sperm acquisition and storage (Arthur, Hauschteck-Jungen, Nöthiger, & Ward, 1998). Two, females from different isogenic lines harbor heritable variation in female sperm storage behavior (Lüpold et al., 2013). And three, a gene that is expressed in the spermathecal ducts called glucose dehydrogenase (*Gld*) is upregulated in a mated female; and the gene product has been shown to be required for sperm storage in St (Iida & Cavener, 2004). Once sperm is successfully stored, it must be released from the SSOs to fertilize the eggs. Although the release of sperm from storage organs is coordinated by female-produced neuromodulators, tyramine, its derivative octopamine (Avila et al., 2012) and *Gld* (Iida & Cavener, 2004), it is unlikely that they can control the specific sperm that is being released. After remating, the sperm from multiple males appears to be mixed (Manier et al., 2010) and thus the preferential use of one male's sperm over the others appears to be highly unlikely.

### 5.2.3  Sperm Ejection

As males transfer much more sperm that can be stored, consequently, the extra sperm is expelled from the female reproductive tract and is referred to as an *ejection* (Lüpold et al., 2013; Manier et al., 2010; Snook & Hosken, 2004). After remating, the ejection is composed of sperm from both the virginal and remating event (Manier et al., 2010). The timing of this ejection can influence the proportion of offspring sired by the first and second male (Manier et al., 2010). Females with longer ejection latencies produced more offspring from the second male compared to those females with shorter ejection latencies (Lüpold et al., 2013) suggesting that quick ejections may bias the offspring to the first male. Therefore, if the female actively controls the timing of this ejection, she could then indirectly manipulate amount of first male sperm within storage and thus the probability of offspring from each male. Furthermore, Manier et al. (2010) show that immediately after mating, the sperm from the first male was concentrated in the distal half of the SR and only slowly mixed over time. Therefore, ejections soon after remating should contain relatively more of the second male's sperm compared to ejections long after mating.

Another form of sperm release from the SSOs of the female occurs during remating. When remating occurs, some females release the sperm of first male into the uterus Snook & Hosken (2004). If females have control over this release, it is possible that by releasing more sperm from SSOs, they can influence how much of each male's sperm is stored after the second mating and ultimately biasing paternity of offspring.

### 5.2.4  Female Reproductive Organ Anatomy

Conformational changes take place in female reproductive tract during and after copulation (Adams & Wolfner, 2007) which require Acps, specifically Acp36DE (Avila & Wolfner, 2009), but not sperm. In a scan for genomic and proteomic changes, postmating found genes involved in muscle activity, contractions, and muscle tissue development indicating that prepared females to manipulate sperm through sperm storage, displacement, or ejection (Mack et al., 2006). Consistent with these findings, Kapelnikov et al. (2008) showed that mating triggers changes in both the genetic expression and proteins abundance that mediates the development of the reproductive tract to deal with the postmating environment (Gioti et al., 2012).

The female reproductive tract is highly innervated by terminal branches of abdominal nerves (Heifetz & Wolfner, 2004). One function of these nerves could be to modulate the responsiveness of the musculature and epithelium. Heifetz and Wolfner (2004) examined the effects of mating on vesicle release in cells throughout the reproductive tract. Sperm, semen, and the mating itself induced changes in vesicle release within the reproductive tract. Initially, the peptidergic nerve termini inntervating the lower reproductive tract including the uterus are found to increase vesicle release immediately following mating and by the higher reproductive tract including the common and lateral oviduct 3 hours after mating (Heifetz & Wolfner, 2004). These results provide evidence that reproductive tract may mediate PMRs to the male ejaculate via neuromodulator release and active female control of these in a male trait-dependent manner would influence offspring production and support cryptic female choice.

### 5.2.5 Oogenesis Stage Manipulation

Once mated, females receive SP from males. In the CA complex, SP increases biosynthesis of JH, which causes eggs to move past the stage 9 block and increase uptake of yolk proteins. However, the presence of JH alone does not increase the transcription of these yolk proteins. Another molecule has shown to be involved in this process and has mated-status-related changes: 20-hydroxyecdysone (20E). As the hypothesized function of this molecule is to reduce ovulation rate, if females have active control over its expression, 20E may represent the female response to the male-derived SP and resulting JH synthesis (Soller, Bownes, & Kubli, 1999).

Findings of a strong genetic basis to ovulation and egg-laying rate (Boulétreau–Merle, Terrier, & Fouillet, 1989; Lüpold et al., 2013) suggest that females may moderate progeny production. If this is possible, females may increase or decrease this process depending on her previous partner(s), potential partner(s) in the environment, or other physical variables such as food availability. Moreover, specific genes have been identified to be involved in egg-production rate through mutagenesis (*lozenge*; Fuyama, 1995) and expression manipulations (*sra*, *norpA*, *broad*, *grapes*, and candidate genes *CG11700*, *CG4612*, *CG30169*, and *CG3961*; Ejima et al., 2004). If females had control over the expression of those genes or the activity of the cells in which those genes are expressed, it could provide a mechanism for influencing her PMR in a context-dependant manner.

Although it is clear that females can influence offspring production, the degree to which this occurs and the mechanism in which she gains this control is still unknown. Females are active participants during the sperm storage and appear to have at least some control over the fate of sperm within her reproductive tract. The identification of the genes involved in cryptic and noncryptic PCMC is the gateway into the exploration into the neuronal network that supports these decisions.

## 6. EGG-LAYING AND OVIPOSITION SITE SELECTION

After successful copulation, ovulation, and fertilization of an egg, females must start the process of oviposition site selection. This choice represents the only form of overt maternal care that *Drosophila* mothers extend to their offspring because once laid, a *Drosophila* egg is left to its own devises. The developmental time from egg to larva is about 22 h at 24 °C (Markow, Beall, & Matzkin, 2009). The larva that will hatch from this egg will therefore experience the microenvironment that its mother selected for it a day earlier: a nice patch of real estate on a fermenting fruit. In a heterogeneous and changing environment, oviposition site selection by females is one of the major factors that will determine offspring survival. In this section, we will see that females demonstrate impressive sophistication in selecting oviposition site.

### 6.1 What is Egg-Laying Behavior?

The process of laying an egg involves a stereotypical behavioral sequence (Yang, Belawat, Hafen, Jan, & Jan, 2008). It begins with a searching phase where females walk over potential substrates and probe with their legs, mouth parts, and ovipositor, which harbor the fly's chemosensory system (Stocker, 1994; Vosshall & Stocker, 2007). Once a site has been found, females begin the ovipositor motor program, which consists in bending their abdomen downward, extruding their ovipositor, and squeezing out an egg. This is followed by a "clean and rest" program, where the female grooms its ovipositor with its hind legs and then stays immobile for a while. The cycle of search, oviposit, clean, and rest nearly always resumes (Yang et al., 2008).

### 6.2 Genetics of Oviposition Site Selection

Females are selective toward egg-laying sites even going as far as withholding eggs when their present environment does not make the cut. For instance, female *D. melanogaster* are repelled by the toxic hexanoic and octanoic acids found at high doses in the fruit of *Morinda citrifolia* and the perception of them inhibits oviposition on acetic acid (AA), a preferred oviposition substrate (Legal, Moulin, & Jallon, 1999). Insights into the genetics of this avoidance comes from a comparison with *Drosophila sechellia*, a close relative of *D. melanogaster* that specializes on the *M. citrifolia* fruit (Lachaise & Cariou, 1988), and is attracted to hexanoic and octanoic acids (Matsuo, Sugaya, Yasukawa, Aigaki, & Fuyama, 2007). Genetic mapping of loci underlying differences between *D. melanogaster* and *D. sechellia* repulsion and attraction to hexanoic acid, respectively, identified two adjacent genes encoding OBPs: Obp57d and Obp57e (Matsuo et al., 2007). These genes were cloned from *D. sechellia* and introduced into a *D. melanogaster* strain lacking Obp57d/ eKO which produced a female *D. melanogaster* expressing the *D. sechellia* OBPs. These females shifted their oviposition site preference and preferred to lay eggs on high concentration of octanoic acid just as *D. sechellia* females (Matsuo et al., 2007). Obp57e is expressed in the lymph bathing taste sensillae of the legs and is transcribed at higher levels in *D. sechellia* than in *D. melanogaster* (Matsuo et al., 2007). It is likely that Obp57e higher expression in *D. sechellia* results in a greater binding of hexanoic and octanoic acid, reducing the response to these acids either by titrating or degrading them.

Genetic variation for oviposition site preference is also observed within species in different strains of *D. melanogaster*. In a laboratory setting, some strains prefer to lay eggs on a piece of paper placed next to food, while others preferring the surface of the food. This oviposition site preference can be altered by artificial selection, revealing the existence of genetic variability for the trait (Takamura & Fuyama, 1980). The two modes of oviposition appear to be connected with the preference of some females to insert eggs in the food versus leaving it on top (Takamura & Fuyama, 1980). These differences are likely due to substrate consistency or chemical composition because amputation of the leg tarsi, which contains both taste and mechanosensory sensillae, reduced the tendency of flies that normally lay egg on paper to do so (Takamura & Fuyama, 1980). *Drosophila melanogaster* lay their eggs on decaying fruits, an environment containing high amount of ethanol and other fermentation products. The preference for inserting eggs in the food versus leaving on the food may be related to the egg resistance to ethanol as eggs inserted in food containing ethanol have a greater survival that those laid on top (Delden & Kamping, 1990; Kamping & van Delden, 1990). In turn, the greater resistance of eggs from some strains to ethanol may have evolved, unexpectedly, from exposure to parasitoid wasps. Parasitoid wasps are the main predator of *D. melanogaster* larvae, using them as egg-laying substrate and a source of food for their larvae. Females seeing parasitoid wasps in their environment switch egg-laying preference toward site containing greater amount of ethanol, a compound which protect larvae against parasitoid infection (Kacsoh, Lynch, Mortimer, & Schlenke, 2013). Populations of flies exposed to greater predation by wasps may be selected for higher ethanol tolerance.

## 6.3 Sensing and Processing of Oviposition Sites

Females take into account a host of sensory information when choosing egg-laying sites. Both visual information, such as the color of the food substrate (del Solar & Palomino, 1966), and textural information of the substrate affects egg-laying choice (Rockwell & Grossfield, 1978). Olfaction is used to sense the microbiota of potential egg-laying substrate and plays a major role in oviposition site selection. For instance, flies do not lay eggs on substrate containing the odorant Geosmin, which is associated with harmful bacteria (Stensmyr et al., 2012). On the other hand, females are attracted to beneficial microbes as indicated by the fact that flies not only fly toward but also lay eggs on substrates containing AA (Eisses, 1997). This substance probably indicates the presence of microbes, such as yeast, which are necessary for the growth of larvae (Baumberger, 1917; Becher et al., 2012).

The means by which females use sensory information to choose egg-laying site are complex. Although females fly toward acetic acid, they avoid walking on substrates containing this acid for longer than the time taken to lay eggs (Joseph, Devineni, King, & Heberlein, 2009). A mutation in the *Poxn* gene prevents the differentiation of taste sensillae specifically in the mouthparts (Boll & Noll, 2002). Such mutant females reduce egg-laying on acetic acid (Joseph et al., 2009), and increase positional avoidance of acetic acid. Surgical removal of the olfactory system leaves egg-laying preference for acetic acid intact but females lose their positional aversion to acetic acid. Acetic acid is thus detected by separate sensory systems to generate distinct behavioral outputs: taste informs egg-laying and olfaction indicates positional avoidance (Joseph et al., 2009).

Lobeline, a bitter-tasting compound, also acts as an attractant for egg-laying (Yang et al., 2008), which flies also avoid walking on (Joseph & Heberlein, 2012). The egg-laying and positional avoidance response to lobeline is abrogated by silencing neurons expressing the bitter taste receptor Gr66a (Joseph & Heberlein, 2012). This indicates that taste of the same compound can both trigger and promote avoidance. Using a mosaic approach, Joseph and Heberlein (2012) showed that expression of Gr66a in the internal mouthparts is responsible for egg-laying attraction to lobeline while Gr66a in gustatory neurons of the legs receive input for the positional aversion to lobeline. It appears that the two opposite signals from Gr66a are integrated in a higher order brain structure called the MB because physiological inactivation of this brain center results in female losing both positional aversion and egg-laying attraction to lobeline (Joseph & Heberlein, 2012).

Females given a choice to lay eggs on lobeline or sugar will lay eggs on lobeline. This preference is in part regulated by the level of expression of the insulin-like peptide *Ilp-7* as overexpression of *Ilp-7* results in egg-laying on both lobeline and sugar (Yang et al., 2008). It is however unclear how this effect is controlled. Egg-laying is under tight neuronal control and its regulatory circuitry is one of the best understood female behaviors. Octopaminergic neurons both contract the ovaries and relax the oviduct which facilitates ovulation (Middleton et al., 2006), and mutations affecting octopamine synthesis result in deficient egg-laying. The muscle fibers of the reproductive tract are regulated by excitatory glutamatergic neurons and inhibitory octopaminergic neurons (Kapelnikov et al., 2008; Middleton et al., 2006; Rodríguez-Valentín et al., 2006). Female-specific *Ilp7*-expressing neuron corresponds to some of the glutamatergic neurons innervating the oviduct, where they are likely to act to stimulate oviduct contraction (Castellanos, Tang, & Allan, 2013). After eggs exit the ovary, they are propelled through the oviduct by somatic-like muscles that ring the oviduct (Hudson, Petrella, Tanaka, & Cooley, 2008). It is possible that *Ilp-7* neurons in the brain are involved in the oviposition site preference and send information to *Ilp-7* motor neuron for egg-laying.

The preference for laying eggs on lobeline vs sugar (Yang et al., 2008) is however not innate but the result of phenotypic plasticity in female oviposition preference. This preference is modulated by the size of the arena in which the egg-laying substrates are located. When hatching on lobeline medium located close to the sugar substrate, larvae can travel to the sugar patch on which they can feed (Schwartz, Zhong, Bellemer, & Tracey, 2012). During the time period required from egg-laying and larval hatching, the sugar patch also has time to diffuse in the arena, further reducing the distance a larva would have to travel to feed. Females seem to be able to predict the foraging cost of their offspring because they lay egg on the sugar-free patch only when there is a nearby sugar patch. However, when the sugar patch is further away from the sugar-free patch, females always lay eggs on the sugar. Experiments show that larvae are unable to find the sugar patch when they are on the opposite of a larger arena (Schwartz et al., 2012). Thus when the foraging costs are high, females lay eggs on the nutritious substrate. These experiments suggest that females measure space when making oviposition–related choices and female egg-laying decisions match not present but expected larval foraging costs (Schwartz et al., 2012).

Another example of how females use their olfactory system to choose an egg-laying site comes from fruit preference. Females prefer citrus fruits as egg-laying sites, such as oranges or grapefruits, over a range of other fruits (Dweck et al., 2013). This preference appears to be due to limonene, a molecule present in citrus fruits. This odorant does not attract females to the fruit but rather acts as an oviposition stimulant once the female has landed on the substrate. Artificial activation of the Or19b ORN, the neuron–sensing limonene, is sufficient to induce oviposition on a substrate devoid of limonene (Dweck et al., 2013). This begs the question of what is the function of laying eggs on citrus fruits? Parasitoid wasps, such as *Leptopilina boulardi*, avoid the smell of citrus fruits and larvae found on media containing citrus smell suffer reduced parasitization rates compared to those on media devoid of citrus smell (Dweck et al., 2013). So parasites might have been a major driver of the evolution of egg-laying substrate preference.

## 6.4 Raising Offspring Together? Social Dimension of Egg-Laying

The complexity of egg-laying site has a major social dimension. Females have a gregarious tendency in choosing the site in which to oviposit, and also prefer to lay their eggs in areas which have already been occupied by larvae (del Solar & Palomino, 1966). Larger adult aggregations on fruit substrates increase offspring fitness at different levels. Increased density of adult flies on substrates, prior to the introduction of larvae, enhanced larval survival and yields larger flies (Wertheim, Marchais, Vet, & Dicke, 2002b). We have seen in Section 3 that larger flies are more attractive. The increased survival is probably linked to increased inoculation rate of wild yeast species by the joining adults and decreased detrimental fungal growth (Hoffmann & Harshman, 1985; Wertheim, 2005; Wertheim et al., 2002b).

We have seen in Section 2 that flies emit aggregation pheromones, which facilitate adult encounters. cVA is also a signal for egg-laying. In outdoor experiments, flies aggregate their eggs on the exact micropatch on which synthetic cVA had been applied (Wertheim et al., 2002a). This effect has also been replicated in the lab where addition of cVA on food substrate makes it more likely to be chosen over a substrate made of the same food but without cVA (Sarin & Dukas, 2009). Males may be able to deposit cVA locally as conditioning food with males also increases stimulation of egg-laying by females (Hoffmann & Harshman, 1985; Wertheim et al., 2002a). The same signal indicating aggregation is also used to indicate egg-laying, showing that these two phenomena evolved in concert.

Females are also capable of exchanging information about egg-laying sites via social interactions. A naïve mated female that experiences two food patches in the presence of females who had laid eggs on one subsequently exhibits a preference for laying eggs on the same patch even though the two are of equal quality (Sarin & Dukas, 2009). Females therefore biased their egg-laying preference based on the choices made by other females. This holds in situations where the naïve female is able to observe other females but also in the absence of demonstration (Battesti, Moreno, Joly, & Mery, 2012). Although the former situation may be established through social copying, the mechanism through which this information is transferred in the absence of demonstration is unclear. The naïve female may for instance smell the micro-environment previously encountered by the experienced females through odors left on her body. No matter the mechanism, these experiments demonstrate that flies exchange social information about egg-laying sites.

These examples make it clear that egg-laying sites are chosen with great care. This importance is also reflected in the ability of flies to remember sites containing aversive substance. When exposed to a food source with an aversive substance perfumed with an olfactory cue, flies will subsequently avoid laying eggs on substrate perfumed with the cue but without aversive substance (Mery & Kawecki, 2002). This demonstrates that they have been conditioned to associate the cue with the toxic substance. An increased ability to avoid unfavorable sites by learning to avoid them would increase the number of eggs laid on favourable substrate thus increasing its contribution to the next generation. Flies selected to learn to modify their substrate preference in response to an aversive cue can evolve better ability to respond to conditioning (Mery & Kawecki, 2002). These flies became faster learner.

The reproductive behavior of *Drosophila* females relates in many ways to the niche in which they deposit their eggs. The care in selecting those sites is an indirect form of parental care. Because of the social dimension of egg-laying, oviposition site selection should be seen as a form of sociality.

# 7. CONCLUSION

In this chapter, we set out to review female reproductive behavior not only considering the ultimate and proximate perspective but also an integration of the two. We compiled the literature regarding the genetic, cellular, and chemical basis of how females make decisions that influence not only herself but also the genotype and environment of her offspring in hope of resolving the evolutionary processes that contribute to species-wide behavior. One apparent theme that occurred but was not necessarily saliently highlighted was the influence of the environment on the fly, namely, the social environment and how it contributes to phenotypic plasticity. As we follow the adult fly through her reproductive decisions in chronological order, the importance of her social life revealed itself.

The specific aspect of the environment that determines the value of a food patch is reflected in the receptors that flies use to locate it. The biological basis of habitat location and navigation are influenced not only by perception of food odors but also by those emitted from flies, together producing the smell of sociality (Section 2). The social group continues to be required as she moves into the next decision of reproduction: choosing the best male for the job. Reproductive decisions concerning the female mate choice (Section 3) clearly require the female to interact with a social group (even if the social group is a single male). Studying mate choice in a well-controlled environment and from a reductionist perspective may have biased researchers to observe single pairs in isolation, while overlooking important aspects of mating behaviors. With the identification of the genetic and neuronal basis of female precopulatory mate choice, including the detection and sensory basis of the male traits and the brain centers responsible for integration of this information, a full realization of the influence of the social environment on mate choice will eventually be achieved. Furthermore, after copulation, the physiological changes that facilitate the transition of the female into the PMR (Section 4), the reproductive decisions including polyandry (Section 5), and egg-laying behavior (Section 6) are all largely determined by the social and physical environment of the female.

In the natural environment, flies are almost always found in groups. The constant sociality of flies must lead us to think more about social influence on female behavior as it is more far-reaching than once thought. The complexity of the social life of *D. melanogaster* goes beyond aggregation as the type and number of flies present on a given food patch has nontrivial effects on the genetics of individual group members. These effects are reviewed in detail by Schneider, Atallah, & Levine (2012). We however want to highlight that group composition affects gene regulation, allowing the genotype of group members to affect each other's phenotype. The social experience influences the transcription of genes controlling circadian timing and pheromone production (Krupp et al., 2008) with correlated effects on rhythmic patterns of locomotor activity and pheromone displays phenotypes (Levine,

2002; Kent et al., 2008; Krupp et al., 2008, 2013). In turn, these effects correlate with changes in reproduction: increasing the genetic diversity of the group also increases female mating frequency, which changes the genotype of offspring they produce (Billeter et al., 2012; Krupp et al., 2008). How information about the genotype of individual group members is transferred remains unknown. Chemosensory transmission is the prime candidate because an intact olfactory system is required for females to detect group composition (Billeter et al., 2012). However, most experiments reduce the social impact on fly genetic, physiological, and behavioral phenotypes. These lab assays may not reveal the full behavioral repertoire of flies in the wild. As we relay on the results from these studies to inform us of the proximate and ultimate processes of reproduction, we may be only getting part of the story.

A final take-home message from this chapter should also be that we really do not know much about female reproductive behavior. Although environmental factors, genetic variants, and physiological changes have been identified, the present understanding of the neuronal circuitry that supports this behavior and the evolutionary processes that shaped it is far from complete. This is not stated to be discouraging or disheartening; it is meant to promote attention to the areas that are lacking and help the advancement of the field. Although we site numerous papers with varying experimental designs, reporting results with varying effects, and concluding relationships with varying strengths, one thing is certain: we actually say so little. In the end, we reviewed not just what we know, but really what we do not know. And there is value in that.

## ACKNOWLEDGMENTS

We thank Bregje Wertheim and Joel Levine for feedback on some of the sections of this review. ML and JCB are funded by an Open Programma grant from the Dutch organization for scientific research (NWO).

## REFERENCES

Adams, E. M., & Wolfner, M. F. (April 2007). Seminal proteins but not sperm induce morphological changes in the *Drosophila melanogaster* female reproductive tract during sperm storage. *Journal of Insect Physiology, 53*(4), 319–331.

Aigaki, T., Fleischmann, I., Chen, P. S., & Kubli, E. (October 1991). Ectopic expression of sex peptide alters reproductive behavior of female *D. melanogaster. Neuron, 7*(4), 557–563.

Amitin, E. G., & Pitnick, S. (January 2007). Influence of developmental environment on male- and female-mediated sperm precedence in *Drosophila melanogaster. Journal of Evolutionary Biology, 20*(1), 381–391.

Anagnostou, C., Dorsch, M., & Rohlfs, M. (May 17, 2010). Influence of dietary yeasts on *Drosophila melanogaster* life-history traits. *Entomologia Experimentalis et Applicata, 136*(1), 1–11.

Antony, C., Davis, T. L., Carlson, D. A., Pechine, J. M., & Jallon, J. M. (December 1985). Compared behavioral responses of male *Drosophila melanogaster* (Canton S) to natural and synthetic aphrodisiacs. *Journal of Chemical Ecology, 11*(12), 1617–1629.

Antony, C., & Jallon, J.-M. (1982). The chemical basis for sex recognition in *Drosophila melanogaster. Journal of Insect Physiology, 28*(10), 873–880.

Arienti, M., Antony, C., Wicker-Thomas, C., Delbecque, J.-P., & Jallon, J.-M. (September 1, 2010). Ontogeny of *Drosophila melanogaster* female sex-appeal and cuticular hydrocarbons. *Integrative Zoology, 5*(3), 272–282.

Arnqvist, G., & Kirkpatrick, M. (May 2005). The evolution of infidelity in socially monogamous passerines: the strength of direct and indirect selection on extrapair copulation behavior in females. *The American Naturalist, 165*(Suppl 5), S26–S37.

Arnqvist, G., & Nilsson, T. (August 2000). The evolution of polyandry: multiple mating and female fitness in insects. *Animal Behaviour, 60*(2), 145–164.

Arthur, B., Hauschteck-Jungen, E., Nöthiger, R., & Ward, P. I. (1998). A female nervous system is necessary for normal sperm storage in *Drosophila melanogaster*: a masculinized nervous system is as good as none. *Proceedings of the Royal Society B: Biological Sciences, 265*(1407), 1749–1753.

Asahina, K., Watanabe, K., Duistermars, B. J., Hoopfer, E., González, C. R., Eyjólfsdóttir, E. A., et al. (January 2014). Tachykinin-expressing neurons control male-specific aggressive arousal in *Drosophila. Cell, 156*(1–2), 221–235.

Averhoff, W. W., & Richardson, R. H. (September 1, 1974). Pheromonal control of mating patterns in *Drosophila melanogaster. Behavior Genetics, 4*(3), 207–225.

Avila, F. W., Bloch Qazi, M. C., Rubinstein, C. D., & Wolfner, M. F. (March 20, 2012). A requirement for the neuromodulators octopamine and tyramine in *Drosophila melanogaster* female sperm storage. *Proceedings of the National Academy of Sciences of the United States of America, 109*(12), 4562–4567.

Avila, F. W., Sirot, L. K., Laflamme, B. A., Rubinstein, C. D., & Wolfner, M. F. (January 7, 2011). Insect seminal fluid proteins: Identification and function. *Annual Review of Entomology, 56*(1), 21–40.

Avila, F. W., & Wolfner, M. F. (September 15, 2009). Acp36DE is required for uterine conformational changes in mated *Drosophila* females. *Proceedings of the National Academy of Sciences of the United States of America, 106*(37), 15796–15800.

Balfanz, S., Strünker, T., Frings, S., & Baumann, A. (2005). A family of octapamine receptors that specifically induce cyclic AMP production or Ca2+ release in *Drosophila melanogaster. Journal of Neurochemistry, 93*(2), 440–451.

Bargmann, C. I. (June 2012). Beyond the connectome: how neuromodulators shape neural circuits. *BioEssays, 34*(6), 458–465.

Barnes, A. I., Boone, J. M., Partridge, L., & Chapman, T. (April 2007). A functioning ovary is not required for sex peptide to reduce receptivity to mating in *D. melanogaster. Journal of Insect Physiology, 53*(4), 343–348.

Barnes, A. I., Wigby, S., Boone, J. M., Partridge, L., & Chapman, T. (July 22, 2008). Feeding, fecundity and lifespan in female *Drosophila melanogaster. Proceedings of the Royal Society B: Biological Sciences, 275*(1643), 1675–1683.

Bartelt, R. J., Schaner, A. M., & Jackson, L. L. (December 1985). cis-Vaccenyl acetate as an aggregation pheromone in *Drosophila melanogaster. Journal of Chemical Ecology, 11*(12), 1747–1756.

Bastock, M., & Manning, A. (January 1, 1955). The courtship of *Drosophila melanogaster. Behaviour, 8*(1), 85–110.

Bateman, A. J. (1948). Intra-sexual selection in *Drosophila. Heredity, 2*(Pt. 3), 349–368.

Battesti, M., Moreno, C., Joly, D., & Mery, F. (February 2012). Spread of social information and dynamics of social transmission within *Drosophila* groups. *Current Biology, 22*(4), 309–313.

Baumberger, J. P. (February 1917). The food of *Drosophila melanogaster* meigen. *Proceedings of the National Academy of Sciences of the United States of America, 3*(2), 122–126.

Becher, P. G., Bengtsson, M., Hansson, B. S., & Witzgall, P. (June 2010). Flying the fly: long-range flight behavior of *Drosophila melanogaster* to attractive odors. *Journal of Chemical Ecology, 36*(6), 599–607.

Becher, P. G., Flick, G., Rozpędowska, E., Schmidt, A., Hagman, A., Lebreton, S., et al. (May 18, 2012). Yeast, not fruit volatiles mediate *Drosophila melanogaster* attraction, oviposition and development. *Functional Ecology, 26*(4), 822–828.

Becnel, J., Johnson, O., Luo, J., Nässel, D. R., & Nichols, C. D. (2011). The serotonin 5-HT7Dro receptor is expressed in the brain of *Drosophila*, and is essential for normal courtship and mating. *PLoS ONE, 6*(6), e20800.

Belgacem, Y. H., & Martin, J.-R. (October 24, 2002). Neuroendocrine control of a sexually dimorphic behavior by a few neurons of the pars intercerebralis in *Drosophila*. *Proceedings of the National Academy of Sciences of the United States of America, 99*(23), 15154–15158.

Belgacem, Y. H., & Martin, J.-R. (2005). Disruption of insulin pathways alters trehalose level and abolishes sexual dimorphism in locomotor activity in *Drosophila*. *Journal of Neurobiology, 66*(1), 19–32.

Belgacem, Y. H., & Martin, J.-R. (2007). Hmgcr in the corpus allatum controls sexual dimorphism of locomotor activity and body size via the insulin pathway in *Drosophila*. *PLoS ONE, 2*(1), e187.

Bellen, H. J., & Kiger, J. A. (January 1987). Sexual hyperactivity and reduced longevity of dunce females of *Drosophila melanogaster*. *Genetics, 115*(1), 153–160.

Benton, R., Vannice, K. S., Gomez-Diaz, C., & Vosshall, L. B. (January 9, 2009). Variant ionotropic glutamate receptors as chemosensory receptors in *Drosophila*. *Cell, 136*(1), 149–162.

Billeter, J.-C., Atallah, J., Krupp, J. J., Millar, J. G., & Levine, J. D. (October 15, 2009). Specialized cells tag sexual and species identity in *Drosophila melanogaster*. *Nature, 461*(7266), 987–991.

Billeter, J.-C., & Goodwin, S. F. (July 19, 2004). Characterization of *Drosophila* fruitless–gal4 transgenes reveals expression in male-specific fruitless neurons and innervation of male reproductive structures. *Journal of Comparative Neurology, 475*(2), 270–287.

Billeter, J.-C., Jagadeesh, S., Stepek, N., Azanchi, R., & Levine, J. D. (June 22, 2012). *Drosophila melanogaster* females change mating behaviour and offspring production based on social context. *Proceedings of the Royal Society B: Biological Sciences, 279*(1737), 2417–2425.

Billeter, J.-C., & Levine, J. D. (February 2013). Who is he and what is he to you? Recognition in *Drosophila melanogaster*. *Current Opinion in Neurobiology, 23*(1), 17–23.

Billeter, J.-C., Rideout, E. J., Dornan, A. J., & Goodwin, S. F. (2006a Sep 5). Control of male sexual behavior in *Drosophila* by the sex determination pathway. *Current Biology, 16*(17), R766–R776.

Billeter, J.-C., Villella, A., Allendorfer, J. B., Dornan, A. J., Richardson, M., Gailey, D. A., et al. (2006b Jun 6). Isoform-specific control of male neuronal differentiation and behavior in *Drosophila* by the fruitless gene. *Current Biology, 16*(11), 1063–1076.

Birkhead, T. R., & Pizzari, T. (April 2002). Postcopulatory sexual selection. *Nature Reviews Genetics, 3*(4), 262–273.

Bloch Qazi, M. C., & Wolfner, M. F. (October 1, 2003). An early role for the *Drosophila melanogaster* male seminal protein Acp36DE in female sperm storage. *Journal of Experimental Biology, 206*(Pt 19), 3521–3528.

Boll, W., & Noll, M. (December 1, 2002). The *Drosophila* Pox neuro gene: control of male courtship behavior and fertility as revealed by a complete dissection of all enhancers. *Development, 129*(24), 5667–5681.

Bono, J. M., Matzkin, L. M., Kelleher, E. S., & Markow, T. A. (May 10, 2011). Postmating transcriptional changes in reproductive tracts of con- and heterospecifically mated *Drosophila mojavensis* females. *Proceedings of the National Academy of Sciences of the United States of America, 108*(19), 7878–7883.

Boulétreau-Merle, J. (1976). Destruction of the pars intercerebralis in *drosophila melanogaster*: Effect on the fecundity and the stimulation through copulation. *Journal of Insect Physiology, 22*(7), 933–940.

Boulétreau-Merle, J., Terrier, O., & Fouillet, P. (April 1989). Chromosomal analysis of initial retention capacity in virgin *Drosophila melanogaster* females. *Heredity, 62*(Pt 2), 145–151.

Bousquet, F., Nojima, T., Houot, B., Chauvel, I., Chaudy, S., Dupas, S., et al. (January 3, 2012). Expression of a desaturase gene, desat1, in neural and nonneural tissues separately affects perception and emission of sex pheromones in *Drosophila*. *Proceedings of the National Academy of Sciences of the United States of America, 109*(1), 249–254.

Bownes, M., Scott, A., & Shirras, A. (May 1988). Dietary components modulate yolk protein gene transcription in *Drosophila melanogaster*. *Development, 103*(1), 119–128.

Branson, K., Robie, A. A., Bender, J., Perona, P., & Dickinson, M. H. (May 3, 2009). High-throughput ethomics in large groups of *Drosophila*. *Nature Methods, 6*(6), 451–457.

Bretman, A., Fricke, C., & Chapman, T. (May 7, 2009). Plastic responses of male *Drosophila melanogaster* to the level of sperm competition increase male reproductive fitness. *Proceedings of the Royal Society B: Biological Sciences, 276*(1662), 1705–1711.

Bretman, A., Lawniczak, M. K. N., Boone, J., & Chapman, T. (January 2010). A mating plug protein reduces early female remating in *Drosophila melanogaster*. *Journal of Insect Physiology, 56*(1), 107–113.

Brieger, G., & Butterworth, F. M. (February 27, 1970). *Drosophila melanogaster*: Identity of male lipid in reproductive system. *Science, 167*(3922). 1262–2.

Brown, W. D., Bjork, A., Schneider, K., & Pitnick, S. (June 2004). No evidence that polyandry benefits females in *Drosophila melanogaster*. *Evolution, 58*(6), 1242–1250.

Butlin, R. K., & Ritchie, M. G. (May 20, 1989). Genetic coupling in mate recognition systems: what is the evidence? *Biological Journal of the Linnean Society, 37*(3), 237–246.

Butterworth, F. M. (March 21, 1969). Lipids of *Drosophila*: a newly detected lipid in the male. *Science, 163*(873), 1356–1357.

Byrne, P. G., & Rice, W. R. (August 25, 2005). Remating in *Drosophila melanogaster*: an examination of the trading-up and intrinsic male-quality hypotheses. *Journal of Evolutionary Biology, 18*(5), 1324–1331.

Byrne, P. G., & Rice, W. R. (April 22, 2006). Evidence for adaptive male mate choice in the fruit fly *Drosophila melanogaster*. *Proceedings of the Royal Society B: Biological Sciences, 273*(1589), 917–922.

Cachero, S., Ostrovsky, A. D., Yu, J. Y., Dickson, B. J., & Jefferis, G. S. X.E. (September 28, 2010). Sexual dimorphism in the fly brain. *Current Biology, 20*(18), 1589–1601.

Cameron, E., Day, T., & Rowe, L. (June 2007). Sperm competition and the evolution of ejaculate composition. *The American Naturalist, 169*(6), E158–E172.

Carhan, A., Allen, F., Armstrong, J. D., Goodwin, S. F., & O'Dell, K. M. C. (November 2005). Female receptivity phenotype of icebox mutants caused by a mutation in the L1-type cell adhesion molecule neuroglian. *Genes, Brain and Behavior, 4*(8), 449–465.

Carlsson, M. A., Diesner, M., Schachtner, J., & Nässel, D. R. (May 20, 2010). Multiple neuropeptides in the *Drosophila* antennal lobe suggest complex modulatory circuits. *Journal of Comparative Neurology, 518*(16), 3359–3380.

Carney, G. E., & Bender, M. (2000). The *Drosophila* ecdysone receptor (EcR) gene is required maternally for normal oogenesis. *Genetics*.

Carvalho, G. B., Kapahi, P., Anderson, D. J., & Benzer, S. (2006). Allocrine modulation of feeding behavior by the sex peptide of *Drosophila*. *Current Biology, 16*(7), 692–696.

Castellanos, M. C., Tang, J. C.Y., & Allan, D.W. (September 2013). Female-biased dimorphism underlies a female-specific role for post-embryonic Ilp7 neurons in *Drosophila* fertility. *Development, 140*(18), 3915–3926.

Chan,Y. B., & Kravitz, E. A. (December 4, 2007). Specific subgroups of FruM neurons control sexually dimorphic patterns of aggression in *Drosophila melanogaster*. *Proceedings of the National Academy of Sciences of the United States of America, 104*(49), 19577–19582.

Chapman, T. (September 5, 2006). Evolutionary conflicts of interest between males and females. *Current Biology, 16*(17), R744–R754.

Chapman, T., Arnqvist, G., Bangham, J., & Rowe, L. (2003a). Sexual conflict. *Trends in Ecology & Evolution, 18*(1), 41–47.

Chapman, T., Bangham, J., Vinti, G., Seifried, B., Lung, O., Wolfner, M. F., et al. (2003b Aug 19). The sex peptide of *Drosophila melanogaster*. Female post-mating responses analyzed by using RNA interference. *Proceedings of the National Academy of Sciences of the United States of America, 100*(17), 9923–9928.

Chapman, T., Choffat,Y., Lucas, W. E., Kubli, E., & Partridge, L. (November 1996). Lack of response to sex-peptide results in increased cost of mating in dunce *Drosophila melanogaster* females. *Journal of Insect Physiology, 42*(11–12), 1007–1015.

Chapman, T., Liddle, L. F., Kalb, J. M., Wolfner, M. F., & Partridge, L. (January 19, 1995). Cost of mating in *Drosophila melanogaster* females is mediated by male accessory gland products. *Nature, 373*(6511), 241–244.

Chapman, T., Neubaum, D. M., Wolfner, M. F., & Partridge, L. (June 7, 2000). The role of male accessory gland protein Acp36DE in sperm competition in *Drosophila melanogaster*. *Proceedings of the Royal Society B: Biological Sciences, 267*(1448), 1097–1105.

Chapman, T., & Partridge, L. (June 22, 1996). Female fitness in *Drosophila melanogaster*: An interaction between the effect of nutrition and of encounter rate with males. *Proceedings of the Royal Society B: Biological Sciences, 263*(1371), 755–759.

Chen, P. S., Stumm-Zollinger, E., Aigaki, T., Balmer, J., Bienz, M., & Böhlen, P. (July 29, 1988). A male accessory gland peptide that regulates reproductive behavior of female *D. melanogaster*. *Cell, 54*(3), 291–298.

Chertemps, T., Duportets, L., Labeur, C., Ueda, R., Takahashi, K., Saigo, K., et al. (March 13, 2007). A female-biased expressed elongase involved in long-chain hydrocarbon biosynthesis and courtship behavior in *Drosophila melanogaster*. *Proceedings of the National Academy of Sciences of the United States of America, 104*(11), 4273–4278.

Chertemps, T., Duportets, L., Labeur, C., Ueyama, M., & Wicker-Thomas, C. (August 2006). A female-specific desaturase gene responsible for diene hydrocarbon biosynthesis and courtship behaviour in *Drosophila melanogaster*. *Insect Molecular Biology, 15*(4), 465–473.

Chertemps, T., François, A., Durand, N., Rosell, G., Dekker, T., Lucas, P., et al. (2012). A carboxylesterase, esterase-6, modulates sensory physiological and behavioral response dynamics to pheromone in *Drosophila*. *BMC Biology, 10*, 56.

Chow, C. Y., Wolfner, M. F., & Clark, A. G. (September 27, 2010). The genetic basis for male × female interactions underlying variation in reproductive phenotypes of *Drosophila*. *Genetics*.

Cirera, S., & Aguadé, M. (September 1997). Evolutionary history of the sex-peptide (Acp70A) gene region in *Drosophila melanogaster*. *Genetics, 147*(1), 189–197.

Civetta, A., Rosing, K. R., & Fisher, J. H. (2008). Differences in sperm competition and sperm competition avoidance in *Drosophila melanogaster*. *Animal Behaviour*.

Clark, A. G., Aguadé, M., Prout, T., Harshman, L. G., & Langley, C. H. (January 1995). Variation in sperm displacement and its association with accessory gland protein loci in *Drosophila melanogaster*. *Genetics, 139*(1), 189–201.

Clark, A. G., & Begun, D. J. (July 1998). Female genotypes affect sperm displacement in *Drosophila*. *Genetics, 149*(3), 1487–1493.

Clark, A. G., Begun, D. J., & Prout, T. (January 8, 1999). Female × male interactions in *Drosophila* sperm competition. *Science, 283*(5399), 217–220.

Clyne, J. D., & Miesenböck, G. (April 2008). Sex-specific control and tuning of the pattern generator for courtship song in *Drosophila*. *Cell, 133*(2), 354–363.

Clyne, P., Grant, A., O'Connell, R., & Carlson, J. R. (1997). Odorant response of individual sensilla on the *Drosophila* antenna. *Invertebrate Neuroscience, 3*(2–3), 127–135.

Cognigni, P., Bailey, A. P., & Miguel-Aliaga, I. (2011). Enteric neurons and systemic signals couple nutritional and reproductive status with intestinal homeostasis. *Cell Metabolism, 13*(1), 92–104.

Cole, S., Carney, G., McClung, C., Willard, S., Taylor, B., Hirsh, J. (2005). Two functional but noncomplementing *Drosophila tyrosine* decarboxylase genes. *Journal of Biological Chemistry, 280*(15), 14948.

Connolly, K., & Cook, R. (1973). Rejection responses by female *Drosophila melanogaster*: their ontogeny, causality and effects upon the behaviour of the courting male behaviour. *JSTOR*, 142–166.

Costa, R. (1989). Esterase-6 and the pheromonal effects of cis-vaccenyl acetate in *Drosophila melanogaster*. *Journal of Evolutionary Biology, 2*(6), 395–407.

Couto, A., Alenius, M., & Dickson, B. J. (September 6, 2005). Molecular, anatomical, and functional organization of the *Drosophila* olfactory system. *Current Biology, 15*(17), 1535–1547.

Cowling, D. E., & Burnet, B. (1981). Courtship songs and genetic control of their acoustic characteristics in sibling species of the *Drosophila melanogaster* subgroup. *Animal Behaviour, 29*(3), 924–935.

Coyne, J., Crittenden, A., & Mah, K. (September 2, 1994). Genetics of a pheromonal difference contributing to reproductive isolation in *Drosophila*. *Science, 265*(5177), 1461–1464.

Crossley, S. A., Bennet-Clark, H. C., & Evert, H. T. (January 1995). Courtship song components affect male and female *Drosophila* differently. *Animal Behaviour, 50*(3), 827–839.

Dallerac, R., Labeur, C., Jallon, J. M., Knipple, D. C., Roelofs, W. L., & Wicker-Thomas, C. (August 15, 2000). A delta 9 desaturase gene with a different substrate specificity is responsible for the cuticular diene hydrocarbon polymorphism in *Drosophila melanogaster*. *Proceedings of the National Academy of Sciences of the United States of America, 97*(17), 9449–9454.

Datta, S. R., Vasconcelos, M. L., Ruta, V., Luo, S., Wong, A., Demir, E., et al. (March 27, 2008). The *Drosophila* pheromone cVA activates a sexually dimorphic neural circuit. *Nature, 452*(7186), 473–477.

de Bruyne, M., Foster, K., & Carlson, J. R. (May 2001). Odor coding in the *Drosophila* antenna. *Neuron, 30*(2), 537–552.

del Solar, E., & Palomino, H. (1966). Choice of oviposition in *Drosophila melanogaster*. *The American Naturalist, 100*(911), 127–133.

Delden, W., & Kamping, A. (September 1990). Genetic variation for oviposition behavior in *Drosophila melanogaster*. II. Oviposition preferences and differential survival. *Behavior Genetics, 20*(5), 661–673.

Demir, E., & Dickson, B. J. (June 3, 2005). Fruitless splicing specifies male courtship behavior in *Drosophila*. *Cell, 121*(5), 785–794.

der Goes van Naters van, W., & Carlson, J. R. (April 2007). Receptors and neurons for fly odors in *Drosophila*. *Current Biology, 17*(7), 606–612.

Dickson, B. J. (November 7, 2008). Wired for sex: the neurobiology of *Drosophila* mating decisions. *Science, 322*(5903), 904–909.

Ding, Z., Haussmann, I., Ottiger, M., & Kubli, E. (June 2003). Sex-peptides bind to two molecularly different targets in *Drosophila melanogaster* females. *Journal of Neurobiology, 55*(3), 372–384.

Ditch, L. M., Shirangi, T., Pitman, J. L., Latham, K. L., Finley, K. D., Edeen, P. T., et al. (December 2, 2004). *Drosophila* retained/dead ringer is necessary for neuronal pathfinding, female receptivity and repression of fruitless independent male courtship behaviors. *Development, 132*(1), 155–164.

Domanitskaya, E. V., Liu, H., Chen, S., & Kubli, E. (November 2007). The hydroxyproline motif of male sex peptide elicits the innate immune response in *Drosophila* females. *FEBS Journal, 274*(21), 5659–5668.

Dow, M. A., & Schilcher von, F. (April 10, 1975). Aggression and mating success in *Drosophila melanogaster. Nature, 254*(5500), 511–512.

Drummond-Barbosa, D., & Spradling, A. C. (March 2001). Stem cells and their progeny respond to nutritional changes during *Drosophila* oogenesis. *Developmental Biology, 231*(1), 265–278.

Dudai, Y., Jan, Y. N., Byers, D., Quinn, W. G., & Benzer, S. (May 1976). dunce, a mutant of *Drosophila* deficient in learning. *Proceedings of the National Academy of Sciences of the United States of America, 73*(5), 1684–1688.

Dukas, R. (2005). Learning affects mate choice in female fruit flies. *Behavioral Ecology.*

Dukas, R., & Jongsma, K. (November 2012). Costs to females and benefits to males from forced copulations in fruit flies. *Animal Behaviour, 84*(5), 1177–1182.

Dweck, H. K. M., Ebrahim, S. A. M., Kromann, S., Bown, D., Hillbur, Y., Sachse, S., et al. (December 16, 2013). Olfactory preference for egg laying on citrus substrates in *Drosophila. Current Biology, 23*(24), 2472–2480.

Eberl, D. F., Duyk, G. M., & Perrimon, N. (December 23, 1997). A genetic screen for mutations that disrupt an auditory response in *Drosophila melanogaster. Proceedings of the National Academy of Sciences of the United States of America, 94*(26), 14837–14842.

Eberl, D. F., Hardy, R. W., & Kernan, M. J. (August 15, 2000). Genetically similar transduction mechanisms for touch and hearing in *Drosophila. The Journal of Neuroscience, 20*(16), 5981–5988.

Eisses, K. T. (1997). The influence of 2-propanol and acetone on oviposition rate and oviposition site preference for acetic acid and ethanol of *Drosophila melanogaster. Behavior Genetics, 27*(3), 171–180.

Ejima, A., & Griffith, L. C. (September 19, 2008). Courtship initiation is stimulated by acoustic signals in *Drosophila melanogaster. PLoS ONE, 3*(9), e3246.

Ejima, A., Smith, B. P. C., Lucas, C., der Goes van Naters van, W., Miller, C. J., Carlson, J. R., et al. (April 3, 2007). Generalization of courtship learning in *Drosophila* is mediated by cis-vaccenyl acetate. *Current Biology, 17*(7), 599–605.

Ejima, A., Tsuda, M., Takeo, S., Ishii, K., Matsuo, T., & Aigaki, T. (December 2004). Expression level of sarah, a homolog of DSCR1, is critical for ovulation and female courtship behavior in *Drosophila melanogaster. Genetics, 168*(4), 2077–2087.

Etienne, R., Wertheim, B., Hemerik, L., Schneider, P., & Powell, J. (February 2002). The interaction between dispersal, the Allee effect and scramble competition affects population dynamics. *Ecological Modelling, 148*(2), 153–168.

Everaerts, C., Farine, J.-P., Cobb, M., & Ferveur, J.-F. (2010). *Drosophila* cuticular hydrocarbons revisited: mating status alters cuticular profiles. *PLoS ONE, 5*(3), e9607.

Ewing, L. S., & Ewing, A. W. (1984). Courtship in *Drosophila melanogaster*: behaviour of mixed-sex groups in large observation chambers. *Behaviour*, 184–202.

Fabre, C. C. G., Hedwig, B., Conduit, G., Lawrence, P. A., Goodwin, S. F., & Casal, J. (November 20, 2012). Substrate-borne vibratory communication during courtship in *Drosophila melanogaster. Current Biology, 22*(22), 2180–2185.

Farine, J.-P., Ferveur, J.-F., & Everaerts, C. (July 11, 2012). Volatile *Drosophila* cuticular pheromones are affected by social but not sexual experience. *PLoS ONE, 7*(7), e40396.

Fedorka, K. M., Linder, J. E., Winterhalter, W., & Promislow, D. (May 7, 2007). Post-mating disparity between potential and realized immune response in *Drosophila melanogaster. Proceedings of the Royal Society B: Biological Sciences, 274*(1614), 1211–1217.

Fedorka, K. M., Winterhalter, W. E., & Ware, B. (February 2011). Perceived sperm competition intensity influences seminal fluid protein production prior to courtship and mating. *Evolution, 65*(2), 584–590.

Fernández, M. de LP., Chan, Y.-B., Yew, J.Y., Billeter, J.-C., Dreisewerd, K., Levine, J. D., et al. (2010). Pheromonal and behavioral cues trigger male-to-female aggression in *Drosophila. PLoS Biology, 8*(11), e1000541.

Ferveur, J.-F. (May 2005). Cuticular hydrocarbons: their evolution and roles in *Drosophila* pheromonal communication. *Behavior Genetics, 35*(3), 279–295.

Ferveur, J.-F. (October 4, 2010). *Drosophila* female courtship and mating behaviors: sensory signals, genes, neural structures and evolution. *Current Opinion in Neurobiology*, 1–6.

Ferveur, J. F., Cobb, M., Boukella, H., & Jallon, J. M. (January 1996). World-wide variation in *Drosophila melanogaster* sex pheromone: behavioural effects, genetic bases and potential evolutionary consequences. *Genetica, 97*(1), 73–80.

Ferveur, J. F., Savarit, F., O'Kane, C. J., Sureau, G., Greenspan, R. J., & Jallon, J. M. (June 6, 1997). Genetic feminization of pheromones and its behavioral consequences in *Drosophila* males. *Science, 276*(5318), 1555–1558.

Findlay, G. D., Sitnik, J. L., Wang, W., Aquadro, C. F., Clark, N. L., & Wolfner, M. F. (January 2014). Evolutionary rate covariation identifies new members of a protein network required for *Drosophila melanogaster* female post-mating responses. *PLoS Genetics, 10*(1), e1004108.

Finley, K. D., Edeen, P. T., Foss, M., Gross, E., Ghbeish, N., Palmer, R. H., et al. (December 1, 1998). Dissatisfaction encodes a tailless-like nuclear receptor expressed in a subset of CNS neurons controlling *Drosophila* sexual behavior. *Neuron, 21*(6), 1363–1374.

Finley, K. D., Taylor, B. J., Milstein, M., & McKeown, M. (1997). Dissatisfaction, a gene involved in sex-specific behavior and neural development of *Drosophila melanogaster. Proceedings of the National Academy of Sciences of the United States of America, 94*(3), 913–918.

Fisher, D. N., Doff, R. J., & Price, T. A. R. (2013). True polyandry and pseudopolyandry: why does a monandrous fly remate? *BMC Evolutionary Biology, 13*, 157.

Fishilevich, E., & Vosshall, L. B. (September 6, 2005). Genetic and functional subdivision of the *Drosophila* antennal lobe. *Current Biology, 15*(17), 1548–1553.

Fiumera, A. C., Dumont, B. L., & Clark, A. G. (January 2005). Sperm competitive ability in *Drosophila melanogaster* associated with variation in male reproductive proteins. *Genetics, 169*(1), 243–257.

Fiumera, A. C., Dumont, B. L., & Clark, A. G. (June 2007). Associations between sperm competition and natural variation in male reproductive genes on the third chromosome of *Drosophila melanogaster. Genetics, 176*(2), 1245–1260.

Flatt, T., Min, K.-J., D'Alterio, C., Villa-Cuesta, E., Cumbers, J., Lehmann, R., et al. (April 29, 2008). *Drosophila* germ-line modulation of insulin signaling and lifespan. *Proceedings of the National Academy of Sciences of the United States of America, 105*(17), 6368–6373.

Fleischmann, I., Cotton, B., Choffat, Y., Spengler, M., & Kubli, E. (2001). Mushroom bodies and post-mating behaviors of *Drosophila melanogaster* females. *Journal of Neurogenetics, 15*(2), 117–144.

Fowler, K., & Partridge, L. (April 27, 1989). A cost of mating in female fruitflies. *Nature, 338*(6218), 760–761.

Friberg, U., & Arnqvist, G. (September 2003). Fitness effects of female mate choice: preferred males are detrimental for *Drosophila melanogaster* females. *Journal of Evolutionary Biology, 16*(5), 797–811.

Fricke, C., Bretman, A., & Chapman, T. (January 2010). Female nutritional status determines the magnitude and sign of responses to a male ejaculate signal in *Drosophila melanogaster. Journal of Evolutionary Biology, 23*(1), 157–165.

Fricke, C., Wigby, S., Hobbs, R., & Chapman, T. (February 2009). The benefits of male ejaculate sex peptide transfer in *Drosophila melanogaster. Journal of Evolutionary Biology, 22*(2), 275–286.

Fujii, S., Krishnan, P., Hardin, P., & Amrein, H. (February 6, 2007). Nocturnal male sex drive in *Drosophila. Current Biology, 17*(3), 244–251.

Fuyama, Y. (November 1995). Genetic evidence that ovulation reduces sexual receptivity in *Drosophila melanogaster* females. *Behavior Genetics, 25*(6), 581–587.

Gailey, D. A., Lacaillade, R. C., & Hall, J. C. (May 1986). Chemosensory elements of courtship in normal and mutant, olfaction-deficient *Drosophila melanogaster*. *Behavior Genetics, 16*(3), 375–405.

Ganter, G. K., Desilets, J. B., Davis-Knowlton, J. A., Panaitiu, A. E., Sweezy, M., Sungail, J., et al. (March 2012). *Drosophila* female precopulatory behavior is modulated by ecdysteroids. *Journal of Insect Physiology, 58*(3), 413–419.

Garbaczewska, M., Billeter, J.-C., & Levine, J. D. (November 20, 2012). *Drosophila melanogaster* males increase the number of sperm in their ejaculate when perceiving rival males. *Journal of Insect Physiology*.

Gatti, S., Ferveur, J.-F., & Martin, J.-R. (June 2000). Genetic identification of neurons controlling a sexually dimorphic behaviour. *Current Biology, 10*(11), 667–670.

Giardina, T. J., Beavis, A., Clark, A. G., & Fiumera, A. C. (October 2011). Female influence on pre- and post-copulatory sexual selection and its genetic basis in *Drosophila melanogaster*. *Molecular Ecology, 20*(19), 4098–4108.

Gioti, A., Wigby, S., Wertheim, B., Schuster, E., Martinez, P., Pennington, C. J., et al. (November 7, 2012). Sex peptide of *Drosophila melanogaster* males is a global regulator of reproductive processes in females. *Proceedings of the Royal Society B: Biological Sciences, 279*(1746), 4423–4432.

Gleason, J. M. (May 2005). Mutations and natural genetic variation in the courtship song of *Drosophila*. *Behavior Genetics, 35*(3), 265–277.

Gomez-Diaz, C., Reina, J. H., Cambillau, C., & Benton, R. (April 2013). Ligands for pheromone-sensing neurons are not conformationally activated odorant binding proteins. *PLoS Biology, 11*(4), e1001546.

Gowaty, P. A. (January 2012). The evolution of multiple mating: Costs and benefits of polyandry to females and of polygyny to males. *Fly, 6*(1), 3–11.

Gowaty, P. A., Kim, Y. K., & Anderson, W. W. (July 17, 2012). No evidence of sexual selection in a repetition of Bateman's classic study of *Drosophila melanogaster*. *Proceedings of the National Academy of Sciences of the United States of America, 109*(29), 11740–11745.

Gowaty, P. A., Kim, Y. K., Rawlings, J., & Anderson, W. W. (August 3, 2010). Polyandry increases offspring viability and mother productivity but does not decrease mother survival in *Drosophila pseudoobscura*. *Proceedings of the National Academy of Sciences of the United States of America, 107*(31), 13771–13776.

Göpfert, M. C., & Robert, D. (June 21, 2001). Biomechanics. Turning the key on *Drosophila* audition. *Nature, 411*(6840), 908.

Greenacre, M. L., Ritchie, M. G., Byrne, B. C., & Kyriacou, C. P. (January 1993). Female song preference and the *period* gene in *Drosophila*. *Behavior Genetics, 23*(1), 85–90.

Grigliatti, T. A., Hall, L., Rosenbluth, R., & Suzuki, D. T. (1973). Temperature-sensitive mutations in *Drosophila melanogaster*. *Molecular Genetics and Genomics. Springer-Verlag, 120*(2), 107–114.

Grillet, M., Dartevelle, L., & Ferveur, J.-F. (February 7, 2006). A *Drosophila* male pheromone affects female sexual receptivity. *Proceedings of the Royal Society B: Biological Sciences, 273*(1584), 315–323.

Grillet, M., Everaerts, C., Houot, B., Ritchie, M. G., Cobb, M., & Ferveur, J.-F. (2012). Incipient speciation in *Drosophila melanogaster* involves chemical signals. *Scientific Reports, 2*, 224.

Gromko, M., & Newport, M. (1988). Genetic-basis for remating in *Drosophila melanogaster*. 2. Response to selection based on the behavior of one sex. *Behavior Genetics, 18*(5), 621–632.

Grosjean, Y., Rytz, R., Farine, J.-P., Abuin, L., Cortot, J., Jefferis, G. S. X.E., et al. (October 13, 2011). An olfactory receptor for food-derived odours promotes male courtship in *Drosophila*. *Nature, 478*(7368), 236–240.

Ha, T. S., & Smith, D. P. (August 23, 2006). A pheromone receptor mediates 11-cis-vaccenyl acetate-induced responses in *Drosophila*. *Journal of Neuroscience, 26*(34), 8727–8733.

Haerty, W., Jallon, J. M., Rouault, J., Bazin, C., & Capy, P. (November 2002). Reproductive isolation in natural populations of *Drosophila melanogaster* from Brazzaville (Congo). *Genetica, 116*(2–3), 215–224.

Hall, J. C. (June 1979). Control of male reproductive behavior by the central nervous system of *Drosophila*: dissection of a courtship pathway by genetic mosaics. *Genetics, 92*(2), 437–457.

Han, K.-A., Millar, N. S., & Davis, R. L. (1998). A novel octopamine receptor with preferential expression in *Drosophila* mushroom bodies. *The Journal of Neuroscience*.

Handler, A. M. (September 1982). Ecdysteroid titers during pupal and adult development in *Drosophila melanogaster*. *Developmental Biology, 93*(1), 73–82.

Hardeland, R. (1971). Lighting conditions and mating behavior in *Drosophila*. *American Naturalist*.

Hardin, P. E. (2011). Molecular genetic analysis of circadian timekeeping in *Drosophila*. *Advances in Genetics*, 141–173.

Harshman, L. G., Hoffmann, A. A., & Prout, T. (1988). Environmental effects on remating in *Drosophila melanogaster*. *Evolution*, 312–321.

Hasemeyer, M., Yapici, N., Heberlein, U., & Dickson, B. J. (February 26, 2009). Sensory neurons in the *Drosophila* genital tract regulate female reproductive behavior. *Neuron, 61*(4), 511–518.

Haussmann, I. U., Hemani, Y., Wijesekera, T., Dauwalder, B., & Soller, M. (2013). Multiple pathways mediate the sex-peptide-regulated switch in female *Drosophila* reproductive behaviours. *Proceedings of the Royal Society B: Biological Sciences, 280*(1771), 20131938.

Heifetz, Y., Lung, O., Frongillo, E. A., & Wolfner, M. F. (January 27, 2000). The *Drosophila* seminal fluid protein Acp26Aa stimulates release of oocytes by the ovary. *Current Biology, 10*(2), 99–102.

Heifetz, Y., Vandenberg, L. N., Cohn, H. I., & Wolfner, M. F. (January 18, 2005). Two cleavage products of the *Drosophila* accessory gland protein ovulin can independently induce ovulation. *Proceedings of the National Academy of Sciences of the United States of America, 102*(3), 743–748.

Heifetz, Y., & Wolfner, M. F. (April 20, 2004). Mating, seminal fluid components, and sperm cause changes in vesicle release in the *Drosophila* female reproductive tract. *Proceedings of the National Academy of Sciences of the United States of America, 101*(16), 6261–6266.

Herndon, L. A., & Wolfner, M. F. (1995). A *Drosophila* seminal fluid protein, Acp26Aa, stimulates egg laying in females for 1 day after mating. *Proceedings of the National Academy of Sciences of the United States of America*.

Hodgson, D. J., & Hosken, D. J. (November 21, 2006). Sperm competition promotes the exploitation of rival ejaculates. *Journal of Theoretical Biology, 243*(2), 230–234.

Hoffmann, A. (1987). Territorial encounters between *Drosophila* males of different sizes. *Animal Behaviour, 35*(6), 1899–1901.

Hoffmann, A. A., & Harshman, L. G. (1985). Male effects on fecundity in *Drosophila melanogaster*. *Evolution*, 638–644.

Holman, L., & Kokko, H. (January 21, 2013). The consequences of polyandry for population viability, extinction risk and conservation. *Philosophical Transactions of the Royal Society B: Biological Sciences, 368*(1613). 20120053–3.

Hudson, A. M., Petrella, L. N., Tanaka, A. J., & Cooley, L. (2008). Mononuclear muscle cells in *Drosophila* ovaries revealed by GFP protein traps. *Developmental Biology*.

Iida, K., & Cavener, D. R. (February 2004). Glucose dehydrogenase is required for normal sperm storage and utilization in female *Drosophila melanogaster*. *Journal of Experimental Biology*, *207*(Pt 4), 675–681.

Ikeya, T., Galic, M., Belawat, P., Nairz, K., & Hafen, E. (August 6, 2002). Nutrient-dependent expression of insulin-like peptides from neuroendocrine cells in the CNS contributes to growth regulation in *Drosophila*. *Current Biology*, *12*(15), 1293–1300.

Imhof, M., Harr, B., Brem, G., & Schlötterer, C. (July 1998). Multiple mating in wild *Drosophila melanogaster* revisited by microsatellite analysis. *Molecular Ecology*, *7*(7), 915–917.

Immonen, E., & Ritchie, M. G. (October 5, 2011). The genomic response to courtship song stimulation in female *Drosophila melanogaster*. *Proceedings of the Royal Society B: Biological Sciences*.

Innocenti, P., & Morrow, E. H. (May 2009). Immunogenic males: a genome-wide analysis of reproduction and the cost of mating in *Drosophila melanogaster* females. *Journal of Evolutionary Biology*, *22*(5), 964–973.

Insel, T. R. (January 3, 2010). The challenge of translation in social neuroscience: A review of oxytocin, vasopressin, and affiliative behavior. *Neuron*, *65*(6), 768–779.

Isaac, R. E., Li, C., Leedale, A. E., & Shirras, A. D. (January 7, 2010). *Drosophila* male sex peptide inhibits siesta sleep and promotes locomotor activity in the post-mated female. *Proceedings of the Royal Society B: Biological Sciences*, *277*(1678), 65–70.

Itskov, P. M., & Ribeiro, C. (2013). The dilemmas of the gourmet fly: the molecular and neuronal mechanisms of feeding and nutrient decision making in *Drosophila*. *Frontiers in Neuroscience*, *7*, 12.

Jackson, L. L., Arnold, M. T., & Blomquist, G. J. (January 1981). Surface lipids of *Drosophila melanogaster*: comparison of the lipids from female and male wild type and sex-linked yellow mutant. *Insect Biochemistry*, *11*(1), 87–91.

Jallon, J.-M. (September 1984). A few chemical words exchanged by *Drosophila* during courtship and mating. *Behavior Genetics*, *14*(5), 441–478.

Jallon, J.-M., & David, J. R. (1987). Variation in cuticular hydrocarbons among the eight species of the *Drosophila melanogaster* subgroup. *Evolution. JSTOR*, 294–302.

Jallon, J.-M., & Hotta, Y. (July 1979). Genetic and behavioral studies of female sex appeal in *Drosophila*. *Behavior Genetics*, *9*(4), 257–275.

Jennions, M., & Petrie, M. (2000). Why do females mate multiply? A review of the genetic benefits. *Biological Reviews*.

Joseph, R. M., Devineni, A. V., King, I. F. G., & Heberlein, U. (July 7, 2009). Oviposition preference for and positional avoidance of acetic acid provide a model for competing behavioral drives in *Drosophila*. , *106*(27), 11352–11357.

Joseph, R. M., & Heberlein, U. (October 1, 2012). Tissue-specific activation of a single gustatory receptor produces opposing behavioral responses in *Drosophila*. *Genetics*, *192*(2), 521–532.

Juni, N., & Yamamoto, D. (2009). Genetic analysis of chaste, a new mutation of *Drosophila melanogaster* characterized by extremely low female sexual receptivity. *Journal of Neurogenetics*, *23*(3), 329–340.

Kacsoh, B. Z., Lynch, Z. R., Mortimer, N. T., & Schlenke, T. A. (February 21, 2013). Fruit flies medicate offspring after seeing parasites. *Science*, *339*(6122), 947–950.

Kamikouchi, A. (July 2013). Auditory neuroscience in fruit flies. *Neuroscience Research*, *76*(3), 113–118.

Kamikouchi, A., Inagaki, H. K., Effertz, T., Hendrich, O., Fiala, A., Göpfert, M. C., et al. (March 12, 2009). The neural basis of *Drosophila* gravity-sensing and hearing. *Nature*, *458*(7235), 165–171.

Kamikouchi, A., Shimada, T., & Ito, K. (November 20, 2006). Comprehensive classification of the auditory sensory projections in the brain of the fruit fly *Drosophila melanogaster*. *Journal of Comparative Neurology*, *499*(3), 317–356.

Kamimura, Y. (August 22, 2007). Twin intromittent organs of *Drosophila* for traumatic insemination. *Biology Letters, 3*(4), 401–404.

Kamping, A., & van Delden, W. (September 1990). Genetic variation for oviposition behavior in *Drosophila melanogaster*. I. Quantitative genetic analysis of insertion behavior. *Behavior Genetics, 20*(5), 645–659.

Kapelnikov, A., Rivlin, P. K., Hoy, R. R., & Heifetz, Y. (2008). Tissue remodeling: a mating-induced differentiation program for the *Drosophila* oviduct. *BMC Developmental Biology, 8*, 114.

Keller, A. (February 6, 2007). *Drosophila melanogaster*'s history as a human commensal. *Current Biology, 17*(3), R77–R81.

Kent, C., Azanchi, R., Smith, B., Formosa, A., & Levine, J. D. (September 23, 2008). Social context influences chemical communication in *D. melanogaster* males. *Current Biology, 18*(18), 1384–1389.

Kerr, C., Ringo, J., Dowse, H., & Johnson, E. (November 1997). Icebox, a recessive X-linked mutation in *Drosophila* causing low sexual receptivity. *Journal of Neurogenetics, 11*(3–4), 213–229.

Kimura, K.-I. (February 2011). Role of cell death in the formation of sexual dimorphism in the *Drosophila* central nervous system. *Development, Growth & Differentiation, 53*(2), 236–244.

Kimura, K.-I., Hachiya, T., Koganezawa, M., Tazawa, T., & Yamamoto, D. (September 2008). Fruitless and doublesex coordinate to generate male-specific neurons that can initiate courtship. *Neuron, 59*(5), 759–769.

Kimura, K.-I., Ote, M., Tazawa, T., & Yamamoto, D. (November 10, 2005). Fruitless specifies sexually dimorphic neural circuitry in the *Drosophila* brain. *Nature, 438*(7065), 229–233.

Koganezawa, M., Haba, D., Matsuo, T., & Yamamoto, D. (January 12, 2010). The shaping of male courtship posture by lateralized gustatory inputs to male-specific interneurons. *Current Biology, 20*(1), 1–8.

Kohl, J., Ostrovsky, A. D., Frechter, S., & Jefferis, G. S. X.E. (December 2013). A bidirectional circuit switch reroutes pheromone signals in male and female brains. *Cell, 155*(7), 1610–1623.

Korol, A., Rashkovetsky, E., Iliadi, K., Michalak, P., Ronin, Y., & Nevo, E. (November 7, 2000). Nonrandom mating in *Drosophila melanogaster* laboratory populations derived from closely adjacent ecologically contrasting slopes at "Evolution Canyon". *Proceedings of the National Academy of Sciences of the United States of America, 97*(23), 12637–12642.

Krishnan, B., Levine, J. D., Lynch, M. K., Dowse, H. B., Funes, P., Hall, J. C., et al. (May 17, 2001). A new role for cryptochrome in a *Drosophila* circadian oscillator. *Nature, 411*(6835), 313–317.

Krishnan, P., Chatterjee, A., Tanoue, S., & Hardin, P. E. (June 3, 2008). Spike amplitude of single-unit responses in antennal sensillae is controlled by the *Drosophila* circadian clock. *Current Biology, 18*(11), 803–807.

Krstic, D., Boll, W., & Noll, M. (February 13, 2009). Sensory integration regulating male courtship behavior in *Drosophila*. *PLoS ONE, 4*(2), e4457.

Krupp, J. J., Billeter, J.-C., Wong, A., Choi, C., Nitabach, M. N., & Levine, J. D. (July 10, 2013). Pigment-dispersing factor modulates pheromone production in clock cells that influence mating in *Drosophila*. *Neuron, 79*(1), 54–68.

Krupp, J. J., Kent, C., Billeter, J.-C., Azanchi, R., So, A. K. -C., Schonfeld, J. A., et al. (September 23, 2008). Social experience modifies pheromone expression and mating behavior in male *Drosophila melanogaster*. *Current Biology, 18*(18), 1373–1383.

Kubli, E. (August 2003). Sex-peptides: seminal peptides of the *Drosophila* male. *Cellular and Molecular Life Sciences, 60*(8), 1689–1704.

Kuijper, B., & Morrow, E. (2009). Direct observation of female mating frequency using time-lapse photography. *Fly, 3*, 1–3.

Kuijper, B., Stewart, A., & Rice, W. (2006). The cost of mating rises nonlinearly with copulation frequency in a laboratory population of *Drosophila melanogaster*. *Journal of Evolutionary Biology, 19*(6), 1795–1802.

Kurtovic, A., Widmer, A., & Dickson, B. J. (March 29, 2007). A single class of olfactory neurons mediates behavioural responses to a *Drosophila* sex pheromone. *Nature, 446*(7135), 542–546.

Kvitsiani, D., & Dickson, B. J. (May 23, 2006). Shared neural circuitry for female and male sexual behaviours in *Drosophila*. *Current Biology, 16*(10), R355–R356.

Kyriacou, C., & Hall, J. (April 25, 1986). Interspecific genetic control of courtship song production and reception in *Drosophila*. *Science, 232*(4749), 494–497.

Kyriacou, C. P., & Hall, J. C. (November 1, 1980). Circadian rhythm mutations in *Drosophila melanogaster* affect short-term fluctuations in the male's courtship song. *Proceedings of the National Academy of Sciences of the United States of America, 77*(11), 6729–6733.

Kyriacou, C. P., & Hall, J. C. (1982). The function of courtship song rhythms in *Drosophila*. *Animal Behaviour*.

Kyriacou, C. P., & Hall, J. C. (March 1984). Learning and memory mutations impair acoustic priming of mating behaviour in *Drosophila*. *Nature, 308*(5954), 62–65.

Lachaise, D., & Cariou, M. (1988). Historical biogeography of the *Drosophila melanogaster* species subgroup. *Evolutionary Biology (USA)*.

Lai, J. S. -Y., Lo, S.-J., Dickson, B. J., & Chiang, A.-S. (February 14, 2012). Auditory circuit in the *Drosophila* brain. *Proceedings of the National Academy of Sciences of the United States of America, 109*(7), 2607–2612.

Lasbleiz, C., Ferveur, J., & Everaerts, C. (2006). Courtship behaviour of *Drosophila melanogaster* revisited. *Animal Behaviour*.

Laughlin, J. D., Ha, T. S., Jones, D. N. M., & Smith, D. P. (June 27, 2008). Activation of pheromone-sensitive neurons is mediated by conformational activation of pheromone-binding protein. *Cell, 133*(7), 1255–1265.

Lawniczak, M. K. N., & Begun, D. J. (October 2005). A QTL analysis of female variation contributing to refractoriness and sperm competition in *Drosophila melanogaster*. *Genetics Research, 86*(2), 107–114.

Lebreton, S., Becher, P. G., Hansson, B. S., & Witzgall, P. (January 2012). Attraction of *Drosophila melanogaster* males to food-related and fly odours. *Journal of Insect Physiology, 58*(1), 125–129.

Lee, G., Foss, M., Goodwin, S. F., & Carlo, T. (2000). Spatial, temporal, and sexually dimorphic expression patterns of the fruitless gene in the *Drosophila* central nervous system. *Journal of Neurobiology*.

Lee, G., Hall, J. C., & Park, J. H. (2002). Doublesex gene expression in the central nervous system of *Drosophila melanogaster*. *Journal of Neurogenetics, 16*(4), 229–248.

Lee, H.-G., Rohila, S., & Han, K.-A. (March 5, 2009). The octopamine receptor OAMB mediates ovulation via $Ca^{2+}$/calmodulin-dependent protein kinase II in the *Drosophila* oviduct epithelium. *PLoS ONE, 4*(3), e4716.

Lee, H.-G., Seong, C.-S., Kim, Y.-C., Davis, R. L., & Han, K.-A. (December 1, 2003). Octopamine receptor OAMB is required for ovulation in *Drosophila melanogaster*. *Developmental Biology, 264*(1), 179–190.

Lefevre, G., & Jonsson, U. B. (December 1962). Sperm transfer, storage, displacement, and utilization in *Drosophila melanogaster*. *Genetics, 47*, 1719–1736.

Legal, L., Moulin, B., & Jallon, J. M. (1999). The relation between structures and toxicity of oxygenated aliphatic compounds homologous to the insecticide octanoic acid and the chemotaxis of two species of *Drosophila*. *Pesticide Biochemistry and Physiology*.

Legendre, A., Miao, X.-X., Da Lage, J.-L., & Wicker-Thomas, C. (February 2008). Evolution of a desaturase involved in female pheromonal cuticular hydrocarbon biosynthesis and courtship behavior in *Drosophila*. *Insect Biochemistry and Molecular Biology, 38*(2), 244–255.

Lehnert, B. P., Baker, A. E., Gaudry, Q., Chiang, A.-S., & Wilson, R. I. (January 2013). Distinct roles of TRP channels in auditory transduction and amplification in *Drosophila*. *Neuron, 77*(1), 115–128.

Levine, J., Billeter, J.-C., Krull, U., & Sodhi, R. (May 27, 2010). The cuticular surface of *D. melanogaster*. ToF-SIMS on the fly. *Surface and Interface Analysis, 43*(1–2), 317–321.

Levine, J. D. (December 6, 2002). Resetting the circadian clock by social experience in *Drosophila melanogaster*. *Science, 298*(5600), 2010–2012.

Lilly, M., Kreber, R., Ganetzky, B., & Carlson, J. R. (March 1994). Evidence that the *Drosophila* olfactory mutant smellblind defines a novel class of sodium channel mutation. *Genetics, 136*(3), 1087–1096.

Linder, J. E., & Rice, W. R. (May 2005). Natural selection and genetic variation for female resistance to harm from males. *Journal of Evolutionary Biology, 18*(3), 568–575.

Liu, H., & Kubli, E. (August 19, 2003). Sex-peptide is the molecular basis of the sperm effect in *Drosophila melanogaster*. *Proceedings of the National Academy of Sciences of the United States of America, 100*(17), 9929–9933.

Lof, M. E., de Gee, M., & Hemerik, L. (June 7, 2009). Odor-mediated aggregation enhances the colonization ability of *Drosophila melanogaster*. *Journal of Theoretical Biology, 258*(3), 363–370.

Lone, S. R., & Sharma, V. K. (April 2012). Or47b receptor neurons mediate sociosexual interactions in the fruit fly *Drosophila melanogaster*. *Journal of Biological Rhythms, 27*(2), 107–116.

Long, T. A. F., Pischedda, A., Stewart, A. D., & Rice, W. R. (December 1, 2009). A cost of sexual attractiveness to high-fitness females. *PLoS Biology, 7*(12), e1000254.

Loyau, A., Blanchet, S., Van Laere, P., Clobert, J., & Danchin, E. (2012a Oct 25). When not to copy: female fruit flies use sophisticated public information to avoid mated males. *Scientific Reports, 2*.

Loyau, A., Cornuau, J. H., Clobert, J., & Danchin, E. (2012b Dec 10). Incestuous sisters: Mate preference for brothers over unrelated males in *Drosophila melanogaster*. *PLoS ONE, 7*(12), e51293.

Lung, O., & Wolfner, M. F. (April 27, 2001). Identification and characterization of the major *Drosophila melanogaster* mating plug protein. *Insect Biochemistry and Molecular Biology, 31*(6–7), 543–551.

Lüpold, S., Pitnick, S., Berben, K. S., Blengini, C. S., Belote, J. M., & Manier, M. K. (June 25, 2013). Female mediation of competitive fertilization success in *Drosophila melanogaster*. *Proceedings of the National Academy of Sciences of the United States of America, 110*(26), 10693–10698.

Mack, P. D., Kapelnikov, A., Heifetz, Y., & Bender, M. (July 5, 2006). Mating-responsive genes in reproductive tissues of female *Drosophila melanogaster*. *Proceedings of the National Academy of Sciences of the United States of America, 103*(27), 10358–10363.

Mack, P. D., Priest, N. K., & Promislow, D. E. L. (January 22, 2003). Female age and sperm competition: last-male precedence declines as female age increases. *Proceedings of the Royal Society B: Biological Sciences, 270*(1511), 159–165.

Maisak, M. S., Haag, J., Ammer, G., Serbe, E., Meier, M., Leonhardt, A., et al. (August 8, 2013). A directional tuning map of *Drosophila* elementary motion detectors. *Nature, 500*(7461), 212–216.

Mane, S., Tompkins, L., & Richmond, R. (October 28, 1983). Male esterase 6 catalyzes the synthesis of a sex pheromone in *Drosophila melanogaster* females. *Science, 222*(4622), 419–421.

Manier, M. K., Belote, J. M., Berben, K. S., Novikov, D., Stuart, W. T., & Pitnick, S. (April 15, 2010). Resolving mechanisms of competitive fertilization success in *Drosophila melanogaster*. *Science, 328*(5976), 354–357.

Manning, A. (April 1967). The control of sexual receptivity in female *Drosophila*. *Animal Behaviour, 15*(2), 239–250.

Manoli, D. S., Foss, M., Villella, A., Taylor, B. J., Hall, J. C., & Baker, B. S. (July 21, 2005). Male-specific fruitless specifies the neural substrates of *Drosophila* courtship behaviour. *Nature, 436*(7049), 395–400.

Marcillac, F., Bousquet, F., Alabouvette, J., Savarit, F., & Ferveur, J.-F. (2005a Dec). A mutation with major effects on *Drosophila melanogaster* sex pheromones. *Genetics, 171*(4), 1617–1628.

Marcillac, F., Grosjean, Y., & Ferveur, J.-F. (2005b Feb 7). A single mutation alters production and discrimination of *Drosophila* sex pheromones. *Proceedings of the Royal Society B: Biological Sciences, 272*(1560), 303–309.

Marican, C., Duportets, L., Birman, S., & Jallon, J.-M. (August 2004). Female-specific regulation of cuticular hydrocarbon biosynthesis by dopamine in *Drosophila melanogaster*. *Insect Biochemistry and Molecular Biology, 34*(8), 823–830.

Markow, T. A. (December 2011). "Cost" of virginity in wild *Drosophila melanogaster* females. *Ecology and Evolution, 1*(4), 596–600.

Markow, T. A., Beall, S., & Matzkin, L. M. (2009). Egg size, embryonic development time and ovoviviparity in *Drosophila* species. *Journal of Evolutionary Biology, 22*(2), 430–434.

Markow, T. A., & Hanson, S. J. (January 1981). Multivariate analysis of *Drosophila* courtship. *Proceedings of the National Academy of Sciences of the United States of America, 78*(1), 430–434.

Markow, T. A., & Manning, M. (June 1980). Mating success of photoreceptor mutants of *Drosophila melanogaster*. *Behavioral and Neural Biology, 29*(2), 276–280.

Markow, T. A., & O'Grady, P. M. (December 2005). Evolutionary genetics of reproductive behaviour in *Drosophila*: connecting the dots. *Annual Review of Genetics, 39*(1), 263–291.

Markow, T. A., & O'Grady, P. (October 2008). Reproductive ecology of *Drosophila*. *Functional Ecology, 22*(5), 747–759.

Martin, J. R., Ernst, R., & Heisenberg, M. (January 1999). Temporal pattern of locomotor activity in *Drosophila melanogaster*. *Journal of Comparative Physiology A. Neuroethology, Sensory, Neural, and Behavioral Physiology, 184*(1), 73–84.

Masse, N. Y., Turner, G. C., & Jefferis, G. S. X.E. (August 25, 2009). Olfactory information processing in *Drosophila*. *Current Biology, 19*(16), R700–R713.

Matsuo, T., Sugaya, S., Yasukawa, J., Aigaki, T., & Fuyama, Y. (2007). Odorant-binding proteins OBP57d and OBP57e affect taste perception and host-plant preference in *Drosophila sechellia*. *PLoS Biology, 5*(5), e118.

McGraw, L. A., Clark, A. G., & Wolfner, M. F. (2008). *Post-mating gene expression profiles of female Drosophila melanogaster in response to time and to four male accessory gland proteins.*

McGraw, L. A., Gibson, G., Clark, A. G., & Wolfner, M. F. (August 2004). Genes regulated by mating, sperm, or seminal proteins in mated female *Drosophila melanogaster*. *Current Biology, 14*(16), 1509–1514.

Mery, F., & Kawecki, T. J. (October 29, 2002). Experimental evolution of learning ability in fruit flies. *Proceedings of the National Academy of Sciences of the United States of America, 99*(22), 14274–14279.

Mery, F., Varela, S. A. M., Danchin, E., Blanchet, S., Parejo, D., Coolen, I., et al. (May 12, 2009). Public versus personal information for mate copying in an invertebrate. *Current Biology, 19*(9), 730–734.

Middleton, C., Nongthomba, U., Parry, K., Sweeney, S., Sparrow, J., & Elliott, C. (2006). Neuromuscular organization and aminergic modulation of contractions in the *Drosophila* ovary. *BMC Biology, 4*(1), 17.

Miller, G. T., & Pitnick, S. (November 8, 2002). Sperm–female coevolution in *Drosophila*. *Science, 298*(5596), 1230–1233.

Miyamoto, T., & Amrein, H. (August 1, 2008). Suppression of male courtship by a *Drosophila* pheromone receptor. *Nature Neuroscience, 11*(8), 874–876.

Monastirioti, M. (December 1, 2003). Distinct octopamine cell population residing in the CNS abdominal ganglion controls ovulation in *Drosophila melanogaster*. *Developmental Biology, 264*(1), 38–49.

Morrow, E. H., & Innocenti, P. (August 2012). Female postmating immune responses, immune system evolution and immunogenic males. *Biological Reviews of the Cambridge Philosophical Society*, 87(3), 631–638.

Moshitzky, P., Fleischmann, I., Chaimov, N., Saudan, P., Klauser, S., Kubli, E., et al. (1996). Sex-peptide activates juvenile hormone biosynthesis in the *Drosophila melanogaster* corpus allatum. *Archives of Insect Biochemistry and Physiology*, 32(3–4), 363–374.

Mueller, J. L., Page, J. L., & Wolfner, M. F. (February 2007). An ectopic expression screen reveals the protective and toxic effects of *Drosophila* seminal fluid proteins. *Genetics*, 175(2), 777–783.

Nakano, Y., Fujitani, K., Kurihara, J., Ragan, J., Usui-Aoki, K., Shimoda, L., et al. (June 2001). Mutations in the novel membrane protein spinster interfere with programmed cell death and cause neural degeneration in *Drosophila melanogaster*. *Molecular and Cellular Biology*, 21(11), 3775–3788.

Nakayama, S., Kaiser, K., & Aigaki, T. (April 28, 1997). Ectopic expression of sex-peptide in a variety of tissues in *Drosophila* females using the P[GAL4] enhancer-trap system. *Molecular Genetics and Genomics*, 254(4), 449–455.

Neckameyer, W. S. (June 15, 1996). Multiple roles for dopamine in *Drosophila* development. *Developmental Biology*, 176(2), 209–219.

Neckameyer, W. S. (March 1998). Dopamine modulates female sexual receptivity in *Drosophila melanogaster*. *Journal of Neurogenetics*, 12(2), 101–114.

Neubaum, D. M., & Wolfner, M. F. (October 1999). Mated *Drosophila melanogaster* females require a seminal fluid protein, Acp36DE, to store sperm efficiently. *Genetics*, 153(2), 845–857.

Nighorn, A., Healy, M. J., & Davis, R. L. (March 1991). The cyclic AMP phosphodiesterase encoded by the *Drosophila* dunce gene is concentrated in the mushroom body neuropil. *Neuron*, 6(3), 455–467.

Nilsen, S. P., Chan, Y.-B., Huber, R., & Kravitz, E. A. (August 17, 2004). Gender-selective patterns of aggressive behavior in *Drosophila melanogaster*. *Proceedings of the National Academy of Sciences of the United States of America*, 101(33), 12342–12347.

O'Dell, K. M. (April 1993). The effect of the inactive mutation on longevity, sex, rhythm and resistance to p-cresol in *Drosophila melanogaster*. *Heredity*, 70(Pt 4), 393–399.

Ottiger, M., Soller, M., Stocker, R. F., & Kubli, E. (July 2000). Binding sites of *Drosophila melanogaster* sex peptide pheromones. *Journal of Neurobiology*, 44(1), 57–71.

Ödeen, A., & Moray, C. M. (September 26, 2007). *Drosophila melanogaster* virgins are more likely to mate with strangers than familiar flies. *Naturwissenschaften*, 95(3), 253–256.

Paillette, M., Ikeda, H., & Jallon, J.-M. (January 1991). A new acoustic signal of the fruitflies *Drosophila simulans* and *D. melanogaster*. *Bioacoustics*, 3(4), 247–254.

Partridge, L., & Fowler, K. (January 1990). Non-mating costs of exposure to males in female *Drosophila melanogaster*. *Journal of Insect Physiology*, 36(6), 419–425.

Partridge, L., Green, A., & Fowler, K. (January 1987). Effects of egg-production and of exposure to males on female survival in *Drosophila melanogaster*. *Journal of Insect Physiology*, 33(10), 745–749.

Pavlou, H. J., & Goodwin, S. F. (February 2013). Courtship behavior in *Drosophila melanogaster*: Towards a 'courtship connectome'. *Current Opinion in Neurobiology*, 23(1), 76–83.

Peng, J., Chen, S., Büsser, S., Liu, H., Honegger, T., & Kubli, E. (2005a Feb 8). Gradual release of sperm bound sex-peptide controls female postmating behavior in *Drosophila*. *Current Biology*, 15(3), 207–213.

Peng, J., Zipperlen, P., & Kubli, E. (2005b Sep 20). *Drosophila* sex-peptide stimulates female innate immune system after mating via the Toll and Imd pathways. *Current Biology*, 15(18), 1690–1694.

Pischedda, A., & Rice, W. R. (February 7, 2012). Partitioning sexual selection into its mating success and fertilization success components. *Proceedings of the National Academy of Sciences of the United States of America*, 109(6), 2049–2053.

Pitnick, S. (1991). Male size influences mate fecundity and remating interval in *Drosophila melanogaster*. *Animal Behaviour*.

Priest, N. K., Roach, D. A., & Galloway, L. F. (February 23, 2008). Cross-generational fitness benefits of mating and male seminal fluid. *Biology Letters*, *4*(1), 6–8.

Prokupek, A. M., Kachman, S. D., & Ladunga, I. (2009). Transcriptional profiling of the sperm storage organs of *Drosophila melanogaster*. *Insect Molecular Biology*.

Pyle, D. W., & Gromko, M. H. (1981). Genetic basis for repeated mating in *Drosophila melanogaster*. *The American Naturalist*, 133–146.

Ram, K., & Wolfner, M. (December 14, 2007). Sustained post-mating response in *Drosophila melanogaster* requires multiple seminal fluid proteins. *PLoS Genetics*, *3*(12), e238.

Ram, K. R., & Wolfner, M. F. (September 8, 2009). A network of interactions among seminal proteins underlies the long-term postmating response in *Drosophila*. *Proceedings of the National Academy of Sciences of the United States of America*, *106*(36), 15384–15389.

Ravi Ram, K., & Wolfner, M. F. (September 2007). Seminal influences: *Drosophila* Acps and the molecular interplay between males and females during reproduction. *Integrative and Comparative Biology*, *47*(3), 427–445.

Rexhepaj, A., Liu, H., Peng, J., Choffat, Y., & Kubli, E. (November 2003). The sex-peptide DUP99B is expressed in the male ejaculatory duct and in the cardia of both sexes. *European Journal of Biochemistry*, *270*(21), 4306–4314.

Rezával, C., Nojima, T., Neville, M C., Lin, A. C., & Goodwin, S. F. (2014). Sexually dimorphic octopaminergic neurons modulate female postmating behaviors in *Drosophila*. *Current Biology: CB*, *24*(7), 725–730. http://dx.doi.org/10.1016/j.cub.2013.12.051.

Rezaval, C., Pavlou, H. J., Dornan, A. J., Chan, Y.-B., Kravitz, E. A., & Goodwin, S. F. (May 30, 2012). Neural circuitry underlying *Drosophila* female postmating behavioral responses. *Current Biology*.

Riabinina, O., Dai, M., Duke, T., & Albert, J. T. (March 29, 2011). Active process mediates species-specific tuning of *Drosophila* ears. *Current Biology*, 1–7.

Ribeiro, C., & Dickson, B. J. (June 8, 2010). Sex peptide receptor and neuronal TOR/S6K signaling modulate nutrient balancing in *Drosophila*. *Current Biology*, *20*(11), 1000–1005.

Rice, W. R., Stewart, A. D., Morrow, E. H., Linder, J. E., Orteiza, N., & Byrne, P. G. (February 28, 2006). Assessing sexual conflict in the *Drosophila melanogaster* laboratory model system. *Philosophical Transactions of the Royal Society B: Biological Sciences*, *361*(1466), 287–299.

Riddiford, L. M. (December 1, 2012). How does juvenile hormone control insect metamorphosis and reproduction? *General and Comparative Endocrinology*, *179*(3), 477–484.

Rideout, E. J., Billeter, J.-C., & Goodwin, S. F. (September 4, 2007). The sex-determination genes fruitless and doublesex specify a neural substrate required for courtship song. *Current Biology*, *17*(17), 1473–1478.

Rideout, E. J., Dornan, A. J., Neville, M. C., Eadie, S., & Goodwin, S. F. (April 2010). Control of sexual differentiation and behavior by the doublesex gene in *Drosophila melanogaster*. *Nature Neuroscience*, *13*(4), 458–466.

Ritchie, M. G., Halsey, E. J., & Gleason, J. M. (1999). *Drosophila* song as a species-specific mating signal and the behavioural importance of Kyriacou & Hall cycles in *D. melanogaster* song. *Animal Behaviour*, *58*(3), 649–657.

Ritchie, M. G., Yate, V. H., & Kyriacou, C. P. (May 1994). Genetic variability of the interpulse interval of courtship song among some European populations of *Drosophila melanogaster*. *Heredity*, *72*(Pt 5), 459–464.

Robinett, C. C., Vaughan, A. G., Knapp, J.-M., & Baker, B. S. (May 2010). Sex and the single cell. II. There is a time and place for sex. *PLoS Biology*, *8*(5), e1000365.

Rockwell, R. F., & Grossfield, J. (1978). *Drosophila*: behavioral cues for oviposition. *American Midland Naturalist*.

Rodríguez-Valentín, R., López-González, I., Jorquera, R., Labarca, P., Zurita, M., & Reynaud, E. (October 2006). Oviduct contraction in *Drosophila* is modulated by a neural network that is both, octopaminergic and glutamatergic. *Journal of Cellular Physiology, 209*(1), 183–198.

Rohlfs, M., & Hoffmeister, T. S. (August 7, 2003). An evolutionary explanation of the aggregation model of species coexistence. *Proceedings of the Royal Society B: Biological Sciences, 270*(Suppl 1), S33–S35.

Ronderos, D. S., & Smith, D. P. (February 17, 2010). Activation of the T1 neuronal circuit is necessary and sufficient to induce sexually dimorphic mating behavior in *Drosophila melanogaster*. *The Journal of Neuroscience, 30*(7), 2595–2599.

Rowe, L. (1992). Convenience polyandry in a water strider: foraging conflicts and female control of copulation frequency and guarding duration. *Animal Behaviour*.

Rubinstein, C. D., & Wolfner, M. F. (October 7, 2013). *Drosophila* seminal protein ovulin mediates ovulation through female octopamine neuronal signaling. *Proceedings of the National Academy of Sciences of the United States of America*.

Ruta, V., Datta, S. R., Vasconcelos, M. L., Freeland, J., Looger, L. L., & Axel, R. (December 2, 2010). A dimorphic pheromone circuit in *Drosophila* from sensory input to descending output. *Nature, 468*(7324), 686–690.

Rybak, F., Sureau, G., & Aubin, T. (April 7, 2002). Functional coupling of acoustic and chemical signals in the courtship behaviour of the male *Drosophila melanogaster*. *Proceedings of the Royal Society B: Biological Sciences, 269*(1492), 695–701.

Ryner, L. C., Goodwin, S. F., Castrillon, D. H., Anand, A., Villella, A., Baker, B. S., et al. (December 13, 1996). Control of male sexual behavior and sexual orientation in *Drosophila* by the fruitless gene. *Cell, 87*(6), 1079–1089.

Rytz, R., Croset, V., & Benton, R. (March 1, 2013). Ionotropic receptors (IRs): Chemosensory ionotropic glutamate receptors in *Drosophila* and beyond. *Insect Biochemistry and Molecular Biology*.

Saad, M., Game, A. Y., Healy, M. J., & Oakeshott, J. G. (1994). Associations of esterase 6 allozyme and activity variation with reproductive fitness in *Drosophila melanogaster*. *Genetica, 94*(1), 43–56.

Sakai, T., & Ishida, N. (July 31, 2001). Circadian rhythms of female mating activity governed by clock genes in *Drosophila*. *Proceedings of the National Academy of Sciences of the United States of America, 98*(16), 9221–9225.

Sakai, T., Kasuya, J., Kitamoto, T., & Aigaki, T. (July 2009). The *Drosophila* TRPA channel, Painless, regulates sexual receptivity in virgin females. *Genes, Brain and Behavior, 8*(5), 546–557.

Sakurai, A., Koganezawa, M., Yasunaga, K.-I., Emoto, K., & Yamamoto, D. (2013). Select interneuron clusters determine female sexual receptivity in *Drosophila*. *Nature Communications, 4*, 1825.

Salmon, A. B., Marx, D. B., & Harshman, L. G. (August 2001). A cost of reproduction in *Drosophila melanogaster*: Stress susceptibility. *Evolution, 55*(8), 1600–1608.

Saltz, J. B., & Foley, B. R. (May 2011). Natural genetic variation in social niche construction: Social effects of aggression drive disruptive sexual selection in *Drosophila melanogaster*. *The American Naturalist, 177*(5), 645–654.

Sarin, S., & Dukas, R. (December 22, 2009). Social learning about egg-laying substrates in fruitflies. *Proceedings of the Royal Society B: Biological Sciences, 276*(1677), 4323–4328.

Saudan, P., Hauck, K., Soller, M., Choffat, Y., Ottiger, M., Spörri, M., et al. (February 2002). Ductus ejaculatorius peptide 99B (DUP99B), a novel *Drosophila melanogaster* sex-peptide pheromone. *European Journal of Biochemistry, 269*(3), 989–997.

Savarit, F., Sureau, G., Cobb, M., & Ferveur, J. F. (August 3, 1999). Genetic elimination of known pheromones reveals the fundamental chemical bases of mating and isolation in *Drosophila*. *Proceedings of the National Academy of Sciences of the United States of America, 96*(16), 9015–9020.

Schlief, M. L., & Wilson, R. I. (May 2007). Olfactory processing and behavior downstream from highly selective receptor neurons. *Nature Neuroscience, 10*(5), 623–630.

Schmidt, T., Choffat, Y., Klauser, S., & Kubli, E. (1993). The *Drosophila melanogaster* sex-peptide: A molecular analysis of structure–function relationships. *Journal of Insect Physiology.*

Schnakenberg, S. L., Siegal, M. L., & Bloch Qazi, M. C. (July 1, 2012). Oh, the places they'll go: Female sperm storage and sperm precedence in *Drosophila melanogaster. Spermatogenesis, 2*(3), 224–235.

Schneider, J., Atallah, J., & Levine, J. D. (2012). One, two, and many–a perspective on what groups of *Drosophila melanogaster* can tell us about social dynamics. *Advances in Genetics, 77*, 59–78.

Schwartz, N. U., Zhong, L., Bellemer, A., & Tracey, W. D. (2012). Egg laying decisions in *Drosophila* are consistent with foraging costs of larval progeny. *PLoS ONE, 7*(5), e37910.

Scott, D. (1994). Genetic variation for female mate discrimination in *Drosophila melanogaster. Evolution, 112*–121.

Scott, D., & Richmond, R. C. (January 1987). Evidence against an antiaphrodisiac role for cis-vaccenyl acetate in *Drosophila melanogaster. Journal of Insect Physiology, 33*(5), 363–369.

Scott, D., Shields, A., Straker, M., Dalrymple, H., Dhillon, P. K., & Harbinder, S. (2011). Variation in the male pheromones and mating success of wild caught *Drosophila melanogaster. PLoS ONE, 6*(8), e23645. http://dx.doi.org/10.1371/journal.pone.0023645.

Seeley, C., & Dukas, R. (March 2011). Teneral matings in fruit flies: male coercion and female response. *Animal Behaviour, 81*(3), 595–601.

Semmelhack, J. L., & Wang, J. W. (May 14, 2009). Select *Drosophila* glomeruli mediate innate olfactory attraction and aversion. *Nature, 459*(7244), 218–223.

Shanbhag, S., Müller, B., & Steinbrecht, R. (1999). Atlas of olfactory organs of *Drosophila melanogaster* 1. Types, external organization, innervation and distribution of olfactory sensilla. *International Journal of Insect Morphology and Embryology, 28*(4), 377–397.

Sharon, G., Segal, D., Ringo, J. M., Hefetz, A., Zilber-Rosenberg, I., & Rosenberg, E. (November 16, 2010). Commensal bacteria play a role in mating preference of *Drosophila melanogaster. Proceedings of the National Academy of Sciences of the United States of America, 107*(46), 20051–20056.

Shirangi, T. R., Dufour, H. D., Williams, T. M., & Carroll, S. B. (2009). Rapid evolution of sex pheromone-producing enzyme expression in *Drosophila. PLoS Biology, 7*(8), e1000168.

Shorey, H. H., & Bartell, R. J. (1970). Role of a volatile female sex pheromone in stimulating male courtship behaviour in *Drosophila melanogaster. Animal Behaviour, 18*, 159–164.

Short, S. M., & Lazzaro, B. P. (2013). Reproductive status alters transcriptomic response to infection in female *Drosophila melanogaster. G3: Genes, Genomes and Genetics.*

Silbering, A. F., Rytz, R., Grosjean, Y., Abuin, L., Ramdya, P., Jefferis, G. S. X.E., et al. (September 21, 2011). Complementary function and integrated wiring of the evolutionarily distinct *Drosophila* olfactory subsystems. *The Journal of Neuroscience, 31*(38), 13357–13375.

Singh, D. S. R., Singh, D. B. N., & Hoenigsberg, D. H. F. (September 1, 2002). Female remating, sperm competition and sexual selection in *Drosophila. Genetics and Molecular Research.*

Sirot, L. K., Wolfner, M. F., & Wigby, S. (June 14, 2011). Protein-specific manipulation of ejaculate composition in response to female mating status in *Drosophila melanogaster. Proceedings of the National Academy of Sciences of the United States of America, 108*(24), 9922–9926.

Siwicki, K. K., & Kravitz, E. A. (April 2009). Fruitless, doublesex and the genetics of social behavior in *Drosophila melanogaster. Current Opinion in Neurobiology, 19*(2), 200–206.

Slatyer, R. A., Mautz, B. S., Backwell, P. R. Y., & Jennions, M. D. (May 5, 2011). Estimating genetic benefits of polyandry from experimental studies: a meta-analysis. *Biological Reviews, 87*(1), 1–33.

Smith, D. T., Hosken, D. J., ffrench-Constant, R. H., & Wedell, N. (August 2009). Variation in sex peptide expression in *D. melanogaster. Genetics Research, 91*(4), 237–242.

Smith, D. T., Sirot, L. K., Wolfner, M. F., Hosken, D. J., & Wedell, N. (October 2012). The consequences of genetic variation in sex peptide expression levels for egg laying and retention in females. *Heredity, 109*(4), 222–225.

Snook, R. R., & Hosken, D. J. (April 29, 2004). Sperm death and dumping in *Drosophila. Nature, 428*(6986), 939–941.

Soller, M., Bownes, M., & Kubli, E. (April 1999). Control of oocyte maturation in sexually mature *Drosophila* females. *Cell, 208*(2), 337–351.

Soller, M., Haussmann, I. U., Hollmann, M., Choffat, Y., White, K., Kubli, E., et al. (September 19, 2006). Sex-peptide-regulated female sexual behavior requires a subset of ascending ventral nerve cord neurons. *Current Biology, 16*(18), 1771–1782.

Söderberg, J. A. E., Carlsson, M. A., & Nässel, D. R. (2012). Insulin-producing cells in the *Drosophila* brain also express satiety-inducing cholecystokinin-like peptide, drosulfakinin. *Frontiers in Endocrinology (Lausanne), 3*, 109.

Spieth, H. T. (January 1974). Courtship behavior in *Drosophila. Annual Review of Entomology, 19*(1), 385–405.

Stensmyr, M. C., Dweck, H. K. M., Farhan, A., Ibba, I., Strutz, A., Mukunda, L., et al. (December 7, 2012). A conserved dedicated olfactory circuit for detecting harmful microbes in *Drosophila. Cell, 151*(6), 1345–1357.

Stocker, R. F. (January 1994). The organization of the chemosensory system in *Drosophila melanogaster*: a rewiew. *Cell and Tissue Research, 275*(1), 3–26.

Stockinger, P., Kvitsiani, D., Rotkopf, S., Tirián, L., & Dickson, B. J. (June 3, 2005). Neural circuitry that governs *Drosophila* male courtship behavior. *Cell, 121*(5), 795–807.

Stökl, J., Strutz, A., Dafni, A., Svatos, A., Doubsky, J., Knaden, M., et al. (October 26, 2010). A deceptive pollination system targeting drosophilids through olfactory mimicry of yeast. *Current Biology, 20*(20), 1846–1852.

Suzuki, K., Juni, N., & Yamamoto, D. (February 25, 1997). Enhanced mate refusal in female *Drosophila* induced by a mutation in the spinster locus. Applied entomology and zoology. *Japanese Society of Applied Entomology and Zoology, 32*(1), 235–243.

Swanson, W. J., Wong, A., Wolfner, M. F., Aquardo, C. F. (November 1, 2004). Evolutionary expressed sequence tag analysis of *Drosophila* female reproductive tracts identifies genes subjected to positive selection. *Genetics, 168*(3), 1457–1465.

Swanson, W. J., Clark, A. G., Waldrip-Dail, H. M., Wolfner, M. F., & Aquadro, C. F. (June 19, 2001). Evolutionary EST analysis identifies rapidly evolving male reproductive proteins in *Drosophila. Proceedings of the National Academy of Sciences of the United States of America, 98*(13), 7375–7379.

Symonds, M. R. E., & Wertheim, B. (September 2005). The mode of evolution of aggregation pheromones in *Drosophila* species. *Journal of Evolutionary Biology, 18*(5), 1253–1263.

Szabad, J., & Fajszi, C. (January 1982). Control of female reproduction in *Drosophila*: genetic dissection using gynandromorphs. *Genetics, 100*(1), 61–78.

Takahashi, A., Tsaur, S. C., Coyne, J. A., & Wu, C. I. (March 27, 2001). The nucleotide changes governing cuticular hydrocarbon variation and their evolution in *Drosophila melanogaster. Proceedings of the National Academy of Sciences of the United States of America, 98*(7), 3920–3925.

Takamura, T., & Fuyama, Y. (January 1980). Behavior genetics of choice of oviposition sites in *Drosophila melanogaster*. I. Genetic variability and analysis of behavior. *Behavior Genetics, 10*(1), 105–120.

Takemori, N., & Yamamoto, M.-T. (May 2009). Proteome mapping of the *Drosophila melanogaster* male reproductive system. *Proteomics, 9*(9), 2484–2493.

Takeo, S., Tsuda, M., Akahori, S., Matsuo, T., Aigaki, T. (Jul 25, 2006). The calcineurin regulator sra plays an essential role in female meiosis in *Drosophila. Current Biology, 16*(14), 1435–1440.

Talyn, B., & Dowse, H. (2004). The role of courtship song in sexual selection and species recognition by female *Drosophila melanogaster. Animal Behaviour, 68*, 1165–1180.

Tan, C. K. W., Lovlie, H., Greenway, E., Goodwin, S. F., Pizzari, T., & Wigby, S. (September 25, 2013). Sex-specific responses to sexual familiarity, and the role of olfaction in *Drosophila*. *Proceedings of the Royal Society B: Biological Sciences, 280*(1771). 20131691–1.

Tan, C. K. W., Løvlie, H., Pizzari, T., & Wigby, S. (June 2012). No evidence for pre-copulatory inbreeding avoidance in *Drosophila melanogaster*. *Animal Behaviour, 83*(6), 1433–1441.

Tauber, E., Roe, H., Costa, R., Hennessy, J. M., & Kyriacou, C. P. (January 21, 2003). Temporal mating isolation driven by a behavioral gene in *Drosophila*. *Current Biology, 13*(2), 140–145.

Terhzaz, S., Rosay, P., Goodwin, S. F., & Veenstra, J. A. (January 12, 2007). The neuropeptide SIFamide modulates sexual behavior in *Drosophila*. *Biochemical and Biophysical Research Communications, 352*(2), 305–310.

Tinette, S., Zhang, L., & Robichon, A. (February 2004). Cooperation between *Drosophila* flies in searching behavior. *Genes, Brain and Behavior, 3*(1), 39–50.

Tompkins, L., Gross, A. C., Hall, J. C., Gailey, D. A., & Siegel, R. W. (1982). The role of female movement in the sexual behavior of *Drosophila melanogaster*. *Behavior Genetics, 12*(3), 295–307.

Tompkins, L., & Hall, J. C. (February 1983). Identification of brain sites controlling female receptivity in mosaics of *Drosophila melanogaster*. *Genetics, 103*(2), 179–195.

Tompkins, L., Hall, J. C., & Hall, L. M. (January 1980). Courtship-stimulating volatile compounds from normal and mutant *Drosophila*. *Journal of Insect Physiology, 26*(10), 689–697.

Tootoonian, S., Coen, P., Kawai, R., & Murthy, M. (January 18, 2012). Neural representations of courtship song in the *Drosophila* brain. *The Journal of Neuroscience, 32*(3), 787–798.

Tracey, W. D., Jr., Wilson, R. I., Laurent, G., & Benzer, S. (April 2003). Painless, a *Drosophila* gene essential for nociception. *Cell, 113*(2), 261–273.

Trevitt, S., Fowler, K., & Partridge, L. (1988). An effect of egg-deposition on the subsequent fertility and remating frequency of female *Drosophila melanogaster*. *Journal of Insect Physiology, 34*(8), 821–828.

Trott, A. R., Donelson, N. C., Griffith, L. C., & Ejima, A. (September 25, 2012). Song choice is modulated by female movement in *Drosophila* males. *PLoS ONE, 7*(9), e46025.

Turiegano, E., Monedero, I., Pita, M., Torroja, L., & Canal, I. (June 1, 2012). Effect of *Drosophila melanogaster* female size on male mating success. *Journal of Insect Behavior, 26*(1), 89–100.

Turner, T. L., Miller, P. M., & Cochrane, V. A. (August 20, 2013). Combining genome-wide methods to investigate the genetic complexity of courtship song variation in *Drosophila melanogaster*. *Molecular Biology and Evolution, 30*(9), 2113–2120.

Ueda, A., & Kidokoro, Y. (March 2002). Aggressive behaviours of female *Drosophila melanogaster* are influenced by their social experience and food resources. *Physiological Entomology, 27*(1), 21–28.

Usui-Aoki, K., Ito, H., Ui-Tei, K., Takahashi, K., Lukacsovich, T., Awano, W., et al. (August 2000). Formation of the male-specific muscle in female *Drosophila* by ectopic fruitless expression. *Nature Cell Biology, 2*(8), 500–506.

Vander Meer, R. K., Obin, M. S., Zawistowski, S., Sheehan, K. B., & Richmond, R. C. (January 1986). A reevaluation of the role of cis-vaccenyl acetate, cis-vaccenol and esterase 6 in the regulation of mated female sexual attractiveness in *Drosophila melanogaster*. *Journal of Insect Physiology, 32*(8), 681–686.

Vargas, M. A., Luo, N., Yamaguchi, A., & Kapahi, P. (June 8, 2010). A role for S6 kinase and serotonin in postmating dietary switch and balance of nutrients in *D. melanogaster*. *Current Biology, 20*(11), 1006–1011.

Villella, A., & Hall, J. C. (2008). Neurogenetics of courtship and mating in *Drosophila*. *Advances in Genetics, 62*, 67–184.

Vosshall, L. B., & Stocker, R. F. (2007). Molecular architecture of smell and taste in *Drosophila*. *Annual Review of Neuroscience, 30*, 505–533.

Vrontou, E., Nilsen, S. P., Demir, E., Kravitz, E. A., & Dickson, B. J. (December 1, 2006). Fruitless regulates aggression and dominance in *Drosophila*. *Nature Neuroscience, 9*(12), 1469–1471.

Wang, L., & Anderson, D. J. (January 14, 2010). Identification of an aggression-promoting pheromone and its receptor neurons in *Drosophila*. *Nature, 463*(7278), 227–231.

Wang, L., Han, X., Mehren, J., Hiroi, M., Billeter, J.-C., Miyamoto, T., et al. (June 2011). Hierarchical chemosensory regulation of male–male social interactions in *Drosophila*. *Nature Neuroscience, 14*(6), 757–762.

Waterbury, J. A., Jackson, L. L., & Schedl, P. (August 1, 1999). Analysis of the doublesex female protein in *Drosophila melanogaster*: role on sexual differentiation and behavior and dependence on intersex. *Genetics, 152*(4), 1653–1667.

Wertheim, B. (2005). Evolutionary ecology of communication signals that induce aggregative behaviour. *Oikos*.

Wertheim, B., Allemand, R., Vet, L. E., & Dicke, M. (2006). Effects of aggregation pheromone on individual behaviour and food web interactions: a field study on *Drosophila*. *Ecological Entomology, 31*(3), 216–226.

Wertheim, B., Dicke, M., & Vet, L. E. (2002a). Behavioural plasticity in support of a benefit for aggregation pheromone use in *Drosophila melanogaster*. *Entomologia Experimentalis et Applicata, 103*(1), 61–71.

Wertheim, B., Marchais, J., Vet, L. E. M., & Dicke, M. (2002b Oct). Allee effect in larval resource exploitation in *Drosophila*: an interaction among density of adults, larvae, and micro-organisms. *Ecological Entomology, 27*(5), 608–617.

Wheeler, D., Kyriacou, C., Greenacre, M., Yu, Q., Rutila, J., Rosbash, M., et al. (March 1, 1991). Molecular transfer of a species-specific behavior from *Drosophila simulans* to *Drosophila melanogaster*. *Science, 251*(4997), 1082–1085.

Wicker, C., & Jallon, J. M. (January 1995). Hormonal control of sex pheromone biosynthesis in *Drosophila melanogaster*. *Journal of Insect Physiology, 41*(1), 65–70.

Wicker-Thomas, C., & Hamann, M. (October 2008). Interaction of dopamine, female pheromones, locomotion and sex behavior in *Drosophila melanogaster*. *Journal of Insect Physiology, 54*(10–11), 1423–1431.

Wigby, S., & Chapman, T. (May 2004). Female resistance to male harm evolves in response to manipulation of sexual conflict. *Evolution, 58*(5), 1028–1037.

Wigby, S., & Chapman, T. (February 22, 2005). Sex peptide causes mating costs in female *Drosophila melanogaster*. *Current Biology, 15*(4), 316–321.

Wigby, S., Sirot, L. K., Linklater, J. R., Buehner, N., Calboli, F. C. F., Bretman, A., et al. (May 12, 2009). Seminal fluid protein allocation and male reproductive success. *Current Biology, 19*(9), 751–757.

Wigby, S., Slack, C., Grönke, S., Martinez, P., Calboli, F. C. F., Chapman, T., et al. (February 7, 2011). Insulin signalling regulates remating in female *Drosophila*. *Proceedings of the Royal Society B: Biological Sciences, 278*(1704), 424–431.

Wilson, T. G., DeMoor, S., & Lei, J. (December 2003). Juvenile hormone involvement in *Drosophila melanogaster* male reproduction as suggested by the Methoprene-tolerant(27) mutant phenotype. *Insect Biochemistry and Molecular Biology, 33*(12), 1167–1175.

Wolfner, M. F. (February 2002). The gifts that keep on giving: physiological functions and evolutionary dynamics of male seminal proteins in *Drosophila*. *Heredity, 88*(2), 85–93.

Wolfner, M. F. (2009). Battle and ballet: molecular interactions between the sexes in *Drosophila*. *Journal of Heredity*, 399–410.

Wolfner, M. F. (November 2011). Precious essences: female secretions promote sperm storage in *Drosophila*. *PLoS Biology, 9*(11), e1001191.

Wong, A., Albright, S. N., Giebel, J. D., Ram, K. R., Ji, S., Fiumera, A. C., et al. (October 2008). A role for Acp29AB, a predicted seminal fluid lectin, in female sperm storage in *Drosophila melanogaster*. *Genetics, 180*(2), 921–931.

Wu, Q., & Brown, M. R. (January 2006). Signaling and function of insulin-like peptides in insects. *Annual Review of Entomology, 51*(1), 1–24.

Xu, P., Atkinson, R., Jones, D. N. M., & Smith, D. P. (January 20, 2005). *Drosophila* OBP LUSH is required for activity of pheromone-sensitive neurons. *Neuron, 45*(2), 193–200.

Xue, L., & Noll, M. (March 28, 2000). *Drosophila* female sexual behavior induced by sterile males showing copulation complementation. *Proceedings of the National Academy of Sciences of the United States of America, 97*(7), 3272–3275.

Yamamoto, D., & Koganezawa, M. (September 20, 2013). Genes and circuits of courtship behaviour in *Drosophila* males. *Nature Reviews Neuroscience, 14*(10), 681–692.

Yang, C.-H., Rumpf, S., Xiang, Y., Gordon, M. D., Song, W., Jan, L. Y., et al. (February 26, 2009). Control of the postmating behavioral switch in *Drosophila* females by internal sensory neurons. *Neuron, 61*(4), 519–526.

Yang, C. H., Belawat, P., Hafen, E., Jan, L. Y., & Jan, Y. N. (March 21, 2008). *Drosophila* egg-laying site selection as a system to study simple decision-making processes. *Science, 319*(5870), 1679–1683.

Yapici, N., Kim, Y.-J., Ribeiro, C., & Dickson, B. J. (July 22, 2008). A receptor that mediates the post-mating switch in *Drosophila* reproductive behaviour. *Nature, 451*(7174), 1675–1683.

Yew, J. Y., Cody, R. B., & Kravitz, E. A. (May 20, 2008). Cuticular hydrocarbon analysis of an awake behaving fly using direct analysis in real-time time-of-flight mass spectrometry. *Proceedings of the National Academy of Sciences of the United States of America, 105*(20), 7135–7140.

Yew, J. Y., Dreisewerd, K., Luftmann, H., Müthing, J., Pohlentz, G., & Kravitz, E. A. (August 11, 2009). A new male sex pheromone and novel cuticular cues for chemical communication in *Drosophila*. *Current Biology, 19*(15), 1245–1254.

Yu, J. Y., Kanai, M. I., Demir, E., Jefferis, G. S. X.E., & Dickson, B. J. (September 28, 2010). Cellular organization of the neural circuit that drives *Drosophila* courtship behavior. *Current Biology, 20*(18), 1602–1614.

Zhu, J., Park, K.-C., & Baker, T. C. (March 14, 2003). Identification of odors from overripe mango that attract vinegar flies, *Drosophila melanogaster*. *Journal of Chemical Ecology, 29*(4), 899–909.

CHAPTER TWO

# Association Mapping in Crop Plants: Opportunities and Challenges

**Pushpendra K. Gupta\* Pawan L. Kulwal[†,1] and Vandana Jaiswal\***
\*Department of Genetics and Plant Breeding, Ch. Charan Singh University, Meerut, UP, India
†State Level Biotechnology Centre, Mahatma Phule Agricultural University, Rahuri, MS, India
[1]Corresponding author: e-mail address: pawankulwal@gmail.com

## Contents

*Advances in Genetics*, Volume 85
ISSN 0065-2660
http://dx.doi.org/10.1016/B978-0-12-800271-1.00002-0

## Abstract

The research area of association mapping (AM) is currently receiving major attention for genetic studies of quantitative traits in all major crops. However, the level of success and utility of AM achieved for crop improvement is not comparable to that in the area of human health care for diagnosis of complex human diseases. These AM studies in plants, as in humans, became possible due to the availability of DNA-based molecular markers and a variety of sophisticated statistical tools that are evolving on a regular basis. In this chapter, we first briefly review the significance of a variety of populations that are used in AM studies, then briefly describe the molecular markers and high-throughput genotyping strategies, and finally describe the approaches used for AM studies. The major part of the chapter is, however, devoted to analysis of reasons why the results of AM have been underutilized in plant breeding. We also examine the opportunities available and challenges faced while using AM for crop improvement programs. This includes a detailed discussion of the issues that have plagued AM studies, and the solutions that have become available to deal with these issues, so that in future, the results of AM studies may prove increasingly fruitful for crop improvement programs.

## 1. INTRODUCTION

In recent years, quantitative genetics assumed a central place in genetic studies for all major crops. This has become possible due to the availability of DNA-based molecular markers and a variety of sophisticated statistical tools that are evolving on a regular basis. Using biparental populations and interval mapping (IM), thousands of markers associated with quantitative trait loci (QTL) have been detected for a variety of quantitative traits in a number of crops. However, it is apparent that the QTL that are detected using biparental populations through IM are relevant only for those breeding programs, which involve parents that differ for the QTL so detected. However, the parents of a proposed crossing/breeding program may or may not differ for this QTL. For this reason alone, a large number of such QTLs detected through IM could not be used in actual breeding programs (Bernardo, 2008), thus making these marker–trait associations (MTAs) only partly useful for marker-assisted selection (MAS) in a wider spectrum of breeding programs. In order to overcome this problem, multiple biparental populations, or alternatively, multiple-line cross QTL mapping have been used (for reviews, see Steinhoff et al., 2011; Wurschum, 2012). Although the power of detection in linkage-based QTL-IM is relatively high, the resolution is poor due to strong linkage disequilibrium (LD), making fine mapping with high resolution difficult. Moreover, QTL-IM involving development and use of a biparental population is cost- and time-intensive, which sometimes discourages the use of QTL-IM for genetic studies (Parisseaux & Bernardo, 2004).

LD-based association mapping (AM) is an alternative approach, which uses a set of genotypes of known/unknown ancestry. This set is designed to carry most of the genetic variability for the trait of interest (including multiple alleles at each locus) and represents the products of hundreds of historic recombination cycles, thus providing higher resolution during QTL mapping (Mackay, Stone, & Ayroles, 2009). However, associations detected in AM are often spurious because associations are based on LD, which depends not only on linkage but also on population stratification and relatedness among individuals. In order to overcome this problem, family-based AM has been proposed (Rosyara, Gonzalez-Hernandez, Glover, Gedye, & Stein, 2009; Stich et al., 2006). Consequently, LD-based AM can be either population-based or family-based (Yu & Buckler, 2006). Efforts have also been made to combine linkage-based QTL-IM with LD-based AM, and conduct joint linkage AM (JLAM) to overcome the limitations and exploit the benefits associated with each of the two approaches (linkage and LD). Despite the above merits of AM in providing high resolution, and use of relatively wider range of genetic variability for one or more traits of interest, there are only few published reports of crop breeding, where MTAs discovered through AM have been put to use in crop improvement programs for the purpose of MAS (there may be some examples of the use of AM, which are not documented).

The purpose of this chapter is to examine the reasons why the results of AM have been underutilized in plant breeding. In addition, we propose to examine the opportunities available and challenges faced while using AM for crop improvement programs. We first briefly describe the material and the genotyping technologies that are available and the different approaches that can be used for conducting AM. For this purpose, genome-wide association scans/studies (GWAS) and candidate gene (CG) approaches will be briefly examined, outlining situations, where each can be profitably utilized.

This will be followed by a detailed discussion of the issues that have plagued AM studies, and the solutions that have become available to deal with these issues. In doing so, we will not discuss any details of the procedures that are followed in conducting either the QTL-IM or the AM studies since these are widely known and discussed in the available literature (Doerge, 2002; Wurschum, 2012; Zhu, Gore, Buckler, & Yu, 2008).

## 2. POPULATIONS USED FOR AM STUDIES (MAPPING POPULATIONS, GERMPLASM AND BREEDING POPULATIONS)

The success of any AM program largely depends on the type of material/population used. Both historic germplasm and family-based populations can be used for AM. Breeding populations can also be used for AM, and offer some advantages, while using AM for crop improvement programs.

### 2.1 Biparental and Breeding Populations

In most QTL-IM/AM studies, one or more family-based mapping populations (biparental or multiparental mapping populations) can be used. However, a set of diverse genotypes constituting an association panel can be used only for AM and not for QTL-IM. Breeding populations have also been used for QTL-IM and AM (Stich et al., 2006; Wurschum, 2012), which is attractive for the following reasons. First, since phenotypic data are routinely generated in breeding programs, the genetic architecture underlying important agronomic traits in crops using breeding populations can be dissected with low cost and with only a little extra effort for genotyping. Second, the genetic base in a breeding population is relatively narrow, but the QTL detected in elite breeding material are of direct relevance for breeders. In contrast, many QTL detected through a population derived from crosses between elite and exotic material are useful only for the introgression of traits from exotic material, but not for the selection within elite breeding material that is being handled in a breeding program (Jannink, Bink, & Jansen, 2001). As there are several biparental populations and their respective linkage maps already available in different crops, one may want to evaluate some of these populations for a given trait under similar environmental conditions. Use of a common set of markers for separate joint linkage analysis and AM or JLAM can also prove rewarding (see section 5.4.1 for some details).

## 2.2 Multiparental Mapping Populations

The use of multiparental populations for QTL mapping was first proposed as early as 1990s (Rebai & Goffinet, 1993). This involved an extension of a biparental population to a quadriparental population, in which four parents were crossed in a half diallel. Statistical methods for the analysis of multiparental populations also became available (Jannink & Wu, 2003; Rebai & Goffinet, 2000; Xu, 1998). Subsequently, different mating designs were recommended for developing populations for QTL-IM and AM (Blanc, Charcosset, Mangin, Gallais, & Moreau, 2006; Buckler et al., 2009; Kover et al., 2009; Paulo, Boer, Huang, Koornneef, & van Eeuwijk, 2008; Stich, 2009; Yu, Holland, McMullen, & Buckler, 2008). These designs differ in complexity of the required crosses. For instance, Liu et al. (2011) used 930 genotypes of maize (292 $F_3$ and 638 DHs) that were crossed to a tester, the tester itself being an elite inbred line from the opposite European heterotic pools. Similarly, in sugarbeet, Reif et al. (2010) developed nine families (S1, S2, and DH) through random crosses among 10 inbreds, and produced test-cross progenies by mating every individual of nine families with the diploid pollinator from the male heterotic pool. Three- and four-way crosses among 49 wheat genotypes were also used for developing 82 families (Rosyara et al., 2009).

Two designs involving multiparental populations which became popular included multiparental advanced generation intercrosses (MAGIC) suggested by Cavanagh, Morell, Mackay, and Powell (2008) and nested association–mapping (NAM) strategy suggested by Ed Buckler (Yu et al., 2008). MAGIC populations involve the use of 4, 8, or 16 parents in biparental crosses, followed by crosses between $F_1$ hybrids. In wheat, a MAGIC population has been developed and used for QTL analysis for plant height and hectoliter weight (Huang et al., 2012). In rice, at IRRI, four different MAGIC populations involving a number of both indica and japonica parents have been developed, which are being used both for precise QTL mapping and for variety development (Bandillo et al., 2013). A MAGIC population may also be derived through intermating of parental inbreds (PIs) for four generations (Kover et al., 2009). In contrast to MAGIC, NAM involves use of crosses between one common PI line with a number of founder PIs (for example, 25 as in case of maize). The important advantage which the NAM population offers is that one can conduct JLAM, which is considered superior to QTL IM and AM for QTL detection. However, it is not necessary to develop a NAM population in all crops. NAM populations using more than one common parent or fewer founder parents have also been developed in *Arabidopsis* (Bentsink et al., 2010; Brachi et al., 2010).

Sometimes, a small number of PIs may also be used for diallel crosses for developing a multiparental population. In *Arabidopsis*, four PIs were used and crossed in a diallel to derive six $F_2$ populations, which were used to obtain $F_3$ and topcross populations that were used for QTL-IM (Rebai & Goffinet, 1993). Sibling mating within a biparental population has also been found to increase the mapping resolution (Lee et al., 2002). The above multiple biparental populations or the multiparental mapping populations differ in the following respects: (i) constitution of the genome of each individual being derived from several parental genomes, (ii) the number of recombination breakpoints, and (iii) the marker/QTL allele frequencies. This in turn will influence the power to detect QTL but the relative contribution of each of the above factors to increasing the power will be unknown in practice, although in theory one can perhaps estimate it. These populations designed for QTL-IM can also be used for AM.

In summary, the family–based populations (Figure 2.1) that have been used successfully for AM include the following: (i) $F_3$ families derived from

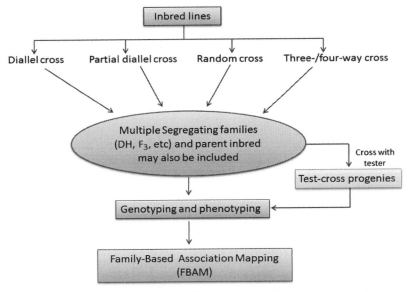

**Figure 2.1** A graphical representation of the protocol for family-based association mapping, where a number of inbred lines can be crossed following any of the different available designs to produce multiple segregating families, which can also be crossed with a tester to produce test-cross progenies. The genotypic and phenotypic data on these multiple segregating families and/or test-cross progenies can be used for association mapping. (For color version of this figure, the reader is referred to the online version of this book.)

a set of diallel crosses using a limited number of inbreds, and used for test crosses by mating every individual of $F_{3.4}$ family and the PIs to one elite inbred tester, unrelated by pedigree. (ii) Families derived from random crosses and a test cross. (iii) Families derived from three-way and four-way crosses (followed by backcrosses and testcrosses). Because of the advantages which plant systems offer over humans in terms of development of populations based on multiple parents, family-based AM is more powerful in these species (Guo, Wang, Guo, & Beavis, 2013). Some of the methods for conducting family-based AM are discussed later in this chapter.

## 3. MOLECULAR MARKERS AND GENOTYPING FOR QTL/AM STUDIES (INCLUDING NEXT GENERATION SEQUENCING)

### 3.1 Single Marker-Based Association Mapping

For QTL-IM and AM studies, a variety of molecular markers have been utilized during the last two decades. More recently, AM studies have also been facilitated by the availability of high-throughput and low-cost next generation sequencing (NGS) platforms, so that much of the genotyping work can now be easily outsourced in a cost-effective manner. These NGS platforms are being extensively utilized for *de novo* development of markers and also for genotyping. In addition to single nucleotide polymorphisms (SNPs) and DArT markers, other newer marker types can also be designed using copy number variations (CNVs), presence and absence variations (PAVs), insertion–deletions (InDels), and insertion-site-based polymorphisms (ISBPs), which are being discovered now in a number of crops (Edwards & Gupta, 2013). The recent development of genotyping by sequencing (GBS) technology at a reasonable cost even in crops, where a reference genome sequence is missing, has also made prior development of markers in a crop unnecessary (Elshire et al., 2011; Poland & Rife, 2012). The technique of GBS has recently been successfully utilized for AM studies in maize (Lipka et al., 2013), sorghum (Morris et al., 2013), and wheat (Saintenac, Jiang, Wang, & Akhunov, 2013).

### 3.2 Haplotype-Based Association Mapping

AM is routinely conducted using one marker at a time. Although, such methods employing only one marker at a time may identify significant associations, the haplotype-based analysis comprising multiple markers may be more effective in providing additional power to the analysis (Liu, Zhang, & Zhao, 2008). With availability of large numbers of markers, particularly

SNPs in many crops and also the recent advances in marker techniques discussed above, haplotype analysis would play a critical role in AM studies. Haplotype-based studies have been successfully carried out mostly in humans due to the availability of data from the HapMap project (Li, Ding, & Abecasis, 2006; Tachmazidou, Verzilli, & Iorio, 2007). Similar efforts in maize and other crops offer further scope for such studies in plant systems (Hao et al., 2012; Inghelandt, Melchinger, Martinant, & Stich, 2012; Lipka et al., 2013; Weber, Zhao, McMullen, & Doebley, 2009).

## 3.3 Imputation of Genotyping Data

Although the recent advances in NGS and GBS techniques made it possible to generate millions of markers, the frequency of missing data is very high and can pose problems during association analysis. The most convenient way to deal with this issue is imputation of genotypes, which can replace missing genotypes with those of the predicted marker alleles. This is done on the basis of observed genotypes at neighboring markers, thereby reducing cost and time in genotyping again for the missing values. Imputation of the missing genotypic data has been a common practice in many genetic studies including AM to boost power, to fine map associations and to facilitate the pooling of results across studies using meta-analysis (for further details, see Asimit & Zeggini, 2010; Balding, 2006; Marchini & Howie, 2010). Importance of data imputation was recently demonstrated in an AM study in rice (Huang et al., 2010), where data imputation reduced the missing genotypes from 61.7 to 2.9% with an accuracy of more than 98%. Although, imputation has increased the utility of these newer marker systems, caution should be exercised while using imputation for the rare variants (Asimit & Zeggini, 2010). For species that do not have an ordered genome sequence, alternative imputation methods have recently been developed that do not require ordered markers (Rutkoski, Poland, Jannink, & Sorrells, 2013). In their study, Random Forest was the best algorithm and up to 50% missing data could be imputed with minimal loss of accuracy.

# 4. APPROACHES USED FOR AM STUDIES (GWAS AND CG APPROACH)

For AM studies, two widely discussed approaches include GWAS and CG approaches. These approaches have been extensively utilized in humans and also in a number of major crops like maize, rice, and wheat.

## 4.1 Genome-Wide Association Studies (GWAS)

GWAS is often utilized, when one is interested in finding all genomic regions that may be involved in controlling the trait of interest. A number of GWAS have actually been conducted in higher plants. However, GWAS can often give statistically significant results at loci unrelated to the trait. These indirect associations are the result of LD between multiple factors, of which only some are physically linked with the trait. Epistasis and population structure can also add to the problem, which can neither be overcome by increasing sampling size nor by increasing the marker density. The problem, however, can be partly resolved through recently proposed multilocus mixed model (MLMM; Segura et al., 2012) and multitrait mixed model (MTMM; Korte et al., 2012). The issues involving GWAS are discussed later in more detail.

## 4.2 Candidate Gene Approach

In cases where some information about the genetics of the target trait is available and CGs can be predicted on the basis of available information, one would like to confirm the genes that control the trait of interest. This may also be used to discover polymorphisms (SNPs or haplotypes) associated with the trait of interest following GWAS. In plants, the first AM studies involving CG approach were conducted for flowering time in maize (Remington et al., 2001; Thorsberry et al., 2001), and were later followed by studies mainly involving genome-wide approach although some additional studies involving CG approach were also conducted in parallel (Ehrenreich et al., 2009; Li et al., 2011a; many more).

CG approach is also useful when after application of multiple testing/false discovery rate (FDR) correction, none of the genome-wide associations are found to be significant for a trait, as observed in a recent study conducted in maize (Cook et al., 2012). In this study, none of the 51,741 SNPs showed significant association with the trait after a multiple test FDR was applied. However, CG association analysis proved effective as they could detect a significant association between oil content and the Phe insertion previously identified to be due to a mutation in the gene *DGAT1-2* for increased oil (Zheng et al., 2008). Two additional SNPs located in the *DGAT1-2* gene were also found to be significantly associated with oil content using CG approach. CG approach can also be used in parallel with GWAS to increase the power and precision of QTL detection as was done in a recent study in maize (Lipka et al., 2013). In still other cases, CG approach is used following GWAS, so that CGs corresponding to some

QTLs identified through GWAS are selected for further study as reported by Weng et al. (2011) in maize.

# 5. MAJOR CONCERNS AND ISSUES IN USING AM

Although a number of AM studies have been carried out involving all major crops, there are hardly any documented examples, where results of AM studies have been used in plant breeding though some undocumented examples may be available. This is partly due to high FDR, and partly due to the difficulty in using markers with rare alleles that may be associated with missing and desirable heritability for the traits of interest. These difficulties are mainly due to a number of major limitations, which we would like to examine in some detail.

## 5.1 Population Structure, Cryptic Familial Relatedness, and Computational Speed

The population used for AM may be structured (with high value of $F_{st}$), consisting of a number of distinct subpopulations. Each of several methods that are available for the study of population structure involves the use of markers (should not be linked with a QTL for the trait) for genotyping the panel to be used for AM. The presence of population structure, if not taken into account, causes confounding in the results leading to inflation in the magnitude of test statistics, thus giving a high frequency of false positives (Breseghello & Sorrells, 2006; Sneller, Mather, & Crepieux, 2009). However, one needs to exercise caution in choosing a method for applying correction that is necessary due to population structure. For instance, the very first study of AM in higher plants involving the gene *Dwarf8* in maize conducted by Thornsberry et al. (2001) was recently found to be flawed, since apparently they applied correction for population structure without realizing that population structure was correlated with the trait of interest (Larsson, Lipka, & Buckler, 2013). The methods developed for dealing with population structure have largely been driven by two major concerns, first for controlling the population structure effectively and second, for reducing the computational demand of the analysis.

### 5.1.1 Approaches for Handling Population Structure

The methods/models that are available for handling the confounding effect of population structure (including kinship) include the following: (i) single locus generalized linear model (GLM; Reeves & Richards, 2009); (ii) single locus mixed linear model (MM or MLM; Yu et al., 2006) that corrects for population structure using either kinship (K) or both stratification and kinship (Q + K), and (iii) multilocus mixed linear model (MLMM) that also corrects for population structure, but simultaneously takes into consideration the background genotype, as also done in composite interval mapping (CIM) of traditional QTL-IM. It is now widely known that MLM and MLMM models are superior to GLM.

(i) *Genomic control (GC), structured association (SA), and principal component analysis (PCA) for population structure:* In GWAS involving any one of the above three methods (GLM, MLM, and MLMM), one can correct for population structure by using any one of the following three approaches: GC (Devlin & Roeder, 1999), SA (Pritchard, Stephens, & Donnelly, 2000), and PCA (Price et al., 2006) or PCoA-MC (Reeves & Richards, 2009). GC was developed mainly for case–control or transmission disequilibrium test (TDT) studies in humans, where Cochran–Armitage Trend tests are applied, and the test statistics are scaled using a multiplication factor (also called *inflation factor*), so that the statistics becomes nearly equal to the expected one. This method, however, lacks power in the presence of strong confounding, and is seldom used for GWAS in plants. In case of SA (e.g., STRUCTURE), random markers are first used to estimate the probable number of sub-populations, whereas in PCA, variation among markers is summarized in the form of number of principal components. The coefficient values of individuals for each principal component describe their relationship to the subpopulations. PCA is considered to be computationally more effective than SA because the latter has been found to overestimate the true number of subpopulations under certain situations. However, both these methods fail to capture the complex pattern of relatedness in the sample (Myles et al., 2009). In this manner, each of the three methods for dealing with the population structure (GC, SA, and PCA) may though prove useful when the nature of population structure is simple, but may fail when the nature of population structure is more complex (Segura et al., 2012).

(ii) *Cryptic familial relatedness and unified mixed linear model:* Correction for population stratification is not always sufficient to avoid false positives as has been shown recently by Larsson et al. (2013) in maize. Therefore, for overcoming the confounding effect due to population stratification (Q matrix) and unequal relatedness (K matrix) among pairs of individuals, mixed models (MLM) proposed by Yu et al. (2006) have been used in many studies and served a useful purpose. In this approach, population structure (Q) is used as fixed effects and relatedness among individuals (K) is used as random effects, which are estimated through variance component analysis (VCA). However, in statistics, the term unrelated individuals convey different meanings to different workers (Bernardo, 1993). Yu et al. (2006) treated random pairs of inbreds as unrelated unless proved otherwise by kinship estimates; they obtained a kinship estimate as the fraction of the genome that is shared between two members of a pair and is identical by descent (as shown by using marker data). According to Zhao et al. (2007), pairs of inbreds that do not share any allele should be treated as unrelated; they also suggested that kinship should simply be estimated as the fraction of shared alleles. However, according to Stich and Melchinger (2009), the definitions (for unrelated individuals) suggested by Yu et al. (2006) and Zhao et al. (2007) are both rather arbitrary; they instead proposed the use of restricted maximum likelihood to estimate the conditional probability of two individuals of a pair being alike on the basis of similar marker alleles given that they are not identical by descent (Lynch, 1988). However, a mixed model can also lead to false negatives by overcompensating (overfitting) for population structure and relatedness [as has been observed by Zhao et al. (2011) for plant height in rice] so that a naive approach (without Q or K) can be useful.

(iii) *Kinship* versus *genome-wide marker* (G) *effect for background QTL in AM:* Bernardo (2013) suggested an AM procedure that utilizes genome-wide markers (G) to account for background QTL, and thus proposed G and QG models. Using simulated data, it was shown that these models show the best adherence to the significance level and give a better balance between many more true QTL detected and fewer false positives. It was argued that by estimating background effect via a genome-wide selection framework (Meuwissen, Hays, & Goddard, 2001) in QG and G models rather than via kinship in the QK or K models (Yu et al., 2006), estimates of marker effect in AM can be improved.

### 5.1.2 Additional Models, where Computational Demand is also Further Reduced (EMMA, FaST-LMM, GRAMMAR, EMMAX, P3D, CMLM, GRAMMAR-Gamma, and GEMMA)

It is known that irrespective of the statistical method chosen, GWAS would require a large sample (both genotypes and markers) to achieve sufficient statistical power, particularly if we wish to detect the small effect polymorphisms that underlie most complex traits (Balding, 2006; Buckler et al., 2009). However, while using MLM approach (described above), a dataset with large sample size creates a heavy computational demand (relative to GLM) because computing time per marker used in AM increases as the cube of the number of genotypes used (Zhang et al., 2010). Improvement in the MLM approach was achieved either through improvement in the VCA, as done in case of Efficient Mixed Model Association or EMMA (Kang et al., 2008), or through the use of a low-rank relatedness matrix (matrix based on a few thousand SNPs instead of all SNPs), as done in Factored Spectrally Transformed Linear Mixed Model or FaST-LMM (Lippert et al., 2011). In particular, EMMA avoids using operations that use a redundant matrix at each iteration during computation of the likelihood function, so that the computation cost of each iteration will reduce from cubic to linear. EMMA has been used in many AM studies (Chan, Rowe, & Kliebenstein, 2010; Li, Huang, Bergelson, Nordborg, & Borevitz, 2010a). FaST-LMM is believed to be a further improvement over EMMA and allows analysis of several thousand individuals with a reasonable number of markers in just a few hours (Lippert et al., 2011).

The three approaches, including MLM, EMMA, and FaST-LMM mentioned above, are described as exact methods, but do not entirely solve the problem of computation demand, so that AM involving the use of a few thousand genotypes with more than half a million markers cannot be undertaken using any of these approaches. In order to overcome this limitation of exact methods, several approximate approaches were proposed and used; these include the following: (i) Genome-wide Rapid Association using Mixed Model and Regression (GRAMMAR; Aulchenko, de Koning, & Haley, 2007), (ii) EMMA eXpedited (EMMAX; Kang et al., 2010), (iii) Population Parameters Previously Determined (P3D), and Compressed MLM (CMLM, both by Zhang et al., 2010) and (iv) GRAMMAR-Gamma (Svishcheva, Axenovich, Belonogova, van Duijn, & Aulchenko, 2012).

GRAMMAR first estimates the residuals adjusted for family effects under the null model (no SNP effect) and then treats these residuals as phenotypes along with marker data for genome-wide pedigree-based association analysis using rapid least-squares methods; this reduces the computation time for each individual SNP (Aulchenko et al., 2007). However, GRAMMAR can be effective only in cases where complete and correct pedigree information is available, thus limiting its application (Amin, van Duijn, & Aulchenko, 2007). An improved version called GRAMMAR-Gamma was later suggested to reduce the computational demand of AM (Svishcheva et al., 2012). In EMMAX, VCA is not repeated for each marker since each marker is assumed to explain only a small fraction of phenotypic effect; instead, heritability estimated from the null model is used for all markers, thus making it possible to perform AM using the vast amount of data in a short time. P3D eliminates the need to estimate population parameters (such as variance components) for testing each marker without compromising the statistical power regardless of the genetic architecture of the phenotypes; this is done by using the preestimated VCs from the null model. The CMLM approach clusters the individuals into fewer groups based on the kinship among the individuals, so that the kinship between pairs of groups replaces the kinship between pairs of individuals for the random effect of an MLM; this reduces the computation demand substantially (Zhang et al., 2010). It has also been shown that the joint use of P3D and CMLM greatly reduces computation demand with increased statistical power (Zhang et al., 2010).

The above mentioned approximate methods work fine, but it is a trade-off since the accuracy is being compromised. Therefore, recently, another exact method called Genome-Wide Efficient Mixed Model (GEMMA) was proposed, which yields accurate $p$ values even in the presence of strong population structure, and even when the marker effect is large, although in terms of computational speed it is only at par with EMMAX (Zhou & Stephens, 2012).

In summary, we now know that several methods that are available to perform association analysis have their own advantages and limitations. The choice really depends on the size of the dataset being analyzed and the information sought. In some recent studies, different methods of AM have also been compared using simulated as well as empirical data (Svishcheva et al., 2012; Zhou & Stephens, 2012).

### 5.1.3 Multilocus AM and Multitrait AM

The simplest model that is most frequently used in GWAS involves testing association between a single-marker locus and a single trait. However, it is widely known that most complex traits are each controlled by a number of QTL/genes, and that many of these genes may be pleiotropic in nature, so that the model involving association of single locus with single trait leads to model misspecification, giving biased results. In linkage-based QTL mapping, the treatment of multiloci and multitraits is common (CIM and multitrait or joint analysis also called as multitrait CIM or MCIM), where we use a number of major QTL as cofactors in CIM and use a number of correlated traits to identify QTL controlling more than one trait. In AM, involving LD, the confounding effect of background loci is even stronger because LD may result due to factors other than linkage also. Conditional analysis on a genome-wide scale involving a number of correlated traits and using genome-wide background loci as cofactors may well lead to higher power and a lower FDR than single locus and single trait approaches (Korte et al., 2012; Segura et al., 2012). However, while doing this, the number of predictors ($p$) is often larger in comparison with number of observations ($n$), so that it becomes a challenge to include all of them as cofactors. LD creates further problems if all the polymorphisms are fitted at a time. In order to overcome this problem, multilocus methods were used in the past, but generally these methods did not include population structure (Cordell, 2002; Hoggart, Whittaker, De Iorio, & Balding, 2008). Recently, two approaches have been proposed to address the issues of "large $p$ small $n$" and "population structure" simultaneously in multilocus AM. *First* is MLMM approach proposed by Segura et al. (2012) and second is linear mixed model–Lasso (LMM–Lasso) approach proposed by Rakitsch, Lippert, Stegle, & Borgwardt (2013). To solve the problem of "large $p$ small $n$," MLMM makes use of stepwise regression with forward inclusion and backward elimination, while LMM–Lasso makes use of Sparse Lasso Regression to deal with the same problem. Recently, MLMM was also used to improve the precision of GWAS for tocochromanol level in maize (Lipka et al., 2013). Similarly MTMM has been suggested to deal with correlated traits in AM (Korte et al., 2012; Liu et al., 2013). However, in both approaches (multilocus and multitrait AM), the mixed model was used for the correction of population structure. The data on correlated traits will help in understanding the mechanism of pleiotropy as well as genotype × environment interaction (Korte et al., 2012). Often in nature, majority of the traits are correlated so that the approaches like MTMM and penalized MTMM will help to understand the genetic complexity of the traits.

The utility of the above methods for GWAS has been demonstrated both in humans and in the model plant species, *Arabidopsis thaliana*. Ideally, one should like to conduct multilocus AM and multitrait AM together, but only "in house" methods for such an analysis have been used so far and no software is available to conduct MLMM and MTMM simultaneously using the same panel for a number of traits.

### 5.1.4 Model Selection for AM

As mentioned above, several models and methods (Table 2.1) with different features are now available to conduct AM. Therefore, one needs to select a model which fits best for the experiment and gives relatively fewer false positives. In AM, it is assumed that markers are generally not associated with the functional polymorphism of the trait of interest and also that the observed $p$-values are normally distributed (Yu et al., 2006). AM approaches

**Table 2.1** Different Methods and Models Used for Association Mapping along with the Different Methods to Control Population Structure, to Generate Kinship Matrix and to Estimate Genome-Wide Marker Effect.

| Method | Statistical Model | Population Structure | Kinship Matrix K | Genome-Wide Marker Effect |
|---|---|---|---|---|
| Simple | $Y = \mu + S\alpha + e$ | — | — | — |
| Q | $Y = \mu + S\alpha + Qv + e$ | Genomic control, Stepwise logistic regression (SLR), STRUCTURE, $\Delta K$ criterion, principal component analysis | — | — |
| K | $Y = \mu + S\alpha + Zu + e$ | — | SPAGeDi; Pedigree information, $K_{\text{unrel}ij} = \dfrac{Sij - 1}{1 - T} + 1;$ $T = S$ entries vs . cultivars | — |
| QK | $Y = \mu + S\alpha + Qv + Zu + e$ | Genomic control, Stepwise logistic regression (SLR), STRUCTURE, $\Delta K$ criterion, principal component analysis | SPAGeDi; Pedigree information, $K_{\text{unrel}ij} = \dfrac{Sij - 1}{1 - T} + 1;$ $T = S$ entries vs . cultivars | — |
| G | $Y = \mu + Mg + e$ | — | — | RR–BLUP[a] |
| QG | $Y = \mu + Qv + Mg + e$ | Genomic control, Stepwise logistic regression (SLR), STRUCTURE, $\Delta K$ criterion, principal component analysis | — | RR–BLUP |

Q, general linear model involving population structure; K, mixed linear model involving kinship; QK, mixed linear model involving population structure and kinship; G, general linear model involving marker effect; QG, general linear model involving marker effect and population structure; Y, $N \times 1$ vector on mean of $N$ individual; $\mu$, grand mean; $\alpha$, effect of marker allele; $S$, incidence vector that relate $\alpha$ to $Y$; $e$, residual; $v$, population structure; Q, incidence matrix that relate $v$ to $Y$; $u$, kinship; Z, incidence matrix that relate $u$ to $Y$; $g$, effect of marker allele from first founder inbred incidence matrix that relate $g$ to $Y$. $S_{ij}$ is the proportion of marker loci with shared variants between inbreds $i$ and $j$.

[a]RR–BLUP is ridge regression-best linear unbiased prediction.

that adhere to the nominal α-level show a uniform distribution of $p$-values. The plot of observed vs expected $p$-values is also known as a quantile–quantile (Q–Q) plot. Another measure called *mean of the squared difference* (MSD) between observed and expected $p$-values (Stich & Melchinger, 2009) of all marker loci may also be used to account for the deviation of the observed $p$-values from the uniform distribution. The high value of MSD indicates a strong deviation of the observed $p$-values from the uniform distribution and suggests that the empirical type I error rate is considerably higher (Stich et al., 2008).

## 5.2 Multiple Testing Problem and FDR

We know that AM studies generally involve the use of more markers than traditional linkage studies. Therefore, there is a need to define an appropriate significance threshold that can take into account the issue of multiple comparisons. In AM, we are dealing with multiple tests, where each of a large number of markers is tested for its association with the trait of interest, and a $p$ value is generated for each test; this is often described as the problem of multiple testing, which also leads to higher frequency of false positives (Brachi et al., 2010). In order to overcome this problem, multiple tests corrections are applied using one of the several methods that are discussed below. Sometimes none of the associations detected using multiple tests are found to be significant (higher level of false negatives), thus making the whole exercise of AM futile, even though genetic variation for the trait of interest might be present.

In AM, each marker is tested separately and a $p$ value is generated for each marker. If the critical threshold $p$ value is 0.05, then there is a 5% error margin for each marker that is subjected to test for significance. If 100 markers are tested, one would expect to find five markers to be significantly associated by chance, irrespective of whether these are associated due to linkage or not. If we test a group of 10,000 markers, 500 markers would exhibit significance due to chance, and only some of them may really be associated. Therefore, it is important to correct the $p$ value of each marker when performing a statistical test on a set of markers. Multiple testing corrections adjust the individual $p$ value for each marker to reduce the overall error rate.

The corrections that have been suggested to overcome the problem of false positive associations due to multiple testing are many; some of the commonly used corrections include the following: (i) Bonferroni correction (Bonferroni, 1936); (ii) Bonferroni step-down correction (or Holm correction; Holm, 1979); (iii) Westfall and Young Permutation (Westfall & Young, 1993); (iv) FDR (Benjamini & Hochberg, 1995); (v) $q$ value (Storey, 2002); and (vi) step-up adaptive method (Benjamini, Krieger, & Yekutieli, 2006). The details of these corrections are available in published literature and the relative level of stringency offered by each method is depicted in Figure 2.2. It may be seen that of the above tests, FDR controlling method proposed by Benjamini & Hochberg (1995) and the $q$ value method are least stringent and are commonly used, since they restrict FDR but allow discovery of significantly associated markers. In contrast, Bonferroni correction is the most stringent, but will disallow many true associations to be detected (high rate of false negatives). For example, if there are 10,000 markers, and cut-off $p$ value is 0.05, then a marker will be declared to be significantly associated only when the actual $p$ value is 0.000005, so that the adjusted value is $0.000005 \times 10,000 = 0.05$. Since there are different methods to account for the false positives, the choice of method depends on the person performing the experiment and also on the costs associated with false positives and false negatives, which may differ from one experiment to the other (Noble, 2009).

Qian & Huang (2005) compared different methods of FDRs for identifying genes with differential expression and concluded that the $q$ value

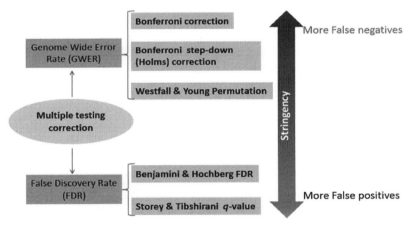

**Figure 2.2** A comparison of methods used for corrections recommended to overcoming the multiple testing problem [genome-wide error rate (GWER) and false discovery rate (FDR)]. Stringency of the results of association mapping involving false negatives and false positives differs in different approaches of GWER and FDR in the order in which they are shown (on the extreme right, the upward arrow indicates the direction of more false negatives and the downward arrow indicates direction of more false positives). (For color version of this figure, the reader is referred to the online version of this book.)

method of Storey (2002) has the highest test power followed by step-up adaptive method of Benjamini et al. (2006) and Benjamini and Hochberg (1995). They also observed that the step-down approaches were most conservative giving results similar to those by Bonferroni method. In a recent AM study on preharvest sprouting in wheat, Kulwal et al. (2012) also reported that a higher number of significant QTLs were detected with the *q* value method than by the Benjamini and Hochberg method. Bonferroni method can be a good choice to control false-positive QTLs in case of human genetic studies, where the risk of identifying a false QTL is a matter of higher concern than ignoring a true QTL. The reasonable choice would be to compare the results obtained using different methods and to see the differences in the number of QTLs identified having biological significance.

## 5.3 Markers with Rare Alleles and Rare Genetic Variants

It has now been widely recognized that the genetic variation of a complex trait is generally controlled by a combination of common and rare genetic variants, and that the contribution of rare variants may be substantial in some cases. In human genetics, it is now known that rare variants generally act along with common variants either independently or in aggregate. In several studies, association of rare variants with complex traits has actually been demonstrated in humans (Nelson et al., 2012; Tennessen et al., 2012). However, in plant systems, the problem of rare variants has not been adequately addressed, and involves two separate issues, namely, the problem due to exclusion of rare marker alleles from the analysis and the problem due to rare genetic variants leading to the so-called "missing heritability".

### 5.3.1 Marker with Rare Alleles

The problem due to exclusion of marker alleles with frequencies lower than 5% from analysis is generally unavoidable in AM studies. This is particularly true when a population of limited size is used, since just a few accessions with rare marker allele do not cover adequate variability to allow their inclusion in statistical analysis. However, we know that in AM panels, the desirable state of a trait may be present at a low frequency, and so will be the marker alleles associated with this desirable state. By eliminating the markers with rare alleles, we lose the chance of detecting markers associated with these rare desirable states of the traits of interest since AM tests are often designed for common variants only.

Several solutions have been suggested to deal with the above problem. *First*, genotypes containing rare variants may be used for development of biparental populations for QTL analysis, thus enhancing the power of mapping (Stich & Melchinger, 2010). *Second,* linkage mapping and LD mapping may be combined for conducting joint-linkage association mapping (JLAM), so that rare marker alleles may be included in the analysis (Brachi et al., 2010; Poland, Bradbury, Buckler, & Nelson, 2011). *Third,* population with large size may be used, so that rare marker alleles may be present in sufficient number for adequate phenotypic variability (associated with rare marker alleles) needed for statistical analysis. *Fourth,* new association analysis methods for rare variants may be used; such new methods are actually being proposed (Asimit & Zeggini, 2010; Bansal, Libiger, Torkamani, & Schork, 2010; Gibson, 2012; Li & Leal, 2008; Zhu et al., 2011).

### 5.3.2 Rare Genetic Variants and Missing Heritability

Rare genetic variants (each with major/minor effect) that may or may not be associated with the common associated markers, often escape detection. It is for this reason that despite the extensive GWAS, the common variants in humans generally explain not more than 5–10% of the heritable component of a trait, although in plant systems, the situation is not so acute (Brachi, Morris, & Borevitz, 2011). This feature is described as "missing heritability," so that there is a reason to search for associations with multiple rare variants (MRV).

The above problem of rare genetic variants has been discussed in many studies (Bansal et al., 2010; Gibson, 2012; Li & Leal, 2008). It has been shown that although the frequency of any single rare or low frequency variant may be low (<5%), collectively, the number of rare variants makes them quite common. There are two hypotheses for explaining the role of these rare variants in controlling human diseases: the old popular common disease common variant (CDCV) hypothesis, and the recent common disease rare variant (CDRV) hypothesis (also called MRV hypothesis). According to the former, the trait is largely controlled by a few common variants with large effect, whereas according to latter, there should be many moderate to large effect rare variants controlling a complex QT. There is increasing evidence to support the CDRV or MRV hypothesis (Li, Byrnes, & Li, 2010b), and the same may be true for plant systems. It has also been shown that the effects of rare variants tend to be larger than those of common variants (Gibson, 2012). This is borne out by the fact that in case–control studies, often the values of odds ratios for common variants are generally <2 (majority fall between 1.1 and 1.4), while those for the rare variants range from 2 to 6 (with a mean approaching 4).

As mentioned above, in many AM studies in humans, the genetic variation explained by the common markers associated with the trait of interest accounts for only 5–10% of the total variation (Eichler et al., 2010; Manolio et al., 2009). In contrast, there are reports in crops like rice, where the genetic variation explained by the markers exceeded even 50% (Huang et al., 2010). The major part of genetic variation which remains unexplained in GWAS (particularly in humans, but also in plants) is often described as "missing heritability," although "missing genetic variability" could be a better expression, since heritability is used to describe proportion of genetic variation in the total phenotypic variation. This aspect has been largely studied in humans and needs to be studied in plants also (Brachi et al., 2011). There are several reasons for the missing heritability. *First* and the most important reason is believed to be the rare variants (discussed above), although according to some, this is not always the case (Gibson, 2012). *Second*, the causal variants, each explains an amount of variation that is too small to reach stringent significance thresholds. *Third*, the causal variants are not in complete LD with the SNPs that are used for genotyping, so that the causal variants are not detected. *Fourth*, the genotyping platforms (e.g., SNP chips) that are used for marker genotyping often carry only the common markers and do not have the markers that are rare and control the targeted trait (for a review, see Zhu et al., 2011). *Fifth*, epistasis and g × e interactions are not examined in majority of GWAS (Makowsky et al., 2011; Manolio et al., 2009), and control part of the genetic variation, thus explaining the missing heritability (Zuk, Hechter, Sunyaev, & Lander, 2012).

The problem of rare variants has also been explained on the basis of synthetic associations, which assume that a number of rare variants may often be associated with a common marker, which is associated with the target trait (Dickson, Wang, Krantz, Hakonarson, & Goldstein, 2010). Experimental evidence has also been provided for the occurrence of synthetic associations (Huff et al., 2012; Kent, 2011), although there are reports, which suggest that synthetic associations cannot fully explain the occurrence of missing heritability (Anderson, Soranzo, Barrett, & Zeggini, 2011; Wray, Purcell, & Visscher, 2011).

Several measures have been suggested for the analysis and interpretation of the rare variant association studies (Bansal et al., 2010; Gibson, 2012; Li & Leal, 2008; Panoutsopoulou, Tachmazidou, & Zeggini, 2013). The issue of rare variants is an active area of research, which will be greatly facilitated with the availability of the facility of targeted genotyping arrays and NGS technologies at the whole-genome and whole-exome scales. This will help in accessing the sequence variation across the full minor allele frequency (MAF) spectrum (Panoutsopoulou et al., 2013).

## 5.4 How to Improve Power for QTL Detection in AM?

In recent years, efforts have been made to develop newer strategies for AM with an aim to improve its power for detecting QTL. These newer approaches include JLAM and family-based AM.

### 5.4.1 Joint Linkage Association Mapping

It has been shown that when compared with QTL interval mapping (QTL-IM), AM has higher resolution, but relatively low power for detecting QTL of interest. Therefore, for a complex trait in a crop of interest, the use of LD-based AM alone is really a tradeoff, unless the experiment is carefully redesigned to combine AM with linkage studies to achieve both high power and high resolution, as in JLAM.

A method combining the strengths of linkage mapping and AM was initially proposed in the field of animal genetics (Churchill et al., 2004; Mott, Talbot, Turri, Collins, & Flint, 2000). In case of plants, JLAM involves either a set of biparental populations or one or more multiparental populations, which can be used for both linkage analysis and AM (Lu et al., 2010; Myles et al., 2009; Reif et al., 2010; Wurschum et al., 2012; Yu et al., 2008); alternatively, one may use two sets of genotypes, one consisting of germplasm and the other consisting of a number of biparental mapping populations, and both these sets are genotyped using the same set of markers, thus permitting JLAM. The use of a number of biparental populations along with a population of inbreds can also allow linkage mapping and LD mapping independently (also called parallel mapping) as well as joint linkage–LD mapping (also called integrated mapping; Lu et al., 2010). This strategy can be helpful in comparing the results of parallel mapping with that of the integrated mapping. In maize, it was shown that integrated mapping could identify many more significant associations than the parallel mapping (Lu et al. 2010). Although, initially JLAM studies focused on available mapping populations or heterogeneous stocks with known pedigrees (Blott et al., 2003; Meuwissen et al., 2002; Mott & Flint, 2002), now with the availability of MAGIC and NAM populations in different crops, it has become possible to perform JLAM effectively although no software package has so far been developed for JLAM.

### 5.4.2 Family-Based AM Studies

Earlier in this chapter, we described a variety of family-based populations for conducting AM. The family-based AM studies like TDT correct for spurious associations that are common in population-based AM involving logistic regression ratio test (Spielman, McGinnes, & Ewens, 1993; Stich et al., 2006). In view of this, family-based AM is recommended to improve precision and power; one such test is quantitative inbred pedigree disequilibrium test (QIPDT), which is an extension of the QPDT originally developed in the context of human genetics (Zhang, Zhang, Li, Sun, & Zhao, 2001). This QIPDT was first applied using data on flowering time in maize inbreds, utilizing the genotypic information of PI lines and genotypic and phenotypic information of their offspring inbreds (Stich et al., 2006). Recently, Guo et al. (2013) has suggested modified models for use of QIPDT in crop species. A list of software that can perform family-based AM is available elsewhere (Ott, Kamatani, & Lathrop, 2011).

## 6. EMERGING RESEARCH AREAS FOR ASSOCIATION MAPPING

### 6.1 Dynamic Trait Mapping and Conditional Genetic Analysis

In majority of QTL mapping studies, only the endpoint phenotype is used as the target trait for analysis. However, several developmental traits such as plant height are dynamic in nature, so that any two genotypes may have the same plant height but different growth trajectories during the development. For this purpose, the data recorded at different developmental stages may be used either independently (data for same stage) or jointly (data for different stages together). In order to identify specific genes and to determine their roles as a function of time for these dynamic traits, a concept of functional mapping has been suggested (Das et al., 2011; Das, Li, Huang, Gai, & Wu, 2012; He, Berg, Li, Vallejos, & Wu, 2010; Wu & Lin, 2006; Zhang, Shi, Li, Chang, & Jing, 2013). In the context of GWAS, it has been called functional genome-wide association mapping (*f*GWAS). Earlier, a relatively similar approach was described as "time-related mapping" (for the data recorded at different time points) as against "time-fixed mapping" for the data recorded only once at the end point (Wu, Li, Tang, Lu, & Worland, 1999).

The advantages of multistage data recording for a dynamic trait are many. *First*, using multistage data, one can identify the developmental stage at which heritability of the trait is highest so that it can be used as the stage for measurement of the trait in subsequent analysis in that population. This can also help in understanding the role of g × e interactions in GWAS. Therefore, the approach of *f*GWAS can maximize our ability to identify the changes affecting only a limited part of development (He et al., 2010) by enabling us to know when and how specific genes/QTLs turn on and turn off for a given trait (Das et al., 2011). In a recent study in wheat, Kulwal et al. (2012) implemented *f*GWAS for preharvest sprouting (PHS) to understand the functionality of the trait by scoring PHS data in response to different periods following seed maturity. *Second*, recording data at different growth stages allows one to estimate the conditional genetic effects using the method proposed by Zhu (1995). By using the data collected during earlier period ($t$-1) to the one collected later ($t$), quantitative genetic effects for the trait can be estimated for each specific stage excluding the effect of earlier developmental stages ($t$-2, etc.). In a recent study of conditional AM conducted in wheat to identify QTLs for plant height, it was suggested that most genes controlling plant height were influenced by the growing environment and none was expressed throughout the entire growth process (Zhang et al., 2011, 2013). Further, the approach of dynamic trait mapping/conditional analysis can also be used in conjunction with that of MTMM described earlier since different developmental stages will be correlated, and the data for different stages may be treated as correlated traits. *Third*, conditional mapping can also be conducted using two different correlated traits to see the effect of one trait on the detection of QTL for the other related trait (for example, grain protein content and thousand grain weight or grain yield in wheat). Conditioning of one trait on the other trait (secondary trait) can allow analysis of the first trait to be done independently of the other trait. The conditional values for the first trait are estimated for the no variation situation in the secondary trait and analysis can be conducted on the conditional as well as unconditional data. The differences in the results of such analysis can provide useful information for improvement of one trait without compromising the improvement of another trait as shown by Wang et al. (2012) for protein content in wheat. Although, such an analysis offers several advantages over the traditional analyses, more empirical studies involving larger datasets are needed to substantiate the importance of this approach.

## 6.2 Bayesian Analysis

Bayesian approaches offer several advantages over the so-called frequentist approaches due to their ability to incorporate background information into the specification of the model (Beaumont & Rannala, 2004). The approach is proving suitable for GWAS (Stephens & Balding, 2009; Wakefield, 2009). The Bayesian approach not only increases the computation speed but also deals with the problems of multiple testing and rare marker alleles (Fernando & Garrick, 2013). Iwata, Uga, Yoshioka, Ebana, and Hayashi (2007) proposed an approach which combines a Bayesian method for mapping multiple QTLs with a regression method that directly incorporates estimates of population structure and has been shown to be effective in controlling both false positives and false negatives using an Asian rice dataset. Similarly, a Bayesian approach has also been used for multilocus mapping and for finding epistatic interactions (Lu, Liu, Wei, & Zhang, 2011; Marttinen & Corander, 2010). With the recent advances in computational analysis, it is expected that the Bayesian approaches will be widely used in future GWAS. However, care needs to be exercised as Bayesian approaches are sensitive to the choice of the prior distribution and could lead to different conclusions if the two studies using the same data use different priors (Shoemaker, Painter, & Weir, 1999).

## 6.3 Epistatic and G × E Interactions

QTL interactions (epistasis and Q × E) have now been identified in many biparental QTL mapping studies. However, there are limited GWAS studies, which involved the analysis of epistatic interactions. One of the reasons that these interactions could not be detected in earlier studies was the lack of suitable time- and cost-effective methods. Several methods proposed now for this purpose include the following: (i) a linear regression approach, that may also include Q and K to control the population structure (Li, Horstman, & Chen, 2011b; Reif et al., 2011; Stracke et al., 2009; Wurschum, Maurer, Schulz, Mohring, & Reif, 2011; Yu et al., 2011); (ii) LASSO penalized logistic regression approach (Wu, Chen, Hastie, Sobel, & Lange, 2009); (iii) Bayesian approach and permutation test (Lu et al., 2011); and (iv) adaptive mixed LASSO (Wang, Eskridge, & Crossa, 2011). Some of these methods have been discussed elsewhere (for details, see Cantor, Lange, & Sinsheimer, 2010; Li et al., 2011b; McKinney & Pajewski, 2012; Wan, Yang, Yang, Zhao, & Yu, 2013). It should also be noted that the contribution of these interacting QTLs, particularly epistatic QTLs, varies from trait to trait, suggesting that these interactions may not be important for all the traits, as shown in maize for leaf architecture and resistance to southern leaf blight (Kump et al., 2011; Tian et al., 2011).

## 6.4 Meta-analysis for GWAS

Meta-analysis combines information from multiple GWAS and can increase the chances of finding true positives among the identified associations (Cantor et al., 2010). Hundreds of studies involving GWAS meta-analysis have been published for humans (Evangelou & Ioannidis, 2013), but there seems to be no published report of meta-analysis for GWAS in plants. Therefore, it will be desirable to conduct meta-analysis using results of several GWAS involving the same trait in the same crop. While doing so, one should recognize that several factors may influence the results of GWAS meta-analysis. *First*, different studies may be based on heterogeneity in data, due to genetic and environmental factors, making interpretation of the results of meta-analysis difficult, although methods have been suggested to deal with this problem (Han & Eskin, 2012). *Second*, sample size and design may be different in different studies included in meta-analysis (Moonesinghe, Khoury, Liu, & Ioannidis, 2008; Spencer, Su, Donnelly, & Marchini, 2009). *Third*, some studies included in a meta-analysis may be based on imputed data, which should be taken into consideration (de Bakker et al., 2008). *Fourth*, meta-analysis for a complex trait involving rare variants may create some problems (Evangelou & Ioannidis, 2013). Some of these issues have also been addressed by Thompson, Attia, and Minelli (2011).

## 6.5 Association Mapping using Transcriptome Data

With recent advances in sequencing techniques, mRNA sequencing is becoming popular for transcriptome analysis in many crops. This can provide information on variation in both gene sequences (as SNP markers) and gene expression (as gene expression markers; GEMs) and can be used effectively for GWAS (Harper et al., 2012). It has also been shown that even for a species without a reference whole genome sequence, these transcriptome-derived markers can be mapped on the basis of the draft genome sequence scaffolds from a related species (Harper et al., 2012).

Recently, mRNAseq data has been utilized for GWAS (described as associative transcriptomics) leading to identification of MTAs in oilseed rape cultivars (Bancroft et al., 2011; Harper et al., 2012). For this purpose, leaf mRNA sequence data from a diversity panel of 84 *Brassica napus* accessions was used to identify more than 100,000 SNPs and more than 125,000 GEMs. Some of these markers were found to be associated with rapeseed quality traits, namely erucic acid and glucosinolate content, when 53 of the above 84 accessions were used for AM; some of these associations also coincided with the genes/QTLs already known to be responsible for these two traits. Such associated SNP and GEM markers can be converted into user-friendly markers for use in molecular breeding. In addition to the detection of MTAs, associative transcriptomics also offers a promise to allow identification of causative genomic DNA regions responsible for trait variation, thus providing insight into the biology of the trait development (Bancroft, 2013).

## 7. SOFTWARE FOR ASSOCIATION ANALYSIS

Several software packages are now available for performing association analysis (Table 2.2). However, many of these packages were originally

**Table 2.2** List of Software Packages Available for Association Mapping (Software which Performs Structure Analysis but not GWAS are Not Listed Here)

| Program | Features | Web Address |
| --- | --- | --- |
| TASSEL | LD statistics, GLM, MLM, CMLM, P3D, genomic selection; graphical interphase, PCA and kinship; free | http://www.maizegenetics.net/tassel/ |
| GAPIT | R-based, CMLM, fast computation, free | http://www.maizegenetics.net/gapit |
| R | Generic, commonly used for programing; free | http://www.r-project.org/ |
| PLINK | Handles virtually unlimited numbers of SNPs; MDS to visualize substructure; free | http://pngu.mgh.harvard.edu/~purcell/plink/ |
| EMMA | Mixed model, corrects for the confounding from population structure and genetic relatedness; free | http://mouse.cs.ucla.edu/emma/ |
| EMMAX | Large-scale association mapping, corrects for the confounding from population structure and genetic relatedness, increased computational speed; free | http://genetics.cs.ucla.edu/emmax/ |
| EIGENSOFT | Uses principal components analysis to explicitly model ancestry differences between cases and controls; free | http://www.hsph.harvard.edu/alkes-price/software/ |
| GGT 2.0 | Graphical genotypes; LD statistics; FDR calculation, does not control for population stratification of its own; free | http://www.wageningenur.nl/en/show/Graphical-GenoTypes-transform-molecular-data-to-colorful-chromosome-drawings.htm |
| GenAMap | Performs automatic structured association mapping (SAM) using different algorithms; good graphical presentation; free | http://sailing.cs.cmu.edu/genamap/ |
| Matapax | GWAS is performed in R environment with EMMA and GAPIT libraries; performs all essential steps for basic GWAS, population structure, fast computation; free | http://matapax.mpimp-golm.mpg.de |
| Merlin | Includes an integrated genotype inference feature for improved analysis when some genotypes are missing, does not control for population stratification of its own; free | http://www.sph.umich.edu/csg/abecasis/merlin/tour/assoc.html |

Continued

**Table 2.2** List of Software Packages Available for Association Mapping (Software which Performs Structure Analysis but not GWAS are Not Listed Here)—cont'd

| Program | Features | Web Address |
|---|---|---|
| ASReml | Handle large data set, calculates population structure and pedigree-based kinship; commercial | http://www.vsni.co.uk/software/asreml |
| SAS | Generic program commonly used in data analysis; commercial | http://www.sas.com |
| JMP Genomics | Calculates population structure and marker-based kinship; commercial | http://www.jmp.com/software/genomics/ |
| SVS | Comprehensive package with better visualization of the results; offers different options; commercial | http://www.goldenhelix.com/SNP_Variation/ |
| GenStat | Performs GLM and MLM, takes care of population structure; commercial | http://www.vsni.co.uk/software/genstat |
| FaST-LMM | For analysis of large data sets (up to 120,000 individuals); free | http://fastlmm.codeplex.com/ |
| GenABEL | Performs GWAS for quantitative as well as binary traits; free | http://www.genabel.org/packages/GenABEL |

designed for the animal systems, so that their suitability for plant system should be examined before using these programs for AM in plants (Zhang, Buckler, Casstevens, & Bradbury, 2009). TASSEL is a software that was designed for plant systems (Bradbury et al., 2007) and has been widely used by the plant community due to its user-friendliness. Another suitable software for plants is GAPIT, which is R based (Lipka et al., 2012), and performs association analysis and genomic prediction based on CMLM. The important feature of this package is that it can handle very large data sets with reduced computation time.

## 8. FUTURE PERSPECTIVES

QTL IM and AM are the two major approaches for the study of genetics of complex traits in all crop plants, so that the associated markers may be used for MAS. Each of these two approaches has its own merits and demerits. While IM has higher power, AM has higher resolution. Also, while results of QTL IM have been extensively utilized for marker–assisted indirect selection for crop improvement, results of AM have been underutilized. However, AM offers several advantages over IM, which has made it a preferred approach in recent years. For instance, it allowed screening of large proportion of genetic variability with higher resolution, so that the results should prove useful for a wider spectrum of breeding programs. In order to make use of these merits of AM, we need to address some of the major issues that have plagued AM and are responsible for its limited use in plant systems; these issues have been discussed in this chapter. We strongly feel that in future, AM studies should be designed keeping these issues in mind. First and foremost is the design of experiment to allow family-based AM or JLAM, so that one makes use of the merits of both linkage-based IM and LD–based AM. In this connection, multiparental populations like NAM and MAGIC and several other mating designs have been tried, and even the use of breeding populations has been recommended for AM. Once we have the right material, we need to address the issues of population structure, model selection, rare marker alleles and rare genetic variants, multiple testing corrections, background genotype, and environmental effects. Decisions are also needed about whether one should use GWAS or CG approach or both (GWAS followed by CG), because the latter overcomes some of the limitations of GWAS.

It is also necessary that one should take decisions about the approaches one would use to bring about precision in both phenotyping and genotyping the population. While phenomic platforms may be used for phenomics (at least in some cases, depending upon the trait being studied), NGS (including GBS) may be used for genotyping, so that one may use large number of sequence-based markers like SNPs, ISBP, and CNV/PAV-based markers, which are the latest markers available now in millions in several crops. While addressing the issue of population structure, one also needs to find out if

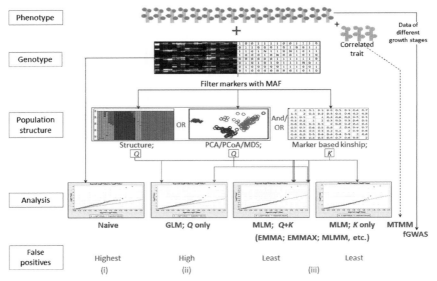

**Figure 2.3** Schematic representations of various steps involved in association analysis. On the extreme left are shown the various steps involved in association mapping including phenotyping, genotyping, study of population structure, association analysis, and the rate of false positives. As shown, association analysis can be conducted either without a study of population structure or using any one or two of the three approaches available for the study of population structure (including study of kinship): (i) When only phenotypic and genotypic data are used without taking care of population structure, the approach is called "naïve" and results in highest rate of false-positive associations. (ii) When phenotypic and genotypic data are used along with the estimates of population structure either using Q values obtained using the software STRUTURE or through the use of principal component analysis (PCA)/principal coordinate analysis (PCoA)/ multidimensional scaling (MDS) following general linear model (GLM), the approach is described as "GLM+Q" and results in a FDR that is lower than the "naive" approach, but is still fairly high. (iii) If estimates of marker-based kinship (K) is added (with or without Q matrix) to GLM approach, the approach is described as mixed linear model (Q+K or K only), which results in the lowest rate of false positives. In addition to the above approaches, the data on correlated traits can also be used for multitrait mixed model (MTMM) association analysis, and the data recorded on a trait at various stages of growth can be used for performing functional mapping (fGWAS), as shown on the extreme right. (For color version of this figure, the reader is referred to the online version of this book.)

the population structure has any direct correlation with the variability for the trait to ensure that it is randomly distributed in each subpopulation, and not associated with the population structure. Although in most of the AM studies, correction for population structure and/or relatedness are applied to reduce the false-positive associations, emphasis should also be given on environmental effect and genetic background as it can cause confounding in AM studies (Vilhjálmsson & Nordborg, 2013). In future, these two factors should also be taken into consideration along with the population structure and relatedness while conducting AM studies.

Another major issue is the elimination of rare marker alleles from analysis, and the detection of only common variants in the analysis, so that a number of rare variants associated with the desirable state of the trait of interest escape detection; this has been described as missing heritability.

We also know that Bonferroni correction or FDR often leads to large proportion of false negatives, and there is no readymade solution available for this limitation. In some cases, one may try CG approach for the associated QTL to eliminate the problem of false negatives due to overcorrection. One may also like to explore the possibility of using Bayesian approaches, which offer reasonable solutions to most of the issues and the concerns related to AM and should receive attention of those planning AM studies. There is no doubt that AM can be more rewarding if the issues associated with it are addressed carefully. A schematic representation showing the steps involved in association analysis are given in Figure 2.3.

## ACKNOWLEDGMENTS

We thank Prof. Mark E. Sorrells, Department of Plant Breeding and Genetics, Cornell University, Ithaca, NY, USA for his useful suggestions and critical comments on the manuscript. PKG held the position of a Senior Scientist of the National Academy of Sciences India (NASI) and VJ held the position of CSIR SRF during the development of this manuscript.

## REFERENCES

Amin, N., van Duijn, C. M., & Aulchenko, Y. S. (2007). A genomic background based method for association analysis in related individuals. *PLoS One, 2*, e1274.

Anderson, C. A., Soranzo, N., Barrett, J. C., & Zeggini, E. (2011). Synthetic associations are unlikely to account for many common disease genome-wide association signals. *PLoS Biology, 9*, e1000580.

Asimit, J., & Zeggini, E. (2010). Rare variant association analysis methods for complex traits. *Annual Review of Genetics, 44*, 293–308.

Aulchenko, Y. S., de Koning, D. J., & Haley, C. (2007). Genome-wide rapid association using mixed model and regression: A fast and simple method for genome-wide pedigree-based quantitative trait loci association analysis. *Genetics, 177*, 577–585.

Balding, D. J. (2006). A tutorial on statistical methods for population association studies. *Nature Reviews Genetics, 7*, 781–791.

Bancroft, I. (2013). Association genetics and more from crop transcriptome sequences. *ISB News Report*, January, 2013. pp. 1–4 .

Bancroft, I., Morgan, C., Fraser, F., Higgins, J., Wells, R., Clissold, L., et al. (2011). Dissecting the genome of the polyploid crop oilseed rape by transcriptome sequencing. *Nature Biotechnology, 29*, 762–766.

Bandillo, N., Raghavan, C., Muyco, P. A., Sevilla, M. A. L., Lobina, I. T., Dilla-Ermita, C. J., et al. (2013). Multi-parent advanced generation inter-cross (MAGIC) populations in rice: Progress and potential for genetics research and breeding. *Rice, 6*, 11.

Bansal, V., Libiger, O., Torkamani, O., & Schork, N. J. (2010). Statistical analysis strategies for association studies involving rare variants. *Nature Reviews Genetics, 11*, 773–785.

Beaumont, M. A., & Rannala, B. (2004). The Bayesian revolution in genetics. *Nature Reviews Genetics, 5*, 251–261.

Benjamini, Y., & Hochberg, Y. (1995). Controlling the false discovery rate: A practical and powerful approach to multiple testing. *Journal of Royal Statistical Society, 57*, 289–300.

Benjamini, Y., Krieger, A., & Yekutieli, D. (2006). Adaptive linear step-up procedures that control the false discovery rate. *Biometrika, 93*, 491–507.

Bentsink, L., Hanson, J., Hanhart, C. J., Blankestijn-de Vries, H., Coltrane, C., Keizer, P., et al. (2010). Natural variation for seed dormancy in *Arabidopsis* is regulated by additive genetic and molecular pathways. *Proceedings of the National Academy of Sciences USA, 107*, 4264–4269.

Bernardo, R. (1993). Estimation of coefficient of coancestry using molecular markers in maize. *Theoretical and Applied Genetics, 85*, 1055–1062.

Bernardo, R. (2008). Molecular markers and selection for complex traits in plants: Learning from the last 20 years. *Crop Science, 48*, 1649–1664.

Bernardo, R. (2013). Genome-wide markers for controlling background variation in association mapping. *The Plant Genome, 6*, 1–9.

Blanc, G., Charcosset, A., Mangin, B., Gallais, A., & Moreau, L. (2006). Connected populations for detecting quantitative trait loci and testing for epistasis: An application in maize. *Theoretical and Applied Genetics, 113*, 206–224.

Blott, S., Kim, J. J., Moisio, S., Schmidt-Kuntzel, A., Cornet, A., Berzi, P., et al. (2003). Molecular dissection of a quantitative trait locus: A phenylalanine-to-tyrosine substitution in the transmembrane domain of the bovine growth hormone receptor is associated with a major effect on milk yield and composition. *Genetics, 163*, 253–266.

Bonferroni, C. E. (1936). Teoria statisti ca delle classi e calcolo delle probabilità. *Pubblicazioni del R Istituto Superiore di Scienze Economiche e Commerciali di Firenze, 8*, 3–62.

Brachi, B., Faure, N., Horton, M., Flahauw, E., Vazquez, A., Nordborg, M., et al. (2010). Linkage and association mapping of *Arabidopsis thaliana* flowering time in nature. *PLoS Genetics, 6*, 1–17.

Brachi, B., Morris, G. P., & Borevitz, J. O. (2011). Genome-wide association studies in plants: The missing heritability is in the field. *Genome Biology, 12*, 232–239.

Bradbury, P. J., Zhang, Z., Kroon, D. E., Casstevens, T. M., Ramdoss, Y., & Buckler, E. S. (2007). TASSEL: Software for association mapping of complex traits in diverse samples. *Bioinformatics, 23*, 2633–2635.

Breseghello, F., & Sorrells, M. E. (2006). Association analysis as a strategy for improvement of quantitative traits in plants. *Crop Science, 46*, 1323–1330.

Buckler, E. S., Holland, J. B., Bradbury, P. J., Acharya, C. B., Brown, P. J., Browne, C., et al. (2009). The genetic architecture of maize flowering time. *Science, 325*, 714–718.

Cantor, R. M., Lange, K., & Sinsheimer, J. S. (2010). Prioritizing GWAS results: A review of statistical methods and recommendations for their application. *American Journal of Human Genetics, 86*, 6–22.

Cavanagh, C., Morell, M., Mackay, I., & Powell, W. (2008). From mutations to MAGIC: Resources for gene discovery, validation and delivery in crop plants. *Current Opinion in Plant Biology, 11*, 215–221.

Chan, E. K. F., Rowe, H. C., & Kliebenstein, D. J. (2010). Understanding the evolution of defense metabolites in *Arabidopsis thaliana* using genome-wide association mapping. *Genetics, 185,* 991–1007.

Churchill, G. A., Airey, D. C., Allayee, H., Angel, J. M., Attie, A. D., Beatty, J., et al. (2004). The collaborative cross, a community resource for the genetic analysis of complex traits. *Nature Genetics, 36,* 1133–1137.

Cook, J. P., McMullen, M. D., Holland, J. B., Tian, F., Bradbury, P., Ross-Ibarra, J., et al. (2012). Genetic architecture of maize kernel composition in the nested association mapping and inbred association panels. *Plant Physiology, 158,* 824–834.

Cordell, H. J. (2002). Epistasis: What it means, what it does not mean, and statistical methods to detect it in humans. *Human Molecular Genetics, 11,* 2463–2468.

Das, K., Li, J., Wang, Z., Tong, C., Fu, G., Li, Y., et al. (2011). A dynamic model for genome-wide association studies. *Human Genetics, 129,* 629–639.

Das, K., Li, R., Huang, Z., Gai, J., & Wu, R. (2012). A Bayesian framework for functional mapping through joint modeling of longitudinal and time-to-event data. *International Journal of Plant Genomics.* http://dx.doi.org/10.1155/2012/680634.

de Bakker, P. I., Ferreira, M. A., Jia, X., Neale, B. M., Raychaudhuri, S., & Voight, B. F. (2008). Practical aspects of imputation-driven meta-analysis of genome-wide association studies. *Human Molecular Genetics, 17,* R122–R128.

Devlin, B., & Roeder, K. (1999). Genomic control for association studies. *Biometrics, 55,* 997–1004.

Dickson, S. P., Wang, K., Krantz, I., Hakonarson, H., & Goldstein, D. B. (2010). Rare variants create synthetic genome-wide associations. *PLoS Biology, 8,* e1000294.

Doerge, R. W. (2002). Mapping and analysis of quantitative trait loci in experimental populations. *Nature Reviews Genetics, 3,* 43–52.

Edwards, D., & Gupta, P. K. (2013). Sequence based DNA markers and genotyping for cereal genomics and breeding. In P. K. Gupta, & R. K. Varshney (Eds.), *Cereal genomics II* (pp. 57–76). Springer.

Ehrenreich, I. M., Hanzawa, Y., Chou, L., Roe, J. L., Kover, P. X., & Purugganan, M. D. (2009). Candidate gene association mapping of *Arabidopsis* flowering time. *Genetics, 183,* 325–335.

Eichler, E. E., Flint, J., Gibson, G., Kong, A., Leal, S. M., Moore, J. H., et al. (2010). Missing heritability and strategies for finding the underlying causes of complex disease. *Nature Reviews Genetics, 11,* 446–450.

Elshire, R. J., Glaubitz, J. C., Sun, Q., Poland, J. A., Kawamoto, K., Buckler, E. S., et al. (2011). A robust, simple genotyping-by-sequencing (GBS) approach for high diversity species. *PLoS One, 6,* e19379.

Evangelou, E., & Ioannidis, J. P. A. (2013). Meta-analysis methods for genome-wide association studies and beyond. *Nature Reviews Genetics, 14,* 379–389.

Fernando, R. L., & Garrick, D. (2013). Bayesian methods applied to GWAS. In Gondro Cedric, et al. (Ed.), *Genome wide association studies and genomic prediction, methods in molecular biology* Vol. 1019. (pp. 237–274). Humana Press.

Gibson, G. (2012). Rare and common variants: Twenty arguments. *Nature Reviews Genetics, 13,* 135–145.

Guo, B., Wang, D., Guo, Z., & Beavis, W. D. (2013). Family-based association mapping in crop species. *Theoretical and Applied Genetics, 126,* 1419–1430.

Han, B., & Eskin, E. (2012). Interpreting meta-analyses of genome-wide association studies. *PLoS Genetics, 8,* e1002555.

Hao, C., Wang, Y., Hou, J., Feuillet, C., Balfourier, F., & Zhang, X. (2012). Association mapping and haplotype analysis of a 3.1-Mb genomic region involved in *Fusarium* head blight resistance on wheat chromosome 3BS. *PLoS One, 7,* e46444.

Harper, A. L., Trick, M., Higgins, J., Clissold, L., Wells, R., Hattori, C., et al. (2012). Associative transcriptomics of traits in the polyploid crop species, *Brassica napus. Nature Biotechnology, 30,* 798–802.

He, Q., Berg, A., Li, Y., Vallejos, C. E., & Wu, R. (2010). Mapping genes for plant structure, development and evolution: Functional mapping meets ontology. *Trends in Genetics, 26*, 39–46.

Hoggart, C. J., Whittaker, J. C., De Iorio, M., & Balding, D. J. (2008). Simultaneous analysis of all SNPs in genome-wide and re-sequencing association studies. *PLoS Genetics, 4*, e1000130.

Holm, S. (1979). A simple sequentially rejective multiple test procedure. *Scandinavian Journal of Statistics, 6*, 65–70.

Huang, X., Wei, X., Sang, T., Zhao, Q., Feng, Q., Zhao, Y., et al. (2010). Genome-wide association studies of 14 agronomic traits in rice landraces. *Nature Genetics, 42*, 961–969.

Huang, B. E., George, A. W., Forrest, K. L., Kilian, A., Hayden, M. J., Morell, M. K., et al. (2012). A multiparent advanced generation inter-cross population for genetic analysis in wheat. *Plant Biotechnology Journal, 10*, 826–839.

Huff, C. D., Witherspoon, D. J., Zhang, Y., Gatenbee, C., Denson, L. A., Kugathasan, S., et al. (2012). Crohn's disease and genetic hitchhiking at IBD5. *Molecular Biology and Evolution, 29*, 101–111.

Inghelandt, D. V., Melchinger, A. E., Martinant, J. P., & Stich, B. (2012). Genome-wide association mapping of flowering time and northern corn leaf blight (*Setosphaeria turcica*) resistance in a vast commercial maize germplasm set. *BMC Plant Biology, 12*, 56.

Iwata, H., Uga, Y., Yoshioka, Y., Ebana, K., & Hayashi, T. (2007). Bayesian association mapping of multiple quantitative trait loci and its application to the analysis of genetic variation among *Oryza sativa* L. germplasms. *Theoretical and Applied Genetics, 114*, 1437–1449.

Jannink, J. L., & Wu, X. L. (2003). Estimating allelic number and identity in state of QTLs in interconnected families. *Genetical Research, 81*, 133–144.

Jannink, J. L., Bink, M. C. A. M., & Jansen, R. C. (2001). Using complex plant pedigrees to map valuable genes. *Trends in Plant Science, 6*, 337–342.

Kang, H. M., Zaitlen, N. A., Wade, C. M., Kirby, A., Heckerman, D., Daly, M. J., et al. (2008). Efficient control of population structure in model organism association mapping. *Genetics, 178*, 1709–1723.

Kang, H. M., Sul, J. H., Service, S. K., Zaitlen, N. A., Kong, S. Y., Freimer, N. B., et al. (2010). Variance component model to account for sample structure in genome-wide association studies. *Nature Genetics, 42*, 348–354.

Kent, J. W. (2011). Rare variants, common markers: Synthetic association and beyond. *Genetic Epidemiology, 35*, S80–S84.

Korte, A., Vilhjalmsson, B. J., Segura, V., Platt, A., Long, Q., & Nordborg, M. (2012). A mixed-model approach for genome-wide association studies of correlated traits in structured populations. *Nature Genetics, 44*, 1066–1071.

Kover, P. X., Valdar, W., Trakalo, J., Scarcelli, N., Ehrenreich, I. M., Purugganan, M. D., et al. (2009). A multiparent advanced generation inter-cross to fine-map quantitative traits in *Arabidopsis thaliana*. *PLoS Genetics, 5*, e1000551.

Kulwal, P., Ishikawa, G., Benscher, D., Feng, Z., Yu, L.-X., Jadhav, A., et al. (2012). Association mapping for pre-harvest sprouting resistance in white winter wheat. *Theoretical and Applied Genetics, 125*, 793–805.

Kump, K. L., Bradbury, P. J., Wisser, R. J., Buckler, E. S., Belcher, A. R., Oropeza-Rosas, M. A., et al. (2011). Genome-wide association study of quantitative resistance to southern leaf blight in the maize nested association mapping population. *Nature Genetics, 43*, 163–168.

Larsson, S. J., Lipka, A. E., & Buckler, E. S. (2013). Lessons from *Dwarf8* on the strengths and weaknesses of structured association mapping. *PLoS Genetics, 9*, e1003246.

Lee, M., Sharopova, N., Beavis, W. D., Grant, D., Katt, M., Blair, D., et al. (2002). Expanding the genetic map of maize with the intermated B73 × Mo17 (IBM) population. *Plant Molecular Biology, 48*, 453–461.

Li, Y., Böck, A., Haseneyer, G., Korzun, V., Wilde, P., Schön, C.-C., et al. (2011a). Association analysis of frost tolerance in rye using candidate genes and phenotypic data from controlled, semi-controlled, and field phenotyping platforms. *BMC Plant Biology, 11*, 146.

Li, Y., Byrnes, A. E., & Li, M. (2010b). To identify associations with rare variants, just WHaIT: Weighted haplotype and imputation-based tests. *American Journal of Human Genetics*, *87*, 728–735.

Li, Y., Ding, J., & Abecasis, G. R. (2006). Mach 1.0: Rapid haplotype reconstruction and missing genotype inference. *American Journal of Human Genetics*, *79*, S2290.

Li, J., Horstman, B., & Chen, Y. (2011b). Detecting epistasis effects in genome-wide association studies based on ensemble approaches. *Bioinformatics*, *27*, 222–229.

Li, Y., Huang, Y., Bergelson, J., Nordborg, M., & Borevitz, J. O. (2010a). Association mapping of local climate-sensitive quantitative trait loci in *Arabidopsis thaliana*. *Proceedings of the National Academy of Sciences USA*, *107*, 21199–21204.

Li, B., & Leal, S. M. (2008). Methods for detecting associations with rare variants for common diseases: Application to analysis of sequence data. *American Journal of Human Genetics*, *83*, 311–321.

Lipka, A. E., Gore, M. A., Lundback, M. M., Mesberg, A., Lin, H., Tiede, T., et al. (2013). Genome-wide association study and pathway-level analysis of tocochromanol level in maize grain. *Genes Genome Genetics*, *3*, 1287–1299.

Lipka, A. E., Tian, F., Wang, Q., Peiffer, J., Li, M., Bradbury, P. J., et al. (2012). GAPIT: Genome Association and Prediction Integrated Tool. *Bioinformatics*, *28*, 2397–2399.

Lippert, C., Listgarten, J., Liu, Y., Kadie, C. M., Davidson, R. I., & Heckerman, D. (2011). FaST linear mixed models for genome-wide association studies. *Nature Methods*, *8*, 833–835.

Liu, W., Gowda, M., Steinhoff, J., Maurer, H. P., Wurschum, T., Londnin, C. F., et al. (2011). Association mapping in an elite maize breeding population. *Theoretical and Applied Genetics*, *123*, 847–858.

Liu, J., Yangy, C., Shi, X., Li, C., Huang, J., Zhao, H., et al. (2013). *A penalized multi-trait mixed model for association mapping in pedigree-based GWAS.* arxiv:1305.4413v1 [stat.ME].

Liu, N., Zhang, K., & Zhao, H. (2008). Haplotype-association analysis. *Advances in Genetics*, *60*, 335–405.

Lu, H. Y., Liu, X. F., Wei, S. P., & Zhang, Y. M. (2011). Epistatic association mapping in homozygous crop cultivars. *PLoS One*, *6*, e17773.

Lu, Y., Zhang, S., Shah, T., Xie, C., Hao, Z., Li, X., et al. (2010). Joint linkage–linkage disequilibrium mapping is a powerful approach to detecting quantitative trait loci underlying drought tolerance in maize. *Proceedings of the National Academy of Sciences USA*, *107*, 19585–19590.

Lynch, M. (1988). Estimation of relatedness by DNA fingerprinting. *Molecular Biology and Evolution*, *5*, 584–599.

Mackay, T. F. C., Stone, E. A., & Ayroles, J. F. (2009). The genetics of quantitative traits: Challenges and prospects. *Nature Reviews Genetics*, *10*, 565–577.

Makowsky, R., Pajewski, N. M., Klimentidis, Y. C., Vazquez, A. I., Duarte, C. W., Allison, D. B., et al. (2011). Beyond missing heritability: Prediction of complex traits. *PLoS Genetics*, *7*, e1002051.

Manolio, T. A., Collins, F. S., Cox, N. J., Goldstein, D. B., Hindorff, L. A., Hunter, D. J., et al. (2009). Finding the missing heritability of complex diseases. *Nature*, *461*, 747–753.

Marchini, J., & Howie, B. (2010). Genotype imputation for genome-wide association studies. *Nature Reviews Genetics*, *11*, 499–511.

Marttinen, P., & Corander, J. (2010). Efficient Bayesian approach for multilocus association mapping including gene-gene interactions. *BMC Bioinformatics*, *11*, 443.

McKinney, B. A., & Pajewski, N. M. (2012). Six degrees of epistasis: Statistical network models for GWAS. *Frontiers in Genetics*, *2*, 1–6.

Meuwissen, T. H. E., Hayes, B. J., & Goddard, M. E. (2001). Prediction of total genetic value using genome-wide dense marker maps. *Genetics*, *157*, 1819–1829.

Meuwissen, T. H. E., Karlsen, A., Lien, S., Olsaker, I., & Goddard, M. E. (2002). Fine mapping of a quantitative trait locus for twinning rate using combined linkage and linkage disequilibrium mapping. *Genetics*, *161*, 373–379.

Moonesinghe, R., Khoury, M. J., Liu, T., & Ioannidis, J. P. (2008). Required sample size and nonreplicability thresholds for heterogeneous genetic associations. *Proceedings of the National Academy of Sciences USA, 105*, 617–622.

Morris, G. P., Ramu, P., Deshpande, S. P., Hash, C.T., Shah,T., Upadhyaya, H. D., et al. (2013). Population genomic and genome-wide association studies of agroclimatic traits in sorghum. *Proceedings of the National Academy of Sciences USA, 110*, 453–458.

Mott, R., & Flint, J. (2002). Simultaneous detection and fine mapping of quantitative trait loci in mice using heterogeneous stocks. *Genetics, 160*, 1609–1618.

Mott, R., Talbot, C. J., Turri, M. J., Collins, A. C., & Flint, J. (2000). A method for fine mapping quantitative trait loci in outbred animal stocks. *Proceedings of the National Academy of Sciences USA, 97*, 12649–12654.

Myles, S., Peiffer, J., Brown, P. J., Ersoz, E. S., Zhang, Z., Costich, D. E., et al. (2009). Association mapping: Critical considerations shift from genotyping to experimental design. *The Plant Cell, 21*, 2194–2202.

Nelson, M. R., Wegmann, D., Ehm, M. G., Kessner, D., StJean, P., Verzilli, C., et al. (2012). An abundance of rare functional variants in 202 drug target genes sequenced in 14,002 people. *Science, 337*, 100–104.

Noble, W. (2009). How does multiple testing correction work? *Nature Biotechnology, 27*, 1135–1137.

Ott, J., Kamatani, Y., & Lathrop, M. (2011). Family-based designs for genome-wide association studies. *Nature Reviews Genetics, 12*, 465–474.

Panoutsopoulou, K., Tachmazidou, I., & Zeggini, E. (2013). In search of low-frequency and rare variants affecting complex traits. *Human Molecular Genetics, 22*, R16–R21.

Parisseaux, B., & Bernardo, R. (2004). *In silico* mapping of quantitative trait loci in maize. *Theoretical and Applied Genetics, 109*, 508–514.

Paulo, M. J., Boer, M., Huang, X., Koornneef, M., & van Eeuwijk, F. A. (2008). A mixed model QTL analysis for a complex cross population consisting of a half diallel of two-way hybrids in *Arabidopsis thaliana*: Analysis of simulated data. *Euphytica, 161*, 107–114.

Poland, J. A., Bradbury, P. J., Buckler, E. S., & Nelson, R. J. (2011). Genome-wide nested association mapping of quantitative resistance to northern leaf blight in maize. *Proceedings of the National Academy of Sciences USA, 108*, 6893–6898.

Poland, J. A., & Rife, T. W. (2012). Genotyping-by-sequencing for plant breeding and genetics. *The Plant Genome, 5*, 92–102.

Price, A. L., Patterson, N. J., Plenge, R. M., Weinblatt, M. E., Shadick, N. A., & Reich, D. (2006). Principal components analysis corrects for stratification in genome-wide association studies. *Nature Genetics, 38*, 904–909.

Pritchard, J. K., Stephens, M., & Donnelly, P. (2000). Inference of population structure using multilocus genotype data. *Genetics, 155*, 945–959.

Qian, H. R., & Huang, S. (2005). Comparison of false discovery rate methods in identifying genes with differential expression. *Genomics, 86*, 495–503.

Rakitsch, B., Lippert, C., Stegle, O., & Borgwardt, K. (2013). A lasso multi-marker mixed model for association mapping with population structure correction. *Bioinformatics, 29*, 206–214.

Rebai, A., & Goffinet, B. (1993). Power of tests for QTL detection using replicated progenies derived from a diallel cross. *Theoretical and Applied Genetics, 86*, 1014–1022.

Rebai, A., & Goffinet, B. (2000). More about quantitative trait locus mapping with diallel designs. *Genetical Research, 75*, 243–247.

Reeves, P., & Richards, C. (2009). Accurate inference of subtle population structure (and other genetic discontinuities) using principal coordinates. *PLoS One, 4*, e4269.

Reif, J. C., Gowda, M., Maurer, H. P., Longin, C. F. H., Korzun, V., Ebmeyer, E., et al. (2011). Association mapping for quality traits in soft winter wheat. *Theoretical and Applied Genetics, 122*, 961–970.

Reif, J. C., Liu, W., Gowda, M., Maurer, H. P., Mohring, J., Fischer, S., et al. (2010). Genetic basis of agronomically important traits in sugar beet (*Beta vulgaris* L.) investigated with joint linkage association mapping. *Theoretical and Applied Genetics, 8*, 1489–1499.

Remington, D. L., Thornsberry, J., Matsuoka, Y., Wilson, L., Rinehart-Whitt, S., Doebley, J., et al. (2001). Structure of linkage disequilibrium and phenotypic associations in the maize genome. *Proceedings of the National Academy of Sciences USA, 98*, 11479–11484.

Rosyara, U. R., Gonzalez-Hernandez, J. L., Glover, K. D., Gedye, K. R., & Stein, J. M. (2009). Family-based mapping of quantitative trait loci in plant breeding populations with resistance to *Fusarium* head blight in wheat as an illustration. *Theoretical and Applied Genetics, 118*, 1617–1631.

Rutkoski, J. E., Poland, J., Jannink, J.-L., & Sorrells, M. E. (2013). Imputation of unordered markers and the impact on genomic selection accuracy. *Genes Genomes Genetics, 3*, 427–439.

Saintenac, C., Jiang, D., Wang, S., & Akhunov, E. (2013). Sequence-based mapping of the polyploid wheat genome. *Genes Genomes Genetics, 3*, 1105–1114.

Segura, V., Vilhjalmsson, B. J., Platt, A., Korte, A., Seren, U., Long, Q., et al. (2012). An efficient multi-locus mixed-model approach for genome-wide association studies in structured populations. *Nature Genetics, 44*, 825–830.

Shoemaker, J. S., Painter, I. S., & Weir, B. S. (1999). Bayesian statistics in genetics. A guide for the uninitiated. *Trends in Genetics, 15*, 354–358.

Sneller, C. H., Mather, D. E., & Crepieux, S. (2009). Analytical approaches and population types for finding and utilizing QTL in complex plant populations. *Crop Science, 49*, 363–380.

Spencer, C. C. A., Su, Z., Donnelly, P., & Marchini, J. (2009). Designing genome-wide association studies: Sample size, power, imputation, and the choice of genotyping chip. *PLoS Genetics, 5*, e1000477.

Spielman, R. S., McGinnes, R. E., & Ewens, W. J. (1993). Transmission test for linkage disequilibrium: The insulin gene region and insulin-dependent diabetes mellitus (IDDM). *American Journal of Human Genetics, 52*, 506–516.

Steinhoff, J., Liu, W., Maurer, H. P., Wurschum, T., Longin, F. H., Rancb, N., et al. (2011). Multiple-line cross QTL-mapping in European elite maize. *Crop Science, 51*, 2505–2516.

Stephens, M., & Balding, D. J. (2009). Bayesian statistical methods for genetic association studies. *Nature Reviews Genetics, 10*, 681–690.

Stich, B. (2009). Comparison of mating designs for establishing nested association mapping populations in maize and *Arabidopsis thaliana*. *Genetics, 183*, 1525–1534.

Stich, B., & Melchinger, A. E. (2009). Comparison of mixed-model approaches for association mapping in rapeseed, potato, sugar beet, maize, and *Arabidopsis*. *BMC Genomics, 10*, 94.

Stich, B., & Melchinger, A. (2010). An introduction to association mapping in plants. *CAB Reviews, 5*, 1–9.

Stich, B., Melchinger, A. E., Piepho, H. P., Heckenberger, M., Maurer, H. P., & Reif, J. C. (2006). A new test for family-based association mapping with inbred lines from plant breeding programs. *Theoretical and Applied Genetics, 113*, 1121–1130.

Stich, B., Mohring, J., Piepho, H. P., Heckenberger, M., Buckler, E. S., & Melchinger, A. E., (2008). Comparison of mixed-model approaches for association mapping. *Genetics, 178*, 1745–1754.

Storey, J. D. (2002). A direct approach to false discovery rates. *Journal of Royal Statistical Society Series B, 64*, 479–498.

Stracke, S., Haseneyer, G., Veyrieras, J. B., Geiger, H. H., Sauer, S., Graner, A., et al. (2009). Association mapping reveals gene action and interactions in the determination of flowering time in barley. *Theoretical and Applied Genetics, 118*, 259–273.

Svishcheva, G. R., Axenovich, T. I., Belonogova, N. M., van Duijn, C. M., & Aulchenko, Y. S. (2012). Rapid variance components-based method for whole-genome association analysis. *Nature Genetics, 44*, 1166–1170.

Tachmazidou, I., Verzilli, C. J., & Iorio, M. D. (2007). Genetic Association mapping via evolution-based clustering of haplotypes. *PLoS Genetics, 3*, e111.

Tennessen, J. A., Bigham, A. W., O'Connor, T. D., Fu, W., Kenny, E. E., Gravel, S., et al. (2012). Evolution and functional impact of rare coding variation from deep sequencing of human exomes. *Science, 64*, 64–69.

Thompson, J. R., Attia, J., & Minelli, C. (2011). The meta-analysis of genome-wide association studies. *Briefings in Bioinformatics, 12*, 259–269.

Thornsberry, J. M., Goodmann, M. M., Doebley, J., Kresovich, S., Nielsen, D., & Buckler, E. S. (2001). *Dwarf8* polymorphisms associate with variation in flowering time. *Nature Genetics, 28*, 286–289.

Tian, F., Bradbury, P., Brown, P., Hung, H., Sun, Q., Flint-Garcia, S., et al. (2011). Genome-wide association study of leaf architecture in the maize nested association mapping population. *Nature Genetics, 43*, 159–162.

Vilhjálmsson, B. J., & Nordborg, M. (2013). The nature of confounding in genome-wide association studies. *Nature Reviews Genetics, 14*, 1–2.

Wakefield, J. (2009). Bayes factors for genome-wide association studies: Comparison with P-values. *Genetic Epidemiology, 33*, 79–86.

Wan, X., Yang, C., Yang, Q., Zhao, H., & Yu, W. (2013). The complete compositional epistasis detection in genome-wide association studies. *BMC Genetics, 14*, 7.

Wang, L., Cui, F., Wang, J., Jun, L., Ding, A., Zhao, C., et al. (2012). Conditional QTL mapping of protein content in wheat with respect to grain yield and its components. *Journal of Genetics, 91*, 303–312.

Wang, D., Eskridge, K. M., & Crossa, J. (2011). Identifying QTLs and epistasis in structured plant populations using adaptive mixed LASSO. *Journal of Agricultural, Biological and Environmental Statistics, 16*, 170–184.

Weber, A. L., Zhao, Q., McMullen, M. D., & Doebley, J. F. (2009). Using association mapping in teosinte to investigate the function of maize selection-candidate genes. *PLoS One, 4*, e8227.

Weng, J., Xie, C., Hao, Z., Wang, J., Liu, C., Li, M., et al. (2011). Genome-wide association study identifies candidate genes that affect plant height in Chinese elite maize (*Zea mays* L.) inbred lines. *PLoS One, 6*, e29229.

Westfall, P. H., & Young, S. S. (1993). *Resampling-based multiple testing: examples and methods for p-value adjustment*. New York: John Wiley.

Wray, N. R., Purcell, S. M., & Visscher, P. M. (2011). Synthetic associations created by rare variants do not explain most GWAS results. *PLoS Biology, 9*, e1000579.

Wu, T. T., Chen, Y. F., Hastie, T., Sobel, E., & Lange, K. (2009). Genome-wide association analysis by lasso penalized logistic regression. *Bioinformatics, 25*, 714–721.

Wu, W. R., Li, W. M., Tang, D. Z., Lu, H. R., & Worland, A. J. (1999). Time related mapping of quantitative trait loci underlying tiller number in rice. *Genetics, 151*, 297–303.

Wu, R., & Lin, M. (2006). Functional mapping – how to map and study the genetic architecture of dynamic complex traits. *Nature Reviews Genetics, 7*, 229–237.

Wurschum, T. (2012). Mapping QTL for agronomic traits in breeding populations. *Theoretical and Applied Genetics, 125*, 201–210.

Wurschum, T., Liu, W., Gowda, M., Maurer, H. P., Fischer, S., Schechert, A., et al. (2012). Comparison of biometrical models for joint linkage association mapping. *Heredity, 108*, 332–340.

Wurschum, T., Maurer, H. P., Schulz, B., Mohring, J., & Reif, J. C. (2011). Genome-wide association mapping reveals epistasis and genetic interaction networks in sugar beet. *Theoretical and Applied Genetics, 123*, 109–118.

Xu, S. (1998). Mapping quantitative trait loci using multiple families of line crosses. *Genetics, 148*, 517–524.

Yu, J., & Buckler, E. S. (2006). Genetic association mapping and genome organization of maize. *Current Opinion in Biotechnology, 17*, 155–160.

Yu, J., Holland, J. B., McMullen, M. D., & Buckler, E. S. (2008). Genetic design and statistical power of nested association mapping in maize. *Genetics, 178*, 539–551.

Yu, L. X., Lorenz, A., Rutkoski, J., Singh, R. P., Bhavani, S., Huerta-Espino, J., et al. (2011). Association mapping and gene–gene interaction for stem rust resistance in spring wheat germplasm. *Theoretical and Applied Genetics, 123*, 1257–1268.

Yu, J., Pressoir, G., Briggs, W. H., Vroh, Bi I., Yamasaki, M., Doebley, J. F., et al. (2006). A unified mixed-model method for association mapping that accounts for multiple levels of relatedness. *Nature Genetics, 38*, 203–208.

Zhang, Z., Buckler, E. S., Casstevens, T. M., & Bradbury, P. J. (2009). Software engineering the mixed model for genome-wide association studies on large samples. *Briefings in Bioinformatics, 10*, 664–675.

Zhang, Z., Ersoz, E., Lai, C. Q., Todhunter, R. J., Tiwari, H. K., Gore, M. A., et al. (2010). Mixed linear model approach adapted for genome-wide association studies. *Nature Genetics, 42*, 355–360.

Zhang, J., Hao, C., Ren, Q., Chang, X., Liu, G., & Jing, R. (2011). Association mapping of dynamic developmental plant height in common wheat. *Planta, 234*, 891–902.

Zhang, B., Shi, W., Li, W., Chang, X., & Jing, R. (2013). Efficacy of pyramiding elite alleles for dynamic development of plant height in common wheat. *Molecular Breeding, 32*, 327–338.

Zhang, S., Zhang, K., Li, J., Sun, F., & Zhao, H. (2001). Test of association for quantitative traits in general pedigrees: The quantitative pedigree disequilibrium test. *Genetic Epidemiology, 21*, 370–375.

Zhao, K., Aranzana, M. J., Kim, S., Lister, C., Shindo, C., Tang, C., et al. (2007). An *Arabidopsis* example of association mapping in structured samples. *PLoS Genetics, 3*, E4.

Zhao, K., Tung, C. W., Eizenga, G. C., Wright, M. H., Ali, M. L., Price, A. H., et al. (2011). Genome-wide association mapping reveals a rich genetic architecture of complex traits in *Oryza sativa. Nature Communications, 2*, 467.

Zheng, P., Allen, W. B., Roesler, K., Williams, M. E., Zhang, S., Li, J., et al. (2008). A phenylalanine in DGAT is a key determinant of oil content and composition in maize. *Nature Genetics, 40*, 367–372.

Zhou, X., & Stephens, M. (2012). Genome-wide efficient mixed-model analysis for association studies. *Nature Genetics, 44*, 821–824.

Zhu, J. (1995). Analysis of conditional genetic effects and variance components in developmental genetics. *Genetics, 141*, 1633–1639.

Zhu, Q., Ge, D., Maia, J. M., Zhu, M., Petrovski, S., Dickson, S. P., et al. (2011). A genome-wide comparison of the functional properties of rare and common genetic variants in humans. *American Journal of Human Genetics, 88*, 458–468.

Zhu, C., Gore, M., Buckler, E. S., & Yu, J. (2008). Status and prospects of association mapping in plants. *The Plant Genome, 1*, 5–20.

Zuk, O., Hechter, E., Sunyaev, S. R., & Lander, E. S. (2012). The mystery of missing heritability: Genetic interactions create phantom heritability. *Proceedings of the National Academy of Sciences USA, 109*, 1193–1198.

# CHAPTER THREE

# The miRNA-Mediated Cross-Talk between Transcripts Provides a Novel Layer of Posttranscriptional Regulation

## Jennifer Y. Tan*,† and Ana C. Marques*,†,1

*MRC Functional Genomics Unit, University of Oxford, Oxford, UK
†Department of Physiology, Anatomy and Genetics, University of Oxford, Oxford, UK
1Corresponding author: e-mail address: ana.marques@dpag.ox.ac.uk

## Contents

## Abstract

Endogenously expressed transcripts that are posttranscriptionally regulated by the same microRNAs (miRNAs) will, in principle, compete for the binding of their shared small noncoding RNA regulators and modulate each other's abundance. Recently, the levels of some coding as well as noncoding transcripts have indeed been found to be regulated in this way. Transcripts that engage in such regulatory interactions are referred to as competitive endogenous RNAs (ceRNAs).

*Advances in Genetics*, Volume 85
ISSN 0065-2660
http://dx.doi.org/10.1016/B978-0-12-800271-1.00003-2

This novel layer of posttranscriptional regulation has been shown to contribute to diverse aspects of organismal and cellular biology, despite the number of functionally characterized ceRNAs being as yet relatively low. Importantly, increasing evidence suggests that the dysregulation of some ceRNA interactions is associated with disease etiology, most preeminently with cancer.

Here we review how posttranscriptional regulation by miRNAs contributes to the cross-talk between transcripts and review examples of known ceRNAs by highlighting the features underlying their interactions and what might be their biological relevance.

## 1. INTRODUCTION

Complex multicellular organisms are made up of hundreds of different cell types, each with distinct morphologies and biological functions. This diversity, which is essential to organism's development and survival, is achieved via context-specific expression from a nearly identical set of genetic instructions containing distinct repertoires of coding and noncoding molecules in different cell types and developmental stages. Such complex spatiotemporal regulation of gene expression requires the intricately and tightly controlled interplay between the genome and its diverse gene products. Posttranscriptional regulation of RNA processing and turnover is essential to fine-tune the cellular concentration of most RNA and protein products (reviewed in Cogni et al., 2000). One preeminent way by which this is achieved is via microRNA (miRNA)-mediated translational repression and/or transcript degradation (Bartel, 2009). MiRNAs are a class of small noncoding RNAs (21–25 nucleotides) that is predicted to target and regulate the abundance of up to 60% of human protein-coding genes (Friedman, Farh, Burge, & Bartel, 2009; Griffiths-Jones, Saini, van Dongen, & Enright, 2008; Lewis, Burge, & Bartel, 2005).

Since the discovery of the first miRNA, *lin-4*, in the early 1990s (Lee, Feinbaum, & Ambros, 1993), thousands of miRNAs have been identified across several eukaryotic species (Griffiths-Jones et al., 2008; Kozomara & Griffiths-Jones, 2011). For example, the number of human miRNAs annotated in miR-Base, a central repository of miRNA annotations (Kozomara & Griffiths-Jones, 2011), is currently over 1870 (miRBase v20, released in June 2013).

Posttranscriptional regulation of gene expression by miRNAs requires their recognition, via partial sequence complementarity, of miRNA response elements (MREs) found within the transcribed sequences of their coding and noncoding targets (Bartel, 2004). A single miRNA can modulate the levels of hundreds of transcripts (Brennecke, Stark, Russell, & Cohen, 2005; Krek et al., 2005; Lewis et al., 2005; Stark, Brennecke, Bushati, Russell, &

Cohen, 2005; Xie et al., 2005) and generally, target recognition is thought to reduce the cellular availability of miRNAs (Ruegger & Grosshans, 2012). These observations suggest that, in principle, transcripts that harbor recognition elements for the same miRNA(s) are capable of modulating each other's expression levels by competing for the binding of shared regulatory miRNAs (Seitz, 2009).

Evidence that targets can effectively modulate the levels of their miRNA regulators came almost simultaneously from the discovery of an endogenously expressed miRNA-sponge in plants (Franco-Zorrilla et al., 2007) and the expression of exogenous MRE-rich artificial transcripts (Ebert, Neilson, & Sharp, 2007). *IPS1* (*INDUCED BY PHOSPHATE STARVATION 1*) is a noncoding RNA whose expression in *Arabidopsis thaliana* is induced upon phosphorous starvation (Franco-Zorrilla et al., 2007). This noncoding transcript contains a conserved MRE for *miR-399*, a miRNA which is also induced upon phosphorous starvation (Franco-Zorrilla et al., 2007). The nonreversible binding of *miR-399* to *IPS1* reduces the cellular availability of this miRNA, which in turn relieves *PHO2*, a protein-coding gene involved in phosphorous response, from *miR-399*-mediated posttranscriptional repression (Franco-Zorrilla et al., 2007) (Figure 3.1). This competition for *miR-399* between *IPS1* and *PHO2* results in their coexpression in vivo (Figure 3.1). Similarly, expression of artificial miRNA sponges, containing multiple binding sites for specific miRNAs, was shown to change the cellular availability of miRNA, thus demonstrating the efficacy of this approach to regulate miRNA levels *in vivo/vitro* (Ebert et al., 2007). Importantly, and as for their endogenous counterparts, the change in miRNA abundance induced by the recognition of artificial transcripts was associated with a significant derepression of these miRNAs endogenous targets (Ebert et al., 2007).

Since then, studies in several organisms, including mammalian and viral species have provided further evidence that reinforce the roles of this newly recognized posttranscriptional mechanism. Coding and noncoding transcripts that are able to regulate each other's abundance through competition for the binding of shared miRNAs are termed *competitive endogenous RNAs* (ceRNAs) (Salmena, Poliseno, Tay, Kats, & Pandolfi, 2011). Given the extensive miRNA regulation in eukaryotic genomes and the large number of loci that each individual miRNA can regulate, virtually all transcripts in the cell can potentially engage and contribute to this layer of miRNA-dependent posttranscriptional regulation.

Here, we describe recent advances in the understanding of this relatively novel layer of posttranscriptional regulation. Given that these interactions

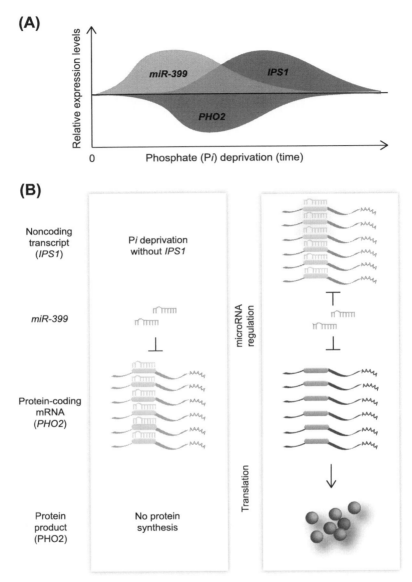

**Figure 3.1** *IPS1* is an endogenous miRNA sponge for *miR-399* and regulates *PHO2*. (A) Illustration of the changes in relative expression levels (*y*-axis) of a microRNA (miRNA), *miR-399* (yellow), a noncoding RNA, *IPS1* (blue), and a protein-coding mRNA, *PHO2* (red), after the induction of phosphate (*Pi*) deprivation (*x*-axis, 0). *IPS1* is a molecular sponge for *miR-399* whose sponging activity reduces *miR-399* abundance and consequently, posttranscriptional derepresses levels of *PHO2*, another gene target of *miR-399*. (B) Upon *Pi* deprivation and prior to *IPS1* induction (left panel), *miR-399* (yellow) posttranscriptionally represses *PHO2* (red) and suppresses the levels of its protein product (red circles). *IPS1* expression (blue, right panel) sequesters *miR-399* and thereby, derepresses *PHO2* from its posttranscriptional regulation and results in increased *PHO2* protein levels. (For interpretation of the references to color in this figure legend, the reader is referred to the online version of this book.)

are dependent on the regulatory roles of miRNAs, we will first describe the principles underlying the posttranscriptional regulation by this class of small noncoding RNAs and illustrate how these rules might contribute to endogenous transcripts' ability to cross-talk. Subsequently, we will review known interactions between ceRNAs by highlighting their similarities and differences in properties and emphasizing their known or predicted contributions to disease and development processes. Finally, we conclude by discussing some of the current challenges encountered in the identification and functional characterization of ceRNAs.

## 2. BIOGENESIS AND MECHANISMS OF MIRNA-MEDIATED REGULATION OF GENE EXPRESSION

### 2.1 Biogenesis and Turnover of miRNAs

Mature miRNAs are the product of a relatively complex biogenesis process (Rodriguez, Griffiths-Jones, Ashurst, & Bradley, 2004; Saini, Griffiths-Jones, & Enright, 2007). Individual primary miRNA transcripts (pri-miRNAs), which can encode one or more mature miRNAs (Cai, Hagedorn, & Cullen, 2004), are generally transcribed by RNA Polymerase II (Pol II). As other Pol II transcripts, pri-miRNAs are capped and polyadenylated (Cai et al., 2004; Lee et al., 2004). Pri-miRNAs can originate at orphan intergenic promoters or overlap the introns or exons of both coding and noncoding transcripts (Cai et al., 2004).

While in the nucleus, pri-miRNAs are cleaved into stem–loop structures of approximately 70 nucleotides in length by Drosha, an RNAse III endonuclease, that exists as part of the protein complex called the Microprocessor complex (Lee et al., 2003) along with its cofactor, DiGeorge Syndrome critical region 8 homolog (DGCR8) (Denli, Tops, Plasterk, Ketting, & Hannon, 2004; Gregory et al., 2004). These hairpin structures, known as the miRNA precursor transcripts (pre-miRNAs), are exported from the nucleus into the cytoplasm (Lund, Guttinger, Calado, Dahlberg, & Kutay, 2004), where they are cleaved by a second RNAse III endonuclease, Dicer, into mature miRNA duplexes of approximately 21–25 base pairs (bp) with 3′ overhangs (Gregory et al., 2004). Following Dicer cleavage, mature single-stranded miRNAs are incorporated into the RNA-induced silencing complex (RISC). Loading of mature miRNAs into RISC requires the recognition of the miRNAs' 3′ overhangs by the Argonaute (AGO) proteins, core protein component of this complex (Hutvagner & Simard, 2008; Meister et al., 2004). In principle, both strands of the miRNA duplex can give rise to functional mature miRNAs (Marco, Macpherson, Ronshaugen, & Griffiths-Jones, 2012). Despite this, for most miRNAs, one strand (the guide strand) is preferentially loaded into the RISC complex. This strand is thought to be selected based on its higher thermodynamic stability (Khvorova, Reynolds, & Jayasena, 2003; Krol et al., 2004; Schwarz et al., 2003). Nonetheless, for a few miRNAs, both strands of the duplex give rise to functional mature miRNAs that are stably incorporated into the RISC complex and often target different transcripts (Okamura, Chung, & Lai, 2008).

MiRNA biogenesis is critical to survival. *Dicer*-null mouse (Bernstein et al., 2003) and zebrafish (Wienholds, Koudijs, van Eeden, Cuppen, & Plasterk, 2003) embryos exhibit development arrest and early embryonic death. Similarly, the deletion of one of the AGO proteins in flies, *dAGO1*, leads to their death during embryogenesis or early larval development stages (Kataoka, Takeichi, & Uemura, 2001; Okamura, Ishizuka, Siomi, & Siomi, 2004), where as the deletion of the two AGO-encoding genes, *alg-1* and *alg-2* in worms, results in reduced levels of mature miRNA levels and lethality early in development (Grishok et al., 2001).

The steady-state level of miRNAs is determined by both their rate of synthesis and decay. However and in contrast to our relatively extensive understanding of the processes underlying miRNA biogenesis, mechanisms that control miRNA decay and destabilization, which are likely to vary for different miRNAs, are less well understood (Gantier et al., 2011; Kai & Pasquinelli, 2010; Liu, 2008). Most mature miRNAs appear to be relatively stable and persist in the cell for several days after their transcription and processing (Bail et al., 2010; Gantier et al., 2011; Kai & Pasquinelli, 2010; Krol, Loedige, & Filipowicz, 2010). Target recognition, particularly through their binding to highly complementary target sequences, can trigger miRNA degradation (Ameres et al., 2010; Baccarini et al., 2011; Pasquinelli, 2012; Ruegger & Grosshans, 2012; Wyman et al., 2011). In addition, target-induced miRNA destabilization is often accompanied by the emergence of modified miRNA species with extended tails or trimmed ends and can induce miRNA degradation (Flynt, Greimann, Chung, Lima, & Lai, 2010). Found in various organisms, the target recognition-mediated regulation is likely be an important contributor to the regulation of miRNA abundance (Ruegger & Grosshans, 2012). However, the specific cellular conditions and molecular mechanisms that trigger the degradation of miRNAs bound to their targets through imperfect complementary are still yet largely unknown (Liu, 2008).

The reciprocal regulation of miRNA and target abundance hints toward a complex and highly connected posttranscriptional regulatory network.

Establishing the extent and complexity of such networks will require the identification of the physiological conditions that control changes in miRNA abundance, and in particular, the mechanisms underlying miRNA turnover following target recognition.

## 2.2 Mechanisms of miRNA-Mediated Regulation of Gene Expression

MiRNA-mediated regulation of transcript abundance requires the recognition of target genes by the mature miRNA-loaded RISC complex (miRISC) (Bartel, 2009; Krol et al., 2010). The association between miRISC and its targets is dependent on Watson–Crick base pairing between the RISC-loaded miRNA and the MREs embedded within the target sequence (Bartel, 2004, 2009; Lewis et al., 2005). The minimum sequence complementarity required for canonical miRNA–target association between an MRE and an miRNA involves at least nucleotides 2–8 from the 5′ end of the mature miRNA (referred to as the "seed" region) (Bartel, 2009; Lewis et al., 2005). Beyond that, the degree of sequence complementarity between the miRNA and its bound target will, to some extent, determine the mechanism of miRNA-mediated silencing (Bartel, 2009).

In plants, most miRNA are highly complementary to their target transcripts, with most interactions often extending beyond the seed sequence (Jones-Rhoades, Bartel, & Bartel, 2006). Target recognition via highly complementary miRNAs triggers the RNA interference (RNAi)-like pathway (Pillai, Bhattacharyya, & Filipowicz, 2007), which leads to transcript cleavage by the AGO protein, induces decapping or deadenylation, and results in target degradation (Behm-Ansmant et al., 2006; Meister et al., 2004; O'Carroll et al., 2007; Wakiyama, Takimoto, Ohara, & Yokoyama, 2007; Wu, Fan, & Belasco, 2006).

In contrast to plant miRNAs, extensive sequence complementary between miRNAs and their targets is rare in animals (Ambros, 2004). Instead, complementarity between the miRNA seed sequence and the target transcript is sufficient for target recognition (Bartel, 2009; Lewis et al., 2005). This type of imperfect sequence complementarity results in bulges between the miRNA:target duplex, which is expected to preclude the AGO-mediated endonucleolytic cleavage of the transcript (Jones-Rhoades et al., 2006) and either induces protein levels exonucleolytic target degradation (Djuranovic, Nahvi, & Green, 2011) or dramatically reduce the protein levels while exerting little impact on the respective mRNA targets (Djuranovic et al., 2011). In addition to transcript degradation, miRNA binding can also induce translational inhibition. This is illustrated by the regulation of *LIN14* and *LIN28* levels by the first identified miRNA, *lin-4*, in *Caenorhabditis elegans* (Lee et al., 1993; Wightman, Ha, & Ruvkun, 1993). Collectively, both translational inhibition and transcriptional repression are common mechanisms of miRNA induced posttranscriptional repression in animals (Djuranovic et al., 2011; Jones-Rhoades et al., 2006).

Transcripts can have more than one response element for the same miRNA (Rajewsky & Socci, 2004) and transcriptome-wide analysis has revealed that the impact of an miRNA on its target product abundance is often directly proportional to the number of harbored MREs (Doench & Sharp, 2004), suggesting that their repeated occurrences have a multiplicative effect. The exception to this effect appear to be when MREs lie within 8 and 40 nucleotides of each other. In these cases, the MREs exert a significantly greater impact on gene expression than what would be expected from independently acting MREs, indicating a cooperative effect from nearby sites (Doench & Sharp, 2004). In addition, the efficacy of miRNA regulation on gene expression is also affected by the relative location of the MREs within its target sequence (Grimson et al., 2007). With a few exceptions, miRNA targeting is more frequent and efficient within the 3′-untranslated regions (UTRs) of protein coding genes than in either the 5′ UTRs or their open-reading frames. This is in part due to the likely displacement of the silencing miRISC complex by the translation machinery as it progresses from the cap-ended side of the transcript (Gu, Jin, Zhang, Sarnow, & Kay, 2009). The lack of translation on long noncoding transcripts has led to the hypothesis that their abundance is more efficiently posttranscriptionally regulated by miRNAs (Ebert & Sharp, 2010a; Hansen et al., 2013).

The ability of any given miRNA to regulate its targets' expression levels is dependent on their subcellular colocalization. The majority of miRNAs are localized in the cytoplasmatic fraction of the cell (Ambros, 2004; Bartel & Chen, 2004), as are the majority of protein-coding mRNAs (Kohler & Hurt, 2007). Similarly, despite being relatively more frequently retained in the nucleus (Derrien et al., 2012), most long noncoding RNAs are also abundantly found in the cytoplasm (Ulitsky & Bartel, 2013). Posttranscriptional regulation outside of the cell's transcriptional core, the nucleus, later in the gene expression pipeline may allow miRNAs to offset inevitable variations that occur during upstream processes, including transcription, splicing, and nuclear export (Bartel, 2004; Ebert & Sharp, 2012).

Furthermore, despite their general cytoplasmatic enrichment, some RISC-associated miRNAs can reenter the nucleus (Hwang, Wentzel, & Mendell, 2007; Meister et al., 2004; Politz, Zhang, & Pederson, 2006; Robb,

Brown, Khurana, & Rana, 2005; Weinmann et al., 2009). These nuclear miRNAs have been proposed to target and modulate the biogenesis and expression levels of primary miRNA transcripts and nuclear-retained RNA transcripts (Chen, Liang, Zhang, & Zen, 2012; Hansen et al., 2011; Liang, Zhang, Zen, Zhang, & Chen, 2013; Tang et al., 2012). Although the prevalence and mechanistic roles of nuclear miRNAs are less well understood than that of their cytoplasmic counterparts, these observations nonetheless expand the extent of miRNA regulation to virtually all levels of gene expression control.

## 2.3 Most miRNAs Confer Robustness to Gene Expression

Some miRNAs can function as molecular switches, regulating the levels of genes that are critical to the transition between distinctive cellular states (switch-like interactions) (Bartel & Chen, 2004; Sotiropoulou, Pampalakis, Lianidou, & Mourelatos, 2009). For instance, the first miRNA to be discovered, *lin-4*, is a switch-like posttranscriptional regulator whose developmentally regulated expression controls the timing of *C. elegans* larval development (Lee et al., 1993). *Lin-4* targets *lin-14*, a gene found to be expressed only at the early stages of larval development. Lin-14 protein inhibits the transcription of genes involved in cell division and differentiation (Wightman et al., 1993). The expression of *lin-4* inhibits *lin-14* later in development and activates stage-specific genes that promote the transitional switch from larval to adult in *C. elegans* (Ambros, 1989; Lee et al., 1993). The mutually exclusive temporal expression of *lin-4* and *lin-14* resembles a larval-to-adult development switch that controls the timing of *C. elegans* stage-specific postembryonic development (Lee et al., 1993). In addition, rather than modulating the expression level of one crucial target, some miRNAs can function as switches by directly or indirectly modulating the expression of several transcripts (Lim et al., 2005). For example, ectopic expression of a brain-specific miRNA, *miR-124*, in cervical cancer-derived HeLa cells induces a global shift in their expression profiles resulting in a repertoire of transcripts that closely resembles those found in brain tissues (Lim et al., 2005).

However, miRNAs capable of inducing switches between different cellular states are rare (Bartel, 2009) and the impact of most miRNAs on target product abundance is typically modest (Baek et al., 2008; Bartel, 2009). This is consistent with most miRNA:target interactions being either neutral (Seitz, 2009) or contributing to fine-tuning of gene expression (Bartel & Chen, 2004; Sevignani, Calin, Siracusa, & Croce, 2006). By fine-tuning transcript levels, most miRNAs likely function to reinforce robust gene expression responses posttranscriptionally (Ebert & Sharp, 2012; Hornstein & Shomron, 2006). Such functions are likely to be nonessential in normal environmental conditions and might only become apparent under stress conditions (Brenner, Jasiewicz, Fahley, Kemp, & Abbott, 2010; van Rooij et al., 2007; Zheng et al., 2011). This might explain, at least in part, why despite that complete loss of miRNA biogenesis being embryonic lethal (Kanellopoulou et al., 2005; Wienholds et al., 2003), knockout of most miRNAs in mice produces no overt morphological or behavioral phenotypes (Miska et al., 2007; Mukherji et al., 2011).

For some miRNAs, their contributions to cellular robustness are reinforced via autoregulatory feedback loops that directly link the output of miRNA actions with their transcriptional regulators (Li, Cassidy, Reinke, Fischboeck, & Carthew, 2009; Linsley et al., 2007; Osella, Bosia, Cora, & Caselle, 2011). For example, the circuitry controlling cell fate decisions in mouse embryonic stem cells (mESCs) is a well-established example of one such circuitry. Transcription of the precursor transcript of *miR-145* is regulated by one of the core transcription factors that control cellular pluripotency, Oct4 (Xu, Papagiannakopoulos, Pan, Thomson, & Kosik, 2009). In turn, mature *miR-145* posttranscriptionally represses the levels of all three core transcription factors that govern ESC pluripotency, including Nanog, Oct4, and Sox2 (Xu et al., 2009). The presence of such feedback circuitries is likely to allow rapid and accurate response to homeostasis disturbance (Li et al., 2009; Osella et al., 2011).

## 2.4 Identification of miRNA–Target Interactions

Several experimental and computational approaches aiming to predict and validate functionally relevant miRNA–target interactions have been developed in recent years.

Computational approaches, designed to identify putative MREs within transcripts, have been extensively used to predict miRNA–target interactions (Chaudhuri & Chatterjee, 2007; Lindow & Gorodkin, 2007; Watanabe, Tomita, & Kanai, 2007; Witkos, Koscianska, & Krzyzosiak, 2011; Zhang & Verbeek, 2010). These tools search for the presence of one or more features known to influence target recognition by miRNAs, namely i) evidence of sequence complementary notably within the miRNA seed sequence; ii) evidence of other sequence features associated with this region of complementary found in (i), such as its distance to the 3′ end of the transcript or target secondary structure; iii) evolutionary conservation of the region of sequence complementary; and iv) the thermodynamic stability of the putative miRNA–target duplex.

TargetScan (Friedman et al., 2009; Garcia et al., 2011; Grimson et al., 2007; Lewis et al., 2005) and miRanda (Enright et al., 2004; John et al., 2004) are among the most commonly used algorithms and are also among the first developed tools to computationally predict MREs. TargetScan (Lewis et al., 2005) relies on perfect (the 8mer site) or near-perfect (the 6mer site and the 7mer site) matches between miRNA and seed regions, along with the thermodynamics and evolutionary conservation of the predicted MREs (Friedman et al., 2009; Garcia et al., 2011; Grimson et al., 2007). miRanda (Enright et al., 2004; John et al., 2004) uses weighted target prediction scores to predict the degree of sequence complementarity, thermodynamics, and conservation. Other popular softwares include PicTar (Krek et al., 2005) that uses combinatorial methods to predict the binding of a single or a group of coexpressed miRNAs, and thus, accounts for potential cooperative effects from the presence and activities of a combination of miRNAs; or TargetBoost (Saetrom, Snove, & Saetrom, 2005) that predicts miRNA–target interaction rules using a machine learning approach built using a large set of validated miRNA targets (Boutla, Delidakis, & Tabler, 2003; Brennecke, Hipfner, Stark, Russell, & Cohen, 2003; Rajewsky & Socci, 2004) and uses a larger set of random sequences as negative training data (Rajewsky & Socci, 2004); or PITA (Hammell et al., 2008; Miranda et al., 2006) that takes into account target site accessibility by factoring mRNA secondary structure.

Despite the large effort and the conceptually different approaches used to develop computational methods to predict miRNA targets, the overlap between predicted and experimentally validated miRNA targets is nevertheless low. This demonstrates our yet incomplete understanding of the mechanisms and properties that underlie miRNA-mediated regulation. For example, on average, 3 in 10 of predicted interactions by the algorithms do not occur *in vivo* (Enright et al., 2004; Kiriakidou et al., 2004; Krek et al., 2005; Lewis, Shih, Jones-Rhoades, Bartel, & Burge, 2003; Lewis et al., 2005). Several studies have compared between the different algorithms (Barbato et al., 2009; Maziere & Enright, 2007) and estimated their relative accuracy and sensitivity (Martin, Schouest, Kovvuru, & Spillane, 2007; Sethupathy, Megraw, & Hatzigeorgiou, 2006) using experimentally validated miRNA targets. By putting the emphasis on matches at the seed regions, miRanda, TargetScan, PicTar, and PITA are all limited by their reliance on the conventional miRNA recognition and binding rules, although the implementation of additional features such as secondary structure predictions and evolutionary conservation has been shown to enhance the accuracy of some of these algorithms (Sethupathy et al., 2006). In most cases, additional cellular context-specific requirements, such as the sequence of tissue specifically isoforms, may vastly improve the accuracy and sensitivity of miRNA–target interaction predictions (Doench & Sharp, 2004; Farh et al., 2005; Grimson et al., 2007). For example, machine learning approaches, similar to Target-Boost (Saetrom et al., 2005), may have an advantage because they do not rely completely on our yet incomplete knowledge of the rules underlying miRNA-target recognition. Surprisingly, simpler approaches seem to have comparable precision to more complex methodologies (Alexiou, Maragkakis, Papadopoulos, Reczko, & Hatzigeorgiou, 2009; Min & Yoon, 2010; Selbach et al., 2008), although their competing times vary substantially with algorithms that incorporate multiple parameters or metrics being more computationally expensive. In addition, all miRNA target prediction algorithms have considerably high false-negative rates (Kumar, Wong, Lizard, Moore, & Lefevre, 2012), demonstrating again the need for enhanced understanding or alternative approaches prior to the development of further miRNA target prediction tools.

In parallel with the growth of computational miRNA target prediction tools, development of experimental techniques for the identification of miRNA–target interactions have also advanced (Rajewsky & Socci, 2004; Rigoutsos, 2009; Sethupathy et al., 2006). Transcriptome–wide analysis has been used to investigate the impact of manipulating the levels of individual miRNAs on global gene expression (Bader, Brown, & Winkler, 2010; Cole et al., 2008; Lim et al., 2005; Liu, Calin, Volinia, & Croce, 2008a; Mallanna & Rizzino, 2010; McLaughlin et al., 2008; Silber et al., 2008; Xiao et al., 2007). Reciprocally, arrays have also been used to pinpoint differentially expressed miRNAs resulting from changes in specific mRNA abundance in distinct cellular context and disease states (Babak, Zhang, Morris, Blencowe, & Hughes, 2004; Barad et al., 2004; Baskerville & Bartel, 2005; Calin et al., 2004; Elkan-Miller et al., 2011; Liang et al., 2005; Liu et al., 2008a; Miska et al., 2004; Nelson et al., 2004; Thomson, Parker, Perou, & Hammond, 2004). More recently, proteomic approaches, such as stable isotope labeling with amino acids in cell culture followed by mass spectrometry, have been employed to evaluate global changes in protein levels following changes in miRNA levels (Baek et al., 2008; Bargaje et al., 2012; Bauer & Hummon, 2012; Ebner & Selbach, 2011; Huang et al., 2012; Kaller et al., 2011; Lossner et al., 2011; Selbach et al., 2008; Yan, Xu, Tan, Liu, & He, 2011; Yang et al., 2010). Despite providing important insights into the global impact on gene product abundances following miRNA perturbation, the fundamental limitation of such analyses is that they do not allow the distinction between direct and indirect effects of miRNA regulation. As a result, these experimental techniques are not directly useful in the identification of miRNA targets (Baek et al., 2008; Johnson et al., 2007; Linsley et al., 2007).

miRNA–target interaction can be identified by detecting transcripts that are physically associated with miRNA-incorporated RISC (Ule et al., 2003). RNA-binding protein immunoprecipitation (RIP) was among one of the first techniques used to map RNA–protein interactions (Niranjanakumari, Lasda, Brazas, & Garcia-Blanco, 2002). This technique requires the cross-linking of AGO to its associated RNA using formaldehyde. Immunoprecipitation of the cross-linked sample using an AGO-specific antibody followed by extensive washes yield AGO-bound RNA (Keene, Komisarow, & Friedersdorf, 2006). Following the digestion of unbound and exposed RNA regions, purified RNA from the AGO-protected regions can be detected by various methods, including microarray (RIP-Chip) (Dolken et al., 2010; Keene et al., 2006; Tan et al., 2009; Wang, Wilfred, Hu, Stromberg, & Nelson, 2010b; Wang et al., 2010c) and high-throughput sequencing (RIP-seq) (Kanematsu, Tanimoto, Suzuki, & Sugano, 2013; Nie et al., 2013).

Similar to RIP, cross-linking immunoprecipitation-high-throughput sequencing (CLIP) also identifies miRNA binding sites by the immunoprecipitation of AGO-associated RISCs (Jensen & Darnell, 2008; Ule et al., 2003; Wang, Tollervey, Briese, Turner, & Ule, 2009). In contrast to RIP, in CLIP protocols, AGO–RISC-associated RNA is cross-linked using UV radiation instead of formaldehyde (Jensen & Darnell, 2008; Ule et al., 2003). UV radiation creates stronger covalent bonds that are irreversible and thus, CLIP allows more stringent purification conditions yielding more specific RNA–AGO interactions (Chi, Zang, Mele, & Darnell, 2009; Hafner et al., 2010; Ule et al., 2003; Zisoulis et al., 2010). Like RIP, AGO-bound RNA from CLIP protocols can be sequenced using high-throughput methods (AGO HITS-CLIP) (Darnell, 2010; Licatalosi et al., 2008). Moreover, photoactivatable ribonucleoside-enhanced CLIP (PAR-CLIP) is another CLIP technique that relies on the incorporation of photoreactive ribonucleoside analogs that allows the efficient cross-linking of photoreactive nucleoside-labeled cellular RNAs to interacting AGO-RISC under UV radiation (Hafner et al., 2010). The isolated MRE-containing RNA can then be deep-sequenced to reveal transcriptome-wide binding sites of miRNAs with relatively high resolution (Ascano, Hafner, Cekan, Gerstberger, & Tuschl, 2012; Chou et al., 2013; Erhard, Dolken, Jaskiewicz, & Zimmer, 2013; Hafner et al., 2010; Hafner, Lianoglou, Tuschl, & Betel, 2012).

Despite the identification of regions within transcripts that are associated with miRISCs, these approaches fail to directly infer the identity of the specific miRNAs that were loaded into target-associated miRISCs (Liu et al., 2013). This limitation can be partially circumvented by exogenously increasing the levels of particular miRNAs, using miRNA mimics, in an attempt to identify targets of the specific miRNAs of interest (Thomson, Bracken, & Goodall, 2011). However, this technique is also limited as it only allows the testing of one miRNA per cellular environment at a time. Moreover, critically, as transfection of miRNA mimics can result in changes in transcriptional programs, this approach might affect the composition of transcript repertoires in the cell. Recently, a new method based on the principles of AGO-CLIP, cross-linking, ligation, and sequencing of hybrids (CLASH) was developed (Helwak, Kudla, Dudnakova, & Tollervey, 2013; Kudla, Granneman, Hahn, Beggs, & Tollervey, 2011); CLASH aims to directly detect functional miRNA–target interactions by introducing an additional ligation step that results in the formation of miRNA–target chimeras (Helwak et al., 2013; Kudla et al., 2011).

The development of these and other high-throughput approaches to map miRNA–target associations allowed an exponentially increase in a number of known bona fide interactions in different cells. Interestingly, all techniques provide evidence that noncanonical (unseeded) miRNA–target interactions are relatively common in animals (Bartel, 2009; Grimson et al., 2007). For example, 18% of the interactions mapped using CLASH suggested miRNA–target pairing that involve bulged or mismatched nucleotides (Helwak et al., 2013). Furthermore, apart from the 3′-UTRs, many putative target–miRNA-seed matches were identified within the coding sequences of genes (Hendrickson, Hogan, Herschlag, Ferrell, & Brown, 2008). Collectively, these observations demonstrate the flexibility of miRNA target recognition and highlight our relative poor understanding of the rules that underlie miRNA–target interactions.

A few databases that compile experimental validated interactions between miRNAs and their targets are available. An example of such a database include starBase (sRNA target base), a database of miRNA–mRNA interactions determined from a comprehensive set of AGO-CLIP-Seq (also known as HITS-CLIP-Seq) and Degradome-Seq data that explored the AGO-binding and cleavage sites in six organisms (Yang, Li, Shao, Zhou, Chen, & Qu, 2011). Furthermore, motivated by the recent findings that many noncoding transcripts function as ceRNAs, including transcribed pseudogenes (Marques et al., 2012; Poliseno et al., 2010), lincRNAs (Cesana et al., 2011), and circRNAs (Hansen et al., 2013; Memczak et al., 2013), several databases were also created with the aim to unravel functional interactions between these transcripts and miRNAs. For instance, the DIANA-LncBase (Paraskevopoulou et al., 2013) integrates putative MREs on the GENCODE-annotated lncRNAs determined using HITS-CLIP and PAR-CLIP experimental data and MREs predicted using miRcode (Jeggari, Marks, & Larsson, 2012).

## 3. MIRNA-MEDIATED CROSS-TALK BETWEEN TRANSCRIPTS

According to current estimates, each human miRNA is able to recognize, and likely to regulate, the abundance of hundreds of gene targets (Lewis et al., 2005; Miranda et al., 2006). Thus, given that miRNA abundance is likely limited, transcripts regulated by the same miRNA(s) have been proposed to compete for the binding of available miRNA(s) and thus be able to modulate each other's expression levels (Seitz, 2009). The first evidence for such cross-talk between endogenously expressed transcripts was described in plants (*A. thaliana*) where *IPS1* was found to compete for the binding of *miR-399* with *PHO2* (Franco-Zorrilla et al., 2007). The *miR-399*-mediated cross-talk between *IPS1* and *PHO2* illustrates what is likely to be a more widespread mechanism of posttranscriptional regulation.

To this day, our literature survey reveals 26 documented examples of endogenous transcripts that modulate the levels of other gene products via this miRNA-mediated mechanism (Table 3.1). Coincidentally, most reports were published following a series of articles (Poliseno et al., 2010; Salmena et al., 2011; Sumazin et al., 2011; Tay et al., 2011) (Figure 3.2) that describe the miRNA-mediated interactions between *PTEN* and several coding and noncoding transcripts in cancerous cells. These works were critical to illustrate how competition for shared miRNAs allows the reciprocal modulation of transcript levels and can lead to changes in cellular physiology and disease.

## 3.1 Competing Exogenous RNAs

Artificial miRNA sponges have been extensively used to manipulate endogenous miRNA levels (Brown & Naldini, 2009; Ebert et al., 2007; Ebert & Sharp, 2010b; Liu, Sall, & Yang, 2008b). The expression of these artificial transcripts, which contain several binding sites for specific miRNAs, can be driven constitutively (Asakawa & Kawakami, 2008; Ebert & Sharp, 2010b; Zhu et al., 2011) or transiently (Bolisetty, Dy, Tam, & Beemon, 2009; Ebert & Sharp, 2010b; Kluiver et al., 2012; Kumar et al., 2008; Penna et al., 2011;

**Table 3.1** Example of Competitive Endogenous RNAs (ceRNAs) Interactions Characterized to Data

| ceRNA | Gene Class | miRNA(s) Implicated | Cross-talking Partner(s) | Condition | Model Organism(s) | References |
|---|---|---|---|---|---|---|
| *IPS1* | Long noncoding RNA | *miR-399* | *PHO2* | Phosphate starvation | *Arabidopsis thaliana* | Franco-Zorrilla et al., 2007 |
| *HSURs* | Small noncoding RNA | *miR-27a/-27b* | *FOXO1* | Viral host invasion | *Herpesvirus saimiri* | Cazalla et al., 2010 |
| *PTENP1* | Transcribed pseudogene | *miR-17-5p/-20/-19* | *PTEN* | Prostate and colon cancers | Mouse/human | Poliseno et al., 2010 |
| *KRAS1P* | Transcribed pseudogene | *let-7/miR-143* | *KRAS* | Cell growth | Mouse/human | Poliseno et al., 2010 |
| *CNOT6L/VAPA* | Protein coding | *miR-17-/19ab/-20ab/-26ab/-93/-106ab* | *PTEN* | Prostate and colon cancers | Human | Tay et al., 2011 |
| *ZEB2* | Protein coding | *miR-181-/200b/-25/-92a* | *PTEN* | BRAF-induced melanoma | Mouse/human | Karreth et al., 2011 |
| *RB1/RUNX1/VEGFA* | Protein coding | Not experimentally validated, full list of predictions can be found in Sumazin et al., 2011 | *PTEN* | Glioblastoma | Human | Sumazin et al., 2011 |
| *DKK1 PCas4/BCAS4* | Protein coding; Unitary transcribed pseudogene/protein coding | *miR-93/-106a miR-185* | *PTEN BCL2/IL17RD/PNPLA3/SHISA7/TAPBP* | Diabetes Neuroblastoma cells | Mouse/human Mouse/human | Ling et al., 2013 Marques et al., 2012 |
| *ciRS-7 (CDR1as)* | Circular RNA | *miR-7* | *SNCA/EGFR/IRS2* | Brain development | Mouse/human | Hansen et al., 2013; Memczak et al., 2013 |

| | | | | | | |
|---|---|---|---|---|---|---|
| *lin-RoR* | Long noncoding RNA | *miR-145* | *Oct4/Nanog/ Sox2* | Pluripotency maintenance/ differentiation | Mouse | Wang et al., 2013 |
| *linc-MD1* | Long noncoding RNA | *miR-133/-135* | *MAML1/ MEF2C* | Muscle differentiation | Human/mouse | Cesana et al., 2011 |
| *H19* | Long noncoding RNA | *let-7* | *Dicer/ Hmga2/ Irs2/Insr/ CypB* | Muscle differentiation | Mouse/human | Kallen et al., 2013 |
| *VCAN* | Protein coding | *miR-133a/- 119a★/-144/-431* | *CD34/FN1* | Hepatocellular carcinoma | Mouse/human | Lee et al., 2009 |
| *CD44* | Protein coding | *miR-216/-330/-608* | *CDC42/ CD44* | Cell cycle | Human | Jeyapalan et al., 2011 |
| *CD44* | Protein coding | *miR-328 miR-512-3p/-491/-671* | *Col1α/ FN1* | Breast cancer | Human | Rutnam & Yang, 2012 |
| *CAV1.2-LTC complex* | Protein coding | *miR-103* | *CAV1.2-LTC complex* | Neuropathic chronic pain | Rat/monkey | Favereaux et al., 2011 |
| *MeCP2* | Protein coding | *miR-483-5p* | *HDAC4/ TBL1X* | Beckwith–Wiedemann syndrome | Human | Han et al., 2013 |
| *Npnt* | Protein coding | *miR-378* | *GalNT7 (GalNAc-T7)* | Osteoblast differentiation | Mouse | Kahai, Lee, Seth, & Yang, 2010 |
| *HULC* | Protein coding | *miR-372* | *PRKACB* | Liver cancer | Human | Wang et al., 2010a |

This list does not contain references to competitive exogenous RNAs or poorly characterized examples of endogenous transcripts.

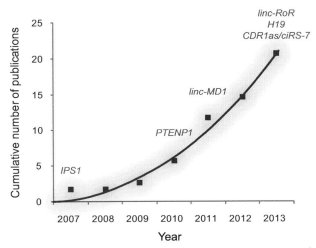

**Figure 3.2** The cumulative number of manuscripts reporting on novel and experimentally characterized ceRNAs published from 2007 to 2013. (For color version of this figure, the reader is referred to the online version of this book.)

Rybak et al., 2008; Otaegi, Pollock, & Sun, 2011) by strong promoter elements. Moreover, miRNA sponges have been used in a variety of different animals (Asakawa & Kawakami, 2008; Kumar et al., 2008; Otaegi et al., 2011; Zhu et al., 2011) and cellular systems (Bolisetty et al., 2009; Kumar et al., 2008; Penna et al., 2011) providing researchers with a tool to investigate miRNA function. Furthermore, these useful systems have also demonstrated that miRNA levels are also regulated by their targets.

Viruses exploit this mechanism of regulation in order to manipulate host miRNA abundances and to increase the efficiency of their pathogenic infection. For example, the genome of *Herpevirus saimiri* virus (HSV) encodes several U-rich noncoding RNAs (*HSURs*), two of which, *HSUR1* and *HSUR2*, contain several response elements for *miR-27* (Cazalla, Yario, & Steitz, 2010). Upon HSV infection of marmoset T cells, viral *HSUR1-2* transcripts are highly expressed and are able to sequester host *miR-27* (Cazalla et al., 2010). *HSUR*-induced reduction in *miR-27* levels leads to the posttranscriptional derepression of the host's endogenous targets, including *FOXO1*, a transcription factor whose silencing promotes cell cycle progression and cell growth of the activated T-cells (Cazalla et al., 2010). Therefore, the *HSUR1-2*-induced *FOXO1* derepression prevents clonal expansion of activated T-cells that is critical in cell-mediated immune response (Mondino, Khoruts, & Jenkins, 1996), likely to benefit efficient viral infection. Other examples of miRNA-mediated interaction between viral and host transcriptome have been recently reviewed in Swaminathan, Martin-Garcia, and Navas–Martin (2013).

## 3.2 Competing Endogenous RNAs

The term competing endogenous RNAs was first used to describe the miRNA–mediated interaction between a retropseudogene, *PTENP1*, and its parental gene, the tumor suppressor, *PTEN* (Poliseno et al., 2010). Poliseno et al. (2010) found that *PTENP1* and *PTEN* harbor several homologous predicted response elements for five different miRNAs—*miR-20a*, *miR-16b*, *miR-21*, *miR-26a*, and *miR-214*—and that the competition for the binding of these shared noncoding regulators allows the two transcripts to reciprocally regulate each other's levels.

Importantly, this work demonstrated that the ability for endogenous transcript to posttranscriptionally regulate gene expression via a miRNA–mediated mechanism is not restricted to noncoding transcripts, as previously shown for *IPS1* (Franco–Zorrilla et al., 2007), *HSUR1-2* (Cazalla et al., 2010), and artificial miRNA sponges (Ebert et al., 2007). Instead, *protein-encoding RNAs*, such as *PTEN*, can also effectively engage in this type of cross-talk and modulate the levels of other transcripts (Poliseno et al., 2010; Salmena et al., 2011). For example, in several tumor tissues, reciprocal ceRNA regulatory interactions were demonstrated between *PTEN* and (i) *PTENP1*, vesicle–associated membrane protein associated protein A (*VAPA*), and CCR4–NOT transcription complex subunit 6–like (*CNOT6L*) in human colon and prostate cancer cells (Tay et al., 2011); (ii) *PTENP1*, *CNOT6L*, and zinc finger E-box binding homeobox 2 (*ZEB2*) in BRAF–induced melanoma cells (Karreth et al., 2011); and (iii) retinoblastoma protein (*RB1*), vascular endothelial growth factor A (*VEGFA*), and runt–related transcription factor 1 (*RUNX1*) in glioblastoma samples (Sumazin et al., 2011) (Figure 3.3). The partially nonoverlapping ceRNA interactions in distinct cellular environments are likely a consequence of, among other things, the cell–specific expression of different miRNA repertoires and gene isoforms.

In addition to *PTENP1*, another *transcribed pseudogene, KRAS1P*, was also demonstrated by Poliseno et al. (2010) to regulate the expression levels of its oncogenic ancestral gene, *KRAS*, by competing for the binding of *let-7* and *miR-143*. Most pseudogenes retain high sequence homology with their ancestors. For instance, 71% of all human transcribed pseudogenes have highly homologous remnant 3′-UTR sequences to their cognate ancestral genes (Kalyana–Sundaram et al., 2012). Therefore, if transcribed, their shared

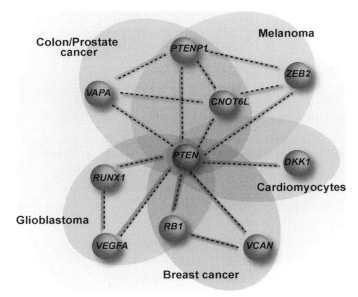

**Figure 3.3** CeRNA network involving the tumor repressor gene *PTEN* (red) and its ceRNAs (blue) in six different types of cancers, including colon/prostate cancer (green shading), melanoma (yellow shading), cardiomyocytes (blue shading), breast cancer (purple shading), and glioblastoma (red shading). Dotted lines represent miRNA-mediated interactions between transcripts. (For interpretation of the references to colour in this figure legend, the reader is referred to the online version of this book.)

sequence similarity might allow them to regulate the expression of their parental genes. Indeed, over 400 pseudogenes were identified to have parental genes that are already implicated in previously proposed ceRNA networks (Sumazin et al., 2011; Tay et al., 2011). Together, these observations suggest that some of the thousands of transcribed pseudogenes found in the mammalian genomes (Ohshima et al., 2003; Torrents, Suyama, Zdobnov, & Bork, 2003; Zhang, Harrison, Liu, & Gerstein, 2003) might also be able to modulate the abundance of their protein-coding paralogs.

In addition to regulating the levels of their cognate ancestral genes, pseudogenes can also modulate the levels of nonhomologous transcripts. This has been shown for *PTENP1*, whose perturbation in levels can affect the abundance of *VAPA*, *CNOT6L*, and *ZEB2* (Figure 3.3). The same was demonstrated more recently for unitary pseudogenes (Marques et al., 2012). In contrast to other types of pseudogenes, unitary pseudogenes do not arise from duplication events, but from the accumulation of loss-of-function mutations within protein-coding genes that lack any apparent homologous loci (Karro et al., 2007; Zhang, Frankish, Hunt, Harrow, & Gerstein, 2010b; Zheng et al., 2005; Zhu et al., 2007). Previously deemed nonfunctional, in mouse, this unique set of transcripts was shown to be transcribed and have preserved their protein-coding ancestors' ability to regulate the levels of nonhomologous genes by competing for shared miRNAs. For example, mouse *Pbcas4* is a transcribed unitary pseudogene of the human *BCAS4* that has conserved its ancestor's response elements for *miR-185*. By retaining these miRNA-binding elements, *Pbcas4* has also preserved *BCAS4*'s miRNA decoying ability and can modulate the levels of orthologous genes within the *BCAS4* ceRNA network in mouse through competition for *miR-185* (Marques et al., 2012). Importantly, unitary pseudogenes demonstrated that the ceRNA functions of some protein–coding transcripts can outlive loss of protein-coding ability (Marques et al., 2012).

Recently, a noncoding *circular RNA* (circRNA) was also shown to be an effective modulator of miRNA levels (Hansen et al., 2011, 2013; Memczak et al., 2013). Cerebellar degeneration–related protein 1 antisense (*CDR1as*) (Memczak et al., 2013) or circRNA sponge for *miR-7* (*ciRS-7*) (Hansen et al., 2013) is a circRNA containing more than 70 conserved *miR-7* response elements (Hansen et al., 2011, 2013; Memczak et al., 2013). *CDR1as/ciRS-7* is relatively highly expressed, with as many as 1400 copies/cell. Similar to *miR-7*, the circRNA is predominantly found in neuronal tissues where it was shown to sequester AGO2-bound *miR-7* (Hansen et al., 2013; Memczak et al., 2013). Consequently, by titrating *miR-7*, *CDR1as/ciRS-7* derepresses endogenous targets of *miR-7*, including *SCNA* (Junn et al., 2009), *EGFR* (Kefas et al., 2008), and *IRS2* (Jiang et al., 2010), which have been implicated in Parkinson's disease, cancer, and diabetes, respectively (Hansen et al., 2013). Interestingly, the linearization of *CDR1as/ciRS-7* leads to at least a 10 fold reduction in the stability of this transcript (Memczak et al., 2013), suggesting that its efficiency as an *miR-7* sponge is likely conferred by its circularity and abundance. In addition to *CDR1as/ciRS-7*, another circRNA, the testis–specific sex-determining region Y, *SRY*, was also proposed to serve as an miRNA sponge for another miRNA, *miR-138* (Hansen et al., 2013). However, aside from the extensively studied *CDR1as/ciRS-7*, nearly all sequences of the other thousands of circRNAs predicted from genome–wide analyses (Danan, Schwartz, Edelheit, & Sorek, 2012; Dixon, Eperon, Hall, & Samani, 2005; Jeck et al., 2013; Salzman, Gawad, Wang, Lacayo, & Brown, 2012; Suzuki et al., 2006) appear to lack the striking functional signature found within *CDR1as/ciRS-7*, namely its extensively high number of predicted MREs for the same miRNA (Hansen et al., 2013; Memczak et al., 2013). Thus, *CDR1as/ciRS-7* might be an exceptional circRNA with miRNA-sponging activity.

Generally, noncoding transcripts, including transcribed pseudogenes, circRNAs, and *long noncoding RNAs* (lncRNAs), have been proposed to act as miRNA sponges that effectively titrate available miRNAs from their endogenous targets, primarily due to the absence of apparent open-reading frames and translation (Su et al., 2013). One such ceRNA is the intergenic lncRNA, *linc-MD1*, that promotes myoblast differentiation by sequestering at least two muscle-specific miRNAs, namely *miR-133* and *miR-135* (Cesana et al., 2011).

## 4. EMERGING ROLES OF COMPETITIVE ENDOGENOUS RNAS

In mammals, most examples of well-characterized ceRNAs have been described in the context of maintenance of pluripotency, development, or disease. In this section, we describe these examples and discuss the important insights they provide to this miRNA-mediated molecular mechanism of posttranscriptional regulation and their potential contributions to different aspects of biology.

### 4.1 Cross-talk between Functionally Related ceRNAs

Most miRNAs function as modulators rather than effectors of gene expression and have a relatively small impact on the levels of their targets (Baek et al., 2008; Bartel, 2009). Interestingly, the targets of most individual miRNAs tend to be functionally related (Chavali et al., 2013; Kim et al., 2009; Merchan, Boualem, Crespi, & Frugier, 2009; Stark, Brennecke, Russell, & Cohen, 2003) suggesting that miRNAs are important for the coordinated modulation of the abundance of the products of genes that are part of the same or related biological pathways. One way this can be achieved is via the competition for shared miRNAs between functionally related transcripts. For instance, a transient decrease in the levels of a given ceRNA in response to a particular environmental change will lead to increased availability of its targeting miRNAs and allow the posttranscriptional repression of these ceRNAs' functionally related targets (Marques, Tan, & Ponting, 2011). Such a mechanism of concerted regulation might be particularly important for multicomponent complexes where the assembly of functional units requires specific relative stoichiometry. For example, *Cav1.2-comprising L-type calcium channel (Cav1.2-LTC)* is a calcium voltage–dependent channel, whose components' expression levels are frequently upregulated in chronic neuropathic pain (Dolmetsch, Pajvani, Fife, Spotts, & Greenberg, 2001; Fossat et al., 2010). One miRNA, *miR-103*, was demonstrated to mediate the posttranscriptional cross-talk between genes encoding all three subunits of Cav1.2-LTC complex, *Cacna1c*, *Cacna2d1*, and *Cacnb1* (Favereaux et al., 2011). *In vivo*, the reduction in *miR-103* levels by intrathecal injections of miRNA inhibitor in naïve rats promoted hypersensitivity to pain in these animals while ectopic overexpression of this miRNA was shown to successfully relieve their pain (Favereaux et al., 2011). *miR-103*-mediated changes in pain levels are likely a consequence of this miRNA's role in coordinating posttranscriptionally the levels of Cav1.2-LTC-encoding mRNAs (Favereaux et al., 2011).

In addition to fine-tuning the relative abundance of components of multiprotein complexes, miRNA-mediated RNA–RNA interactions may also conceivably facilitate the relay of signaling information between genes involved in the same biological pathway, components of the same cellular structures, and even the communication between neighboring cells (van Niel, Porto-Carreiro, Simoes, & Raposo, 2006; Valadi et al., 2007). Some of the examples below illustrate such possibilities.

## 4.2 Contributions of ceRNAs to the Maintenance of Pluripotency and Development

In mice, loss-of-function mutations in *Dicer* lead to developmental arrest and early embryonic lethality (Bernstein et al., 2003). Despite retaining generally normal cellular morphology and expressing all markers of pluripotency, *Dicer*-null mESCs fail to undergo differentiation, suggesting the importance of miRNA–dependent gene expression regulation in development (Hatfield et al., 2005; Kanellopoulou et al., 2005; Muljo et al., 2005). This is, at least in part, a consequence of improper cell-lineage specification, as illustrated by the tissue-specific morphological defects observed in conditional cell-specific *Dicer* knockout mice (Chen et al., 2008; Damiani et al., 2008; Davis et al., 2008; Harfe, McManus, Mansfield, Hornstein, & Tabin, 2005; Frezzetti et al., 2011; Zhang et al., 2010a).

Several of the most highly expressed miRNAs in ESCs, including *miR-145* and *miR-34*, have been shown to posttranscriptionally regulate mRNAs encoding some of the transcription factors that control the regulatory circuitries underlying pluripotency and differentiation (reviewed in Jia, Chen, & Kang, 2013; Judson, Babiarz, Venere, & Blelloch, 2009; Kashyap et al., 2009; Pauli, Rinn & Schier, 2011). Strikingly, the overexpression of a single miRNA, *miR-302*, is sufficient to direct the reprogramming of terminally differentiated cells likely by indirectly regulating the abundance of core transcription factors that control ESC self-renewal and pluripotency (Hu et al., 2013; Lin et al., 2008). On top of *miR-302*, additional miRNAs, including the *miR-290* cluster (Judson et al., 2009), the *miR-17-92* cluster (Mendell, 2008), and *miR-106* clusters (Marson et al., 2008), were also recognized to promote the reprogramming of somatic cells into induced pluripotent stem cells (iPSCs) (Judson et al., 2009; Mallanna & Rizzino, 2010).

Collectively, these studies established the importance of miRNAs in early mammalian development, stem cell renewal, and differentiation processes (Jia et al., 2013; Judson et al., 2009; Kashyap et al., 2009; Pauli et al., 2011). More recently, a number of ceRNAs, including a few intergenic long noncoding RNAs, have also been described to contribute to the regulation of this circuitry (Guttman et al., 2011; Kallen et al., 2013; Wang et al., 2013).

The expression profile of some intergenic lncRNAs (lincRNAs) changes upon reprogramming of human lineage-committed cells, such as fibroblasts, into iPSCs (Anguera et al., 2012; Dinger et al., 2008; Loewer et al., 2010; Mohamed, Gaughwin, Lim, Robson, & Lipovich, 2010). One lincRNA, termed *linc-RoR* for Regulator of Reprogramming, is particularly highly expressed during iPSCs reprogramming and in undifferentiated ESCs (Loewer et al., 2010; Wang et al., 2013) (Figure 3.4). *linc-RoR* overexpression increased the number of iPSC colonies, where its repression reduced the success of somatic cell reprogramming (Loewer et al., 2010). Recently, *linc-RoR* was shown to compete for *miR-145* binding with key self-renewal transcription factors, including *Nanog*, *Oct4*, and *Sox2* (Wang et al., 2013). *linc-RoR* overexpression was associated with increased levels of core pluripotency transcription factors. In contrast, the overexpression of a *linc-RoR* mutant, containing only mutated response elements for *miR-145*, had no impact on the abundance of these transcription factors (Wang et al., 2013). This is consistent with the hypothesis that *linc-RoR* regulates these transcripts via an miR-145–dependent mechanism. Interestingly, *linc-RoR* transcription is in turn regulated by Nanog, Oct4, and Sox2 proteins (Loewer et al. 2010), thus illustrating an autoregulatory feedback loop composed of transcription factors, a lincRNA and a miRNA to the regulatory circuitry that underlies cell fate decisions in ES cells (Wang et al., 2013). In addition to functioning as a ceRNA, *linc-RoR* was also proposed to promote cellular reprogramming by functioning as a molecular scaffold for chromatin-modifying complexes (Zhang et al., 2013), suggesting potential bifunctional roles of this intergenic lncRNA.

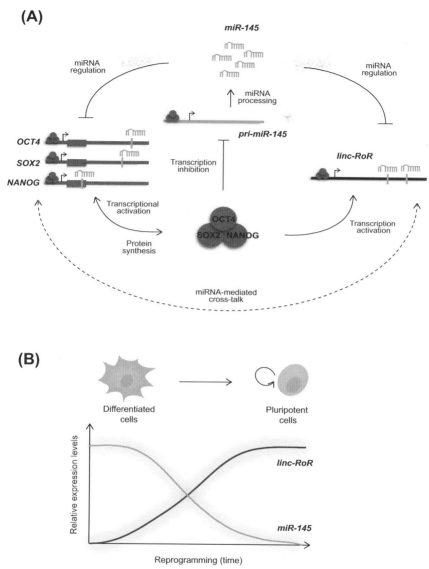

**Figure 3.4** Contributions of *linc-RoR* to cellular reprogramming. A) *MiR-145* (yellow) recognizes *OCT4*, *SOX2*, and *NANOG* (red) mRNAs and *linc-RoR* (blue) via *miR-145* MREs within their sequences (yellow vertical lines) and posttranscriptionally regulates the abundances of OCT4, SOX2, and NANOG's protein products (red circles). These three core transcription factors are essential to transcriptionally activate their own mRNAs and *linc-RoR*, as well as to inhibit *pri-miR145* (red circles bound near the gene transcriptional start sites—represented as forward black arrows on transcripts). Recently, *linc-RoR*, *OCT4*, *SOX2*, and *NANOG* were shown to cross-talk (dashed black line with double arrows) via the competition for *miR-145* binding. B) This miRNA-mediated interaction was proposed to contribute to the reprogramming of terminally differentiated cells. (For interpretation of the references to colour in this figure legend, the reader is referred to the online version of this book.)

As *linc-RoR*, *H19* is also a multifunctional intergenic lncRNA. The imprinted *H19* belongs to a highly conserved cluster of imprinted genes (Brannan, Dees, Ingram, & Tilghman, 1990; Rachmilewitz et al., 1992; Zhang & Tycko, 1992). This gene cluster also contains the paternally expressed insulin-like growth factor 2 (*Igf2*), whose reciprocal expression with the maternally expressed *H19* controls the timing of embryogenesis and postembryonic development (Arima, Matsuda, Takagi, & Wake, 1997; Elkin et al., 1995; Leighton, Ingram, Eggenschwiler, Efstratiadis, & Tilghman, 1995). Interestingly, the *H19* RNA also encodes a conserved placenta-specific miRNA, *miR-675*. This miRNA appears to regulate placental development as it is exclusively expressed when placental growth halts, whereas *H19*-null placentas fail to stop and continue to grow (Keniry et al., 2012).

Although this intergenic lncRNA has been implicated in various human genetic disorders and cancers (Gabory, Ripoche, Yoshimizu, & Dandolo, 2006; Matouk et al., 2007; Yoshimizu et al., 2008), its role and molecular mechanism of function during early embryonic development remain poorly understood. Recently, human *H19* was demonstrated to modulate the levels of the *let-7* miRNA family by acting as an endogenous sponge (Kallen et al., 2013). Similar to *lin-4*, *let-7* is an essential miRNA that regulates early developmental. Specifically, in *C. elegans*, the upregulation of *let-7* significantly represses heterochronic proteins that promote larval-specific cell fates and the developmental transition from larval to adulthood (Moss, 2007; Pasquinelli et al., 2000; Reinhart et al., 2000; Rougvie, 2005; Thummel, 2001). The binding of *let-7* miRNAs to *H19* inhibited the transcript's activities in early embryo development (Kallen et al., 2013). Consistently, the knockdown of *H19* resulted in precocious muscle differentiation, a similar phenotype to that observed following *let-7* overexpression (Kallen et al., 2013).

The accelerated muscle differentiation observed upon *H19* depletion suggests its role in controlling timing of muscle differentiation, similar to that reported for the muscle-specific lincRNA, *linc-MD1*, which governs myoblast differentiation (Cesana et al., 2011). This muscle-specific intergenic long noncoding RNA contributes to the timing of muscle differentiation by acting as a molecular sponge for two muscle-specific miRNAs, *miR-133* and *miR-135* (Cesana et al., 2011) (Figure 3.5). Expressed exclusively during early stages of mouse myoblast differentiation, *linc-MD1* was found to compete for the binding of *miR-133* and *miR-135* with two key myogenic factors, *Maml1* and *Mef2c*, whose protein products activate muscle-specific gene expression (Cesana et al., 2011). Interestingly, the precursor transcript

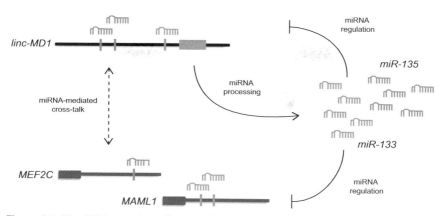

**Figure 3.5** *Linc-MD1* encodes *miR-133* and is an miRNA sponge for this miRNA and *miR-135*. *Linc-MD1* (blue) is a muscle-specific lincRNA that competes with two protein-coding genes, *MAML1* and *MEF2C* (red), involved in the control of timing in myoblast differentiation for the binding (dashed black line with double arrows) of two muscle-specific miRNAs, *miR-133* (yellow) and *miR-135* (green). Not only is *linc-MD1* regulated by *miR-133*, it also encodes for the precursor of *miR-133b* (yellow box). (For interpretation of the references to colour in this figure legend, the reader is referred to the online version of this book.)

of *miR-133b* is derived from *linc-MD1*, thus suggesting an interesting, yet poorly understood, interplay between miRNA processing and modulation (Cesana et al., 2011). The function of mouse *linc-MD1* in myoblast differentiation is also conserved to its human ortholog and its levels are significantly reduced in human myoblasts from Duchene muscular dystrophy (DMD) patients (Cesana et al., 2011), the most common form of muscular dystrophy (Blake, Weir, Newey, & Davies, 2002). Interestingly, the ectopic expression of *linc-MD1* in DMD myoblasts rescued both *MAML1* and *MEF2C* synthesis and improved its delayed myoblast differentiation process compared to human primary myoblast controls (Cesana et al., 2011).

In addition to lincRNAs, the miRNA-mediated cross-talk between protein-coding transcripts has also been found to contribute to mechanisms underlying cellular differentiation. For example, Nephronectin (Npnt), an extracellular matrix (ECM) protein, is critical to osteoblast differentiation (Brandenberger et al., 2001). The levels of *Npnt* in mouse osteoblastic progenitor cells (MC3T3-E1) are modulated by *miR-378* (Kahai et al., 2009), where ectopic expression of *Npnt* enhances osteoblast differentiation and bone nodule formation, while the reciprocal is observed following the overexpression of *miR-378* (Kahai et al., 2009). UDP-*N*-acetyl-alpha-D-galactosamine:polypeptide *N*-acetylgalactosaminyl-transfease 7 (*GalNT-7*), an enzyme that promotes Npnt glycosylation and secretion, is another target of *miR-378* (Kahai et al., 2009).

High levels of both *Npnt* and *GalNT7* are thought to be required for osteo-blast proliferation by promoting Npnt glycosylation and secretion (Kahai et al., 2009). Competition for the binding of *miR-378* between *Npnt* and *GAlNT-7* was proposed to coordinately modulate the relative levels of these tran-scripts during this developmental process (Kahai et al., 2009). In particular, high expression of either *Npnt* or *GalNT-7* reduces the availability of *miR-378* and consequently, posttranscriptionally derepresses each other, as well as other target(s) of this miRNA (Kahai et al., 2009).

In addition to *miR-378*, *Npnt* was also recently proposed to compete for the binding of other miRNAs, including *miR-23a*, *miR-101a*, *miR-296-5p*, *miR-328*, *miR-340-3p*, and *miR-425\**, with two other genes involved in osteoblast differentiation (Lee et al., 2011), namely *β-Catenin* and *GSK3β* (Baron, & Rawadi, 2007; Krishnan, Bryant, & Macdougald, 2006); both genes were found to be significantly upregulated upon the overexpression of *Npnt* 3′-UTR in MC3T3-E1s (Lee et al., 2011). This competition for the binding of shared miRNAs between *Npnt*, *β-Catenin*, and *GSK3β* was proposed to contribute to increased osteoblast differentiation by reducing EGFR and ERK phosphorylation (Lee et al., 2011), two protein kinases whose suppressed phosphorylation activities were shown to promote cel-lular proliferation and osteoblast differentiation (Roberts & Der, 2007).

*Methyl CpG-binding protein 2 (MeCP2)* is a transcription factor essen-tial to the development and maturation of the fetal central nervous system (Chahrour & Zoghbi, 2007; Chen, Akbarian, Tudor, & Jaenisch, 2001; Guy, Cheval, Selfridge, & Bird, 2011; Na & Monteggia, 2011). In humans, *MeCP2* encodes several alternatively spliced isoforms, whose differential expression is tightly regulated during development (Coy, Sedlacek, Bachner, Delius, & Poustka, 1999). In particular, the predominant isoform of *MeCP2* expressed in the human fetal brain has an unusually long 3′-UTR and is specifically expressed during fetal development (Han et al., 2013). Interestingly, despite relatively constant level of its mRNA transcripts found throughout brain development, MeCP2 protein expression is suppressed in the fetus and its product levels increase gradually only post birth (Balmer, Goldstine, Rao, & LaSalle, 2003). This led to the suggestion that the prevalent fetal isoform of *MeCP2* with extended 3′-UTR might facilitate the mechanism underlying the fetal-to-post-natal switch in vertebrate development (Shahbazian, Anta-lffy, Armstrong, & Zoghbi, 2002). This hypothesis was recently validated when the long 3′-UTR of the fetal expressed *MeCP2* isoform was found to harbor two response elements for *miR-483-5p*, an miRNA that is also highly expressed during fetal development (Han et al., 2013). The selective recognition of the fetal expressed *MeCP2* isoform by this miRNA repressed *MeCP2* translation, illustrating how posttranscriptional regulation of alternatively spliced and differentially expressed transcript isoforms allows the fine-tuning of gene product levels (Han et al., 2013). Interestingly, *miR-483-5p* also targets genes that encode two other proteins, Histone deacetylase 4 (HDAC4) and transducin (beta)-like 1 X-linked (TBL1X). These genes physically interact with MeCP2 to form the MeCP2-interacting corepressor complex (Fischle et al., 2002; Han et al., 2013; Yoon et al., 2003). Although competition for the binding of *miR-483-5p* between *MeCP2*, *HDAC4*, and *TBL1X* has not yet been experimentally demonstrated, it is likely that these functionally related proteins can regulate each others' mRNA levels by competing for the binding of shared miRNA(s).

These examples illustrate how miRNA-mediated cross-talk between coding and noncoding transcripts can contribute to the complex circuitries that underlie cell pluripotency and differentiation. Importantly, this extra layer of posttranscriptional regulation might be critical for determining transitions between developmental stages (Ebert & Sharp, 2010b). The contributions of ceRNAs to the control of developmental timing might be particularly impactful when they involve 1) transcripts whose levels are regulated through autoregulatory loops, as that illustrated for *linc-RoR* (Wang et al., 2013), or 2) when functionally related genes are coordinately and cooperatively regulated by a common set of miRNAs, such as that demonstrated by *MeCP2* (Han et al., 2013).

## 4.3 Disease-Associated ceRNAs

Impaired miRNA-mediated cross-talk between transcripts has been associated with diseases, particularly in various cancers, thus highlighting the medical relevance of this layer of posttranscriptional regulation.

As discussed previously, *PTEN* is an important tumor suppressor gene. Along with *PTENP1*, a transcribed *PTEN* retropseudogene and the first well-characterized mammalian ceRNA for *PTEN*, the dysregulation of *PTEN* has been implicated in multiple human cancers (Li et al., 1997). In cancer cells, mutations leading to the loss of *PTENP1* were found to result in decreased levels of *PTEN* mRNA and its encoded protein, likely due to the increased availability of their shared miRNAs (Poliseno et al., 2010). Cells have been shown to be ultrasensitive to changes in PTEN abundance (Alimonti et al., 2010). As deletions in *PTENP1* led to similar magnitude of changes in PTEN as that observed in cancers, this suggests *PTENP1* might contribute to cancer susceptibility and tumor growth (Poliseno et al., 2010).

In addition to *PTENP1*, the levels of *PTEN* in cancer cells are also modulated by other protein-coding ceRNAs (Figure 3.3), including *VAPA* and *CNOT6L* in prostate and colon cancer (Tay et al., 2011), *ZEB2* in melanoma (Karreth et al., 2011), and *RB1*, *VEGFA*, and *RUNX1* in glioblastoma (Sumazin et al., 2011). The complexity of the emerging ceRNA network underlying the posttranscriptional regulation of *PTEN* might be important to ensure robust and coordinated gene expression and to maintain homeostasis (Ebert & Sharp, 2012; Mukherji et al., 2011; Stark et al., 2005). However, this might also represent a source of increased fragility and propensity for disease, as the loss of function in any of PTEN modulators may initiate a cascade of changes that eventually lead to the dysregulation of this tumor suppressor and may promote tumorigenisis.

Recently, *PTEN* was also shown to interact with *Dickkopf-1* (*DKK1*) in diabetic *cardiomyocytes* (Ling et al., 2013). These two transcripts are able to modulate each other's abundances by competing for shared miRNAs, namely *miR-93* and *miR-106* (Ling et al., 2013). Interesting, *DKK1* has also been proposed as a putative ceRNA for *PTEN* in glioblastoma (Sumazin et al., 2011), suggesting this interaction could be relevant to several disease states.

Versican, encoded by the *VCAN* gene, is a large ECM proteoglycan protein (Margolis & Margolis, 1994) whose increased abundance has been associated with tumor growth in breast, brain, ovary, prostate, melanoma, and hepatocellular carcinomas (reviewed in Ricciardelli, Sakko, Ween, Russell, & Horsfall, 2009). The *VCAN* locus encodes four alternatively spliced isoforms (V0-4) (Ito, Shinomura, Zako, Ujita, & Kimata, 1995). Recently, ectopic overexpression of the 3'-UTR of *VCAN*, which is common to all isoforms, was found to enhance the levels of endogenous *VCAN* isoforms V0 and V1 in hepatocellular carcinoma cells (HCC), as well as the levels of two other ECM proteins, *CD34* and fibronectin (*FN1*) (Fang et al., 2013). These ECM proteins and *VCAN* are posttranscriptionally regulated by a number of shared miRNAs, *miR-133a*, *miR-199a*\*, *miR-144*, and *miR-431*, and the upregulation of *CD31* and *FN1* are associated with increased cell proliferation, colony formation, and decreased rate of apoptosis (Fang et al., 2013). Consistently, coordinated upregulation of these ECM proteins in a transgenic mouse model expressing *VCAN*'s 3'-UTR resulted in enhanced cell–cell and organ adhesion and increase propensity to develop hepatocellular carcinoma (Fang et al., 2013).

Additionally, in a mouse breast carcinoma cell line, another *Versican*-encoding isoform (V2) was found to compete for the binding of *miR-199a-3p* with a cell cycle regulator, *Rb1*, and the tumor suppressor gene, *Pten*

(Lee et al., 2010). Interestingly, in an independent study, the human orthologs of *Pten* and *Rb1* were also identified as part of a ceRNA network in gliobastoma that was described earlier (Sumazin et al., 2011) (Figure 3.3), suggesting these interactions contribute to tumorigenesis in different cellular environments. In breast carcinoma cells, the overexpression of the *Vcan* 3′-UTR results in the upregulation of *VCAN* isoform V2, *Rb1*, and *Pten* as a consequence of reduced *miR-199a-3p* availability and results in formation of small tumors–like cell colonies (Lee et al., 2010). Together this suggests *Vcan* might cross-talk via competition for shared miRNAs with other genes involved in cell proliferation, survival, and adhesion, illustrating what may be a complex regulatory mechanism important for cell survival, as proposed by Lee et al. (2010).

Similar to what was found for *VCAN* (Lee et al., 2009, 2011), the overexpression of *CD44* 3′-UTR, a transmembrane glycoprotein involved in cell–cell and cell–matrix interactions, tumor metastasis, and cell migration (Fox et al., 1994; Mantenhorst et al., 1995; Weber, Ashkar, Glimcher, & Cantor, 1996), was found to be sufficient to regulate the cell cycle (Jeyapalan et al., 2011). In a human breast cancer cell line, MT-1, *CD44* 3′-UTR antagonized *miR-216a*, *miR-330*, and *miR-608* levels, and in turn posttranscriptionally derepressed endogenous *CD44* and *CDC42* from miRNA-mediated repression, resulting in the inhibition of cell proliferation and tumor growth (Jeyapalan et al., 2011).

In a separate study, *CD44* 3′-UTR overexpression enhanced metastasis in a human metastatic breast carcinoma cell line (Rutnam & Yang, 2012). *CD44* acted as a competitor for the binding of several miRNAs, namely *miR-328*, *miR-512-3p*, *miR-491*, and *miR-671*, and prevented them from targeting other matrix–encoding mRNAs, including *collagen type 1α1* (*Col1α*) and *FN1* (Rutnam & Yang, 2012). Consequently, derepressed matrix–encoding genes resulted in the enhanced cell motility, invasion, and cell adhesion of the metastatic breast cancer cells (Rutnam & Yang, 2012). The contrasting roles of *CD44* in different tumor cell lines, as an inhibitor (Jeyapalan et al., 2011) and an enhancer (Rutnam & Yang, 2012) of tumor proliferation in human breast carcinoma, suggests context-specific ECM activities are at least partially regulated by the posttranscriptional regulation of different repertoires of cell-specific miRNAs.

Interestingly, overlapping interactions were also reported between *CD44* and *VCAN*, which was previously demonstrated to regulate *CD34* and *FN1* abundances via the competition for a set of miRNAs in human breast carcinoma (Lee et al., 2009). Taken together, these studies revealed an interconnected ceRNA network that includes at least *CD44, CDC42, VCAN, Rb1*, and *PTEN*, which regulates ECM properties via miRNA-mediated cross-talk between matrix protein encoding mRNAs in a human breast carcinoma cell line (Jeyapalan et al., 2011; Lee et al., 2010; Rutnam & Yang, 2012).

Finally, intergenic lncRNAs have also been found to contribute to cancer progression by participating in autoregulatory feedback loops within ceRNA networks. *Highly Up-regulated in Liver Cancer (HULC)* was found to be highly abundant in hepatocellular carcinoma (Panzitt et al., 2007). This intergenic lncRNA is predominantly localized in the cytoplasm and competes for the binding of *miR-372* with *PRKACB*, which encodes a protein kinase that phosphorylates cAMP response binding protein (CREB) (Wang et al., 2010a). Phosphorylation is required for CREB activation, which in turn promotes the transcription activation of *HULC* mRNA (Wang et al., 2010a). In addition to being highly expressed in liver cancer, *HULC* is also abundant in metastasized colorectal carcinomas (Matouk et al., 2009). Overall, the presence of high levels of *HULC* in different cancers may suggest its wider contribution to tumorigenesis.

## 5. INFERENCE OF MIRNA-MEDIATED GENETIC INTERACTIONS

Approaches that integrate information on miRNA target predictions with the expression profiles of the miRNAs and their targets are useful tools to identify genetic interactions between putative ceRNAs. Mutually targeted MRE enrichment (MuTaME) (Tay et al., 2011) and Hermes (Sumazin et al., 2011) are examples of recently developed algorithms.

Mutually targeted MRE enrichment (MuTaMe) was designed to identify transcripts that interact with a gene of interest based on the hypothesis that transcripts sharing more miRNAs will more efficiently cross-talk and regulate each others' expression levels (Tay et al., 2011). This application uses transcriptome-wide predictions of MREs to compute a ceRNA likelihood score for each pair of transcripts based on the number, density, and distribution of the miRNA binding sites. This score is used to rank the likelihood of cross-talking transcripts with the gene of interest (Tay et al., 2011).

The coexpression between predicted ceRNAs can also be used as a biological readout of miRNA-mediated genetic interactions between transcripts. Hermes (Sumazin et al., 2011) integrates information on genome-wide transcript coexpression with the expression profile of miRNAs to infer miRNA-mediated cross-talk between transcripts that share a high number of miRNAs, with emphasis placed on the relatively abundant miRNAs (Sumazin et al., 2011). Based on a similar principle, a web-based tool designed to identify ceRNAs interactions for a target gene named TraceRNA, has also been recently developed (Flores & Huang, 2012; Flores et al., 2013).

Furthermore, several databases that store predicted ceRNA interactions have also recently been made available. For example, an extension of star-Base, starBase v2.0, now incorporates miRNA-mediated interactions with potential ceRNA-acting lncRNAs, circRNAs, and pseudogenes to decode regulatory networks using 108 CLIP-Seq data from 37 independent studies (http://starbase.sysu.edu.cn/). A comprehensive transcriptome-wide analysis of pseudogene expression from approximately 300 samples across 13 cancer and normal tissue types also provides a resource to investigate putative ceRNA cross-talk between coordinately expressed pseudogenes and their protein-coding homologs (Kalyana-Sundaram et al., 2012).

Computational prediction tools that integrate the expression profiles of RNA transcripts and expressed miRNAs, the predicted miRNA:target interactions, as well as databases that store putative or validated ceRNAs will facilitate the identification and further functional characterization of ceRNA networks.

## 6. PERSPECTIVES

The discovery that endogenously expressed transcripts can cross-talk and regulate each others' levels by competing for the binding of shared miRNAs has revealed a new layer of posttranscriptional regulation. The prospective roles of ceRNAs are widespread across all spectrums of cellular and organismal biology and are potentially associated with diseases (Fang et al., 2013; Jeyapalan et al., 2011; Karreth et al., 2011; Lee et al., 2010; Ling et al., 2013; Poliseno et al., 2010; Rutnam & Yang, 2012; Sumazin et al., 2011; Wang et al., 2010a). These associations provide exciting opportunities for the development of new disease biomarkers, as well as cutting edge avenues for novel RNA-based therapeutic interventions.

Despite these encouraging findings, our understanding of the molecular rules that govern this miRNA-mediated cross-talk between transcripts, as well as their prevalence and global relevance to normal physiology and disease remain relatively poorly characterized. The ability of a transcript to impact the levels of other transcripts, with which it shares miRNAs, is dependent primarily on its relative expression levels, as well as the number of functional recognition elements for miRNAs these transcripts share. Mathematical modeling of the dynamics of miRNA–mediated cross-talks revealed that these interactions can be reciprocal (bidirectional) or nonreciprocal (unidirectional) (Ala et al., 2013; Artmann, Jung, Bleckmann, & Beissbarth, 2012; Bosia, Pagnani, & Zecchina, 2013; Figliuzzi, Marinari, & De Martino, 2013), and that the directionality of the cross-talk will depend on the stoichiometry and abundance of the cross-talking transcripts and their shared MREs (Figliuzzi et al., 2013). However, in vivo, some of these properties might influence each other's effects and the relative number of MREs and expression levels of ceRNAs that allow their effective cross-talk, beyond these qualitative inferences, remain unknown.

For example, in melanoma cells, the downregulation of eight nonhomologous ceRNA transcripts led to a significant reduction in *PTEN* levels, while comparable changes in *PTEN* transcript abundance affected only four out of its eight predicted ceRNAs (Karreth et al., 2011), suggesting that not all ceRNA interactions are reciprocally regulated. Contrary to expectations, among all eight *PTEN* ceRNAs, the four transcripts whose cellular abundances was altered in response to *PTEN* perturbation were not those with the lowest expression levels or those with the highest number of *PTEN*-shared MREs (Karreth et al., 2011). Therefore, it seems that the ability of a putative ceRNA to regulate the levels of other transcripts cannot simply be determined using either the expression level or the number of shared MREs with the transcript, but rather, a combination of factors and transcript properties needs to be considered.

Along with the number of shared MREs, their sequence complementarity to the targeting miRNA may also affect the efficiency of a posttranscriptional modulator. For example, the first miRNA sponge identified, *IPS4*, harbors a single MRE for the decoyed miRNA, *miR-399* (Franco–Zorrilla et al., 2007). This particular MRE contains a mismatched nucleotide between the target and positions 10–11 of *miR-399* that is thought to allow *IPS1* to efficiently sequester *miR-399* and prevent its degradation (Franco–Zorrilla et al., 2007). The sequestration and titration of miRNAs can also be achieved by the overexpression (hundreds to thousands of copies per cell) of artificial transcripts (Ebert et al., 2007) or endogenous RNAs that are not easily accessible to RNA exonucleases-mediated degradation, such as the circular RNA, *CDR1as*/*ciRS-7* (Hansen et al., 2013; Memczak et al., 2013). However, and unlike these transcripts, not all effective ceRNAs harbor a disproportionally high numbers of MREs, or are expressed at levels similar to or exceeding their target transcripts. Unlike *IPS1*, ceRNAs such as *linc-MD1* (Cesana et al., 2011), *PTENP1* (Poliseno et al., 2010), and *linc-RoR* (Wang et al., 2013) do not appear to bind miRNAs without undergoing miRNA-mediated repression and do not harbor a remarkably high number of functional MREs, nor are they expressed at exceptionally high levels relative to their targets. Therefore, it is essential to understand how the relative abundance of miRNAs, the number and type of MREs, and the relative levels of interacting transcripts, can affect the ability of transcripts to cross-talk.

Some of these outstanding challenges in the identification and prediction of genetic interactions between ceRNAs are tied to our incomplete understanding of miRNA regulation, such as the transcript properties that determine target recognition by miRNAs and how do the strength of various miRNA–transcript interactions affect the levels of their respective gene products. With the rapid advances in the development of tools to investigate the biology underlying miRNA regulation, we will soon be able to address the currently undetermined features of miRNA target recognition and increase the reliability of ceRNA predictions.

Probably more intriguing, but also more challenging to address, is what are the contributions of the ceRNA posttranscriptional regulatory mechanism to traits and diseases. For example, despite its low expression levels relative to its cross-talking targets, *linc-RoR* was demonstrated to contribute to cellular state determination in ES cells by competing for the binding of *miR-145* with highly expressed core stem cell transcription factors, including *Nanog*, *Oct4*, and *Sox2* (Wang et al., 2013). Surprisingly, *linc-RoR* contains only two *miR145* response elements (Wang et al., 2013). How can this relatively lowly expressed transcript with only two shared MREs modulate the expression levels of multiple abundantly expressed mRNAs and contribute to cell fate decisions? One answer could be that lowly expressed ceRNAs, or those with less effective MREs, can only efficiently influence cellular phenotypes if they cross-talk with (i) highly dose sensitive transcripts, such as *PTEN*, or (ii) genes involved in controlling switch-like decisions, such as the ones underlying the maintenance of pluripotent and differentiation (de Giorgio, Krell, Harding, Stebbing, & Castellano, 2013). Alternatively, some ceRNAs can also amplify changes in transcript levels by acting as part of autoregulatory feedback loops.

This novel mechanism of posttranscriptional regulation undoubtedly has added a new layer of complexity to genetic networks. A full understanding of the biological relevance of this mechanism, however, will require deciphering the rules that determine the interactions between transcripts and their relative contributions to transcript level regulation. Thus far, the impact of endogenously expressed ceRNAs on target expression is relatively modest, suggesting that their individual contribution likely remains unnoticed in circuitries that tolerate large variations in gene expression levels. Nevertheless, they may be critical regulators in highly intricate networks or in governing transitional shifts between cellular states.

## ACKNOWLEDGMENTS

We thank members of Prof Chris Ponting's group for insightful comments and suggestions. ACM is funded by a Dorothy Hodgkin's Fellowship, Royal Society. JYT is funded by the Clarendon Fund and the Natural Sciences Engineering Research Council of Canada. We apologize to the authors of the works we inavertedly failed to refer to in this manuscript.

## REFERENCES

Ala, U., Karreth, F. A., Bosia, C., Pagnani, A., Taulli, R., Leopold, V., et al. (2013). Integrated transcriptional and competitive endogenous RNA networks are cross-regulated in permissive molecular environments. *Proceedings of the National Academy of Sciences of the United States of America, 110*(18), 7154–7159.

Alexiou, P., Maragkakis, M., Papadopoulos, G. L., Reczko, M., & Hatzigeorgiou, A. G. (2009). Lost in translation: An assessment and perspective for computational microRNA target identification. *Bioinformatics, 25*(23), 3049–3055.

Alimonti, A., Carracedo, A., Clohessy, J. G., Trotman, L. C., Nardella, C., Egia, A., et al. (2010). Subtle variations in Pten dose determine cancer susceptibility. *Nature Genetics, 42*(5). 454–U136.

Ambros, V. (1989). A hierarchy of regulatory genes controls a larva-to-adult developmental switch in *C. elegans. Cell, 57*(1), 49–57.

Ambros, V. (2004). The functions of animal microRNAs. *Nature, 431*(7006), 350–355.

Ameres, S. L., Horwich, M. D., Hung, J. H., Xu, J., Ghildiyal, M., Weng, Z. P., et al. (2010). Target RNA-directed trimming and tailing of small silencing RNAs. *Science, 328*(5985), 1534–1539.

Anguera, M. C., Sadreyev, R., Zhang, Z. Q., Szanto, A., Payer, B., Sheridan, S. D., et al. (2012). Molecular signatures of human induced pluripotent stem cells highlight sex differences and cancer genes. *Cell Stem Cell, 11*(1), 75–90.

Arima, T., Matsuda, T., Takagi, N., & Wake, N. (1997). Association of IGF2 and H19 imprinting with choriocarcinoma development. *Cancer Genetics and Cytogenetics, 93*(1), 39–47.

Artmann, S., Jung, K., Bleckmann, A., & Beissbarth, T. (2012). Detection of simultaneous group effects in microRNA expression and related target gene sets. *PLoS One, 7*(6).

Asakawa, K., & Kawakami, K. (2008). Targeted gene expression by the Gal4-UAS system in zebrafish. *Development Growth & Differentiation, 50*(6), 391–399.

Ascano, M., Hafner, M., Cekan, P., Gerstberger, S., & Tuschl, T. (2012). Identification of RNA-protein interaction networks using PAR-CLIP. *Wiley Interdisciplinary Reviews - RNA, 3*(2), 159–177.

Babak, T., Zhang, W., Morris, Q., Blencowe, B. J., & Hughes, T. R. (2004). Probing microRNAs with microarrays: Tissue specificity and functional inference. *RNA, 10*(11), 1813–1819.

Baccarini, A., Chauhan, H., Gardner, T. J., Jayaprakash, A. D., Sachidanandam, R., & Brown, B. D. (2011). Kinetic analysis reveals the fate of a microRNA following target regulation in mammalian cells. *Current Biology, 21*(5), 369–376.

Bader, A. G., Brown, D., & Winkler, M. (2010). The promise of microRNA replacement therapy. *Cancer Research, 70*(18), 7027–7030.

Baek, D., Villen, J., Shin, C., Camargo, F. D., Gygi, S. P., & Bartel, D. P. (2008). The impact of microRNAs on protein output. *Nature, 455*(7209). 64–U38.

Bail, S., Swerdel, M., Liu, H. D., Jiao, X. F., Goff, L. A., Hart, R. P., et al. (2010). Differential regulation of microRNA stability. *RNA - A Publication of the RNA Society, 16*(5), 1032–1039.

Balmer, D., Goldstine, J., Rao, Y. M., & LaSalle, J. M. (2003). Elevated methyl-CpG-binding protein 2 expression is acquired during postnatal human brain development and is correlated with alternative polyadenylation. *Journal of Molecular Medicine-jmm, 81*(1), 61–68.

Barad, O., Meiri, E., Avniel, A., Aharonov, R., Barzilai, A., Bentwich, I., et al. (2004). MicroRNA expression detected by oligonucleotide microarrays: System establishment and expression profiling in human tissues. *Genome Research, 14*(12), 2486–2494.

Barbato, C., Arisi, I., Frizzo, M. E., Brandi, R., Da Sacco, L., & Masotti, A. (2009). Computational challenges in miRNA target predictions: To be or not to be a true target? *Journal of Biomedicine and Biotechnology.*

Bargaje, R., Gupta, S., Sarkeshik, A., Park, R., Xu, T., Sarkar, M., et al. (2012). Identification of novel targets for miR-29a using miRNA proteomics. *PLoS One, 7*(8).

Baron, R., & Rawadi, G. (2007). Targeting the Wnt/beta-catenin pathway to regulate bone formation in the adult skeleton. *Endocrinology, 148*(6), 2635–2643.

Bartel, D. P. (2004). MicroRNAs: Genomics, biogenesis, mechanism, and function. *Cell, 116*(2), 281–297.

Bartel, D. P. (2009). MicroRNAs: Target recognition and regulatory functions. *Cell, 136*(2), 215–233.

Bartel, D. P., & Chen, C. Z. (2004). Micromanagers of gene expression: The potentially widespread influence of metazoan microRNAs. *Nature Reviews Genetics, 5*(5), 396–400.

Baskerville, S., & Bartel, D. P. (2005). Microarray profiling of microRNAs reveals frequent coexpression with neighboring miRNAs and host genes. *RNA - A Publication of the RNA Society, 11*(3), 241–247.

Bauer, K. M., & Hummon, A. B. (2012). Effects of the miR-143/-145 microRNA cluster on the colon cancer proteome and transcriptome. *Journal of Proteome Research, 11*(9), 4744–4754.

Behm-Ansmant, I., Rehwinkel, J., Doerks, T., Stark, A., Bork, P., & Izaurralde, E. (2006). MRNA degradation by miRNAs and GW182 requires both CCR4:NOT deadenylase and DCP1:DCP2 decapping complexes. *Genes & Development, 20*(14), 1885–1898.

Bernstein, E., Kim, S. Y., Carmell, M. A., Murchison, E. P., Alcorn, H., Li, M. Z., et al. (2003). Dicer is essential for mouse development. *Nature Genetics, 35*(3), 215–217.

Blake, D. J., Weir, A., Newey, S. E., & Davies, K. E. (2002). Function and genetics of dystrophin and dystrophin-related proteins in muscle. *Physiological Reviews, 82*(2), 291–329.

Bolisetty, M. T., Dy, G., Tam, W., & Beemon, K. L. (2009). Reticuloendotheliosis virus strain T induces miR-155, which targets JARID2 and promotes cell survival. *Journal of Virology, 83*(23), 12009–12017.

Bosia, C., Pagnani, A., & Zecchina, R. (2013). Modelling competing endogenous RNA networks. *PLoS One, 8*(6), e66609.

Boutla, A., Delidakis, C., & Tabler, M. (2003). Developmental defects by antisense-mediated inactivation of micro-RNAs 2 and 13 in *Drosophila* and the identification of putative target genes. *Nucleic Acids Research, 31*(17), 4973–4980.

Brannan, C. I., Dees, E. C., Ingram, R. S., & Tilghman, S. M. (1990). The product of the H19 gene may function as an RNA. *Molecular and Cellular Biology, 10*(1), 28–36.

Brandenberger, R., Schmidt, A., Linton, J., Wang, D., Backus, C., Denda, S., Muller, U, & Reichardt, L. F. (2001). Identification and characterization of a novel extracellular matrix protein nephronectin that is associated with integrin alpha8beta1 in the embryonic kidney. *Journal of Cell Biology, 154*(2), 447–458.

Brennecke, J., Hipfner, D. R., Stark, A., Russell, R. B., & Cohen, S. M. (2003). bantam encodes a developmentally regulated microRNA that controls cell proliferation and regulates the proapoptotic gene hid in *Drosophila. Cell, 113*(1), 25–36.

Brennecke, J., Stark, A., Russell, R. B., & Cohen, S. M. (2005). Principles of microRNA-target recognition. *PLoS Biology, 3*(3), 404–418.

Brenner, J. L., Jasiewicz, K. L., Fahley, A. F., Kemp, B. J., & Abbott, A. L. (2010). Loss of individual microRNAs causes mutant phenotypes in sensitized genetic backgrounds in *C. elegans. Current Biology, 20*(14), 1321–1325.

Brown, B. D., & Naldini, L. (2009). INNOVATION exploiting and antagonizing microRNA regulation for therapeutic and experimental applications. *Nature Reviews Genetics, 10*(8), 578–585.

Cai, X. Z., Hagedorn, C. H., & Cullen, B. R. (2004). Human microRNAs are processed from capped, polyadenylated transcripts that can also function as mRNAs. *RNA - A Publication of the RNA Society, 10*(12), 1957–1966.

Calin, G. A., Liu, C. G., Sevignani, C., Ferracin, M., Felli, N., Dumitru, C. D., et al. (2004). MicroRNA profiling reveals distinct signatures in B cell chronic lymphocytic leukemias. *Proceedings of the National Academy of Sciences of the United States of America, 101*(32), 11755–11760.

Cazalla, D., Yario, T., & Steitz, J. (2010). Down-regulation of a host microRNA by a *Herpesvirus saimiri* noncoding RNA. *Science, 328*(5985), 1563–1566.

Cesana, M., Cacchiarelli, D., Legnini, I., Santini, T., Sthandier, O., Chinappi, M., et al. (2011). A long noncoding RNA controls muscle differentiation by functioning as a competing endogenous RNA. *Cell, 147*(4). 947–947.

Chahrour, M., & Zoghbi, H. Y. (2007). The story of Rett syndrome: From clinic to neurobiology. *Neuron, 56*(3), 422–437.

Chaudhuri, K., & Chatterjee, R. (2007). MicroRNA detection and target prediction: Integration of computational and experimental approaches. *DNA and Cell Biology, 26*(5), 321–337.

Chavali, S., Bruhn, S., Tiemann, K., Saetrom, P., Barrenas, F., Saito, T., et al. (2013). MicroRNAs act complementarily to regulate disease-related mRNA modules in human diseases. *RNA, 19*(11), 1552–1562.

Chen, J. F., Murchison, E. P., Tang, R., Callis, T. E., Tatsuguchi, M., Deng, Z., et al. (2008). Targeted deletion of Dicer in the heart leads to dilated cardiomyopathy and heart failure. *Proceedings of the National Academy of Sciences of the United States of America, 105*(6), 2111–2116.

Chen, R. Z., Akbarian, S., Tudor, M., & Jaenisch, R. (2001). Deficiency of methyl-CpG binding protein-2 in CNS neurons results in a Rett-like phenotype in mice. *Nature Genetics, 27*(3), 327–331.

Chen, X., Liang, H., Zhang, C. Y., & Zen, K. (2012). miRNA regulates noncoding RNA: A noncanonical function model. *Trends in Biochemical Sciences, 37*(11), 457–459.

Chi, S. W., Zang, J. B., Mele, A., & Darnell, R. B. (2009). Argonaute HITS-CLIP decodes microRNA–mRNA interaction maps. *Nature, 460*(7254), 479–486.

Chou, C. H., Lin, F. M., Chou, M. T., Hsu, S. D., Chang, T. H., Weng, S. L., et al. (2013). A computational approach for identifying microRNA–target interactions using high-throughput CLIP and PAR-CLIP sequencing. *BMC Genomics, 14.*

Cogoni, C., & Macino, G. (2000). Post-transcriptional gene silencing across kingdoms. *Current Opinion in Genetics & Development, 10*(6), 638–643.

Cole, K. A., Attiyeh, E. F., Mosse, Y. P., Laquaglia, M. J., Diskin, S. J., Brodeur, G. M., et al. (2008). A functional screen identifies miR-34a as a candidate neuroblastoma tumor suppressor gene. *Molecular Cancer Research, 6*(5), 735–742.

Coy, J. F., Sedlacek, Z., Bachner, D., Delius, H., & Poustka, A. (1999). A complex pattern of evolutionary conservation and alternative polyadenylation within the long 3′-untranslated region of the methyl-CpG-binding protein 2 gene (MeCP2) suggests a regulatory role in gene expression. *Human Molecular Genetics, 8*(7), 1253–1262.

Damiani, D., Alexander, J. J., O'Rourke, J. R., McManus, M., Jadhav, A. P., Cepko, C. L., et al. (2008). Dicer inactivation leads to progressive functional and structural degeneration of the mouse retina. *Journal of Neuroscience, 28*(19), 4878–4887.

Danan, M., Schwartz, S., Edelheit, S., & Sorek, R. (2012). Transcriptome-wide discovery of circular RNAs in Archaea. *Nucleic Acids Research, 40*(7), 3131–3142.

Darnell, R. B. (2010). HITS-CLIP: Panoramic views of protein-RNA regulation in living cells. *Wiley Interdisciplinary Reviews - RNA, 1*(2), 266–286.

Davis, T. H., Cuellar, T. L., Koch, S. M., Barker, A. J., Harfe, B. D., McManus, M. T., et al. (2008). Conditional loss of Dicer disrupts cellular and tissue morphogenesis in the cortex and hippocampus. *Journal of Neuroscience, 28*(17), 4322–4330.

de Giorgio, A., Krell, J., Harding, V., Stebbing, J., & Castellano, L. (2013). Emerging roles of competing endogenous RNAs in cancer: Insights from the regulation of PTEN. *Molecular and Cellular Biology, 33*(20), 3976–3982.

Denli, A. M., Tops, B. B. J., Plasterk, R. H. A., Ketting, R. F., & Hannon, G. J. (2004). Processing of primary microRNAs by the microprocessor complex. *Nature, 432*(7014), 231–235.

Derrien, T., Johnson, R., Bussotti, G., Tanzer, A., Djebali, S., Tilgner, H., et al. (2012). The GENCODE v7 catalog of human long noncoding RNAs: Analysis of their gene structure, evolution, and expression. *Genome Research, 22*(9), 1775–1789.

Dinger, M. E., Amaral, P. P., Mercer, T. R., Pang, K. C., Bruce, S. J., Gardiner, B. B., et al. (2008). Long noncoding RNAs in mouse embryonic stem cell pluripotency and differentiation. *Genome Research, 18*(9), 1433–1445.

Dixon, R. J., Eperon, I. C., Hall, L., & Samani, N. J. (2005). A genome-wide survey demonstrates widespread non-linear mRNA in expressed sequences from multiple species. *Nucleic Acids Research, 33*(18), 5904–5913.

Djuranovic, S., Nahvi, A., & Green, R. (2011). A parsimonious model for gene regulation by miRNAs. *Science, 331*(6017), 550–553.

Doench, J. G., & Sharp, P. A. (2004). Specificity of microRNA target selection in translational repression. *Genes & Development, 18*(5), 504–511.

Dolken, L., Malterer, G., Erhard, F., Kothe, S., Friedel, C. C., Suffert, G., et al. (2010). Systematic analysis of viral and cellular microRNA targets in cells latently infected with human gamma-Herpesviruses by RISC immunoprecipitation assay. *Cell Host & Microbe, 7*(4), 324–334.

Dolmetsch, R. E., Pajvani, U., Fife, K., Spotts, J. M., & Greenberg, M. E. (2001). Signaling to the nucleus by an L-type calcium channel–calmodulin complex through the MAP kinase pathway. *Science, 294*(5541), 333–339.

Ebert, M. S., Neilson, J. R., & Sharp, P. A. (2007). MicroRNA sponges: Competitive inhibitors of small RNAs in mammalian cells. *Nature Methods, 4*(9), 721–726.

Ebert, M. S., & Sharp, P. A. (2010a). Emerging roles for natural microRNA sponges. *Curr Biology, 20*(19), R858–R861.

Ebert, M. S., & Sharp, P. A. (2010b). MicroRNA sponges: Progress and possibilities. *RNA, 16*(11), 2043–2050.

Ebert, M. S., & Sharp, P. A. (2012). Roles for microRNAs in conferring robustness to biological processes. *Cell, 149*(3), 515–524.

Ebner, O. A., & Selbach, M. (2011). Whole cell proteome regulation by microRNAs captured in a pulsed SILAC mass spectrometry approach. *Methods in Molecular Biology, 725*, 315–331.

Elkan-Miller, T., Ulitsky, I., Hertzano, R., Rudnicki, A., Dror, A. A., Lenz, D. R., et al. (2011). Integration of transcriptomics, proteomics, and microRNA analyses reveals novel microRNA regulation of targets in the mammalian inner ear. *PLoS One, 6*(4).

Elkin, M., Shevelev, A., Schulze, E., Tykocinsky, M., Cooper, M., Ariel, I., et al. (1995). The expression of the imprinted H19 and Igf-2 genes in human bladder-carcinoma. *FEBS Letters, 374*(1), 57–61.

Enright, A. J., John, B., Gaul, U., Tuschl, T., Sander, C., & Marks, D. S. (2004). MicroRNA targets in *Drosophila*. *Genome Biology*, *5*(1).

Erhard, F., Dolken, L., Jaskiewicz, L., & Zimmer, R. (2013). PARma: Identification of microRNA target sites in AGO–PAR–CLIP data. *Genome Biology*, *14*(7), R79.

Fang, L., Du, W. W., Yang, X. L., Chen, K., Ghanekar, A., Levy, G., et al. (2013). Versican 3′-untranslated region (3′-UTR) functions as a ceRNA in inducing the development of hepatocellular carcinoma by regulating miRNA activity. *FASEB Journal*, *27*(3), 907–919.

Farh, K. K. H., Grimson, A., Jan, C., Lewis, B. P., Johnston, W. K., Lim, L. P., et al. (2005). The widespread impact of mammalian microRNAs on mRNA repression and evolution. *Science*, *310*(5755), 1817–1821.

Favereaux, A., Thoumine, O., Bouali-Benazzouz, R., Roques, V., Papon, M. A., Salam, S. A., et al. (2011). Bidirectional integrative regulation of Cav1.2 calcium channel by microRNA miR-103: Role in pain. *EMBO Journal*, *30*(18), 3830–3841.

Figliuzzi, M., Marinari, E., & De Martino, A. (2013). MicroRNAs as a selective channel of communication between competing RNAs: A steady-state theory. *Biophysical Journal*, *104*(5), 1203–1213.

Fischle, W., Dequiedt, F., Hendzel, M. J., Guenther, M. G., Lazar, M. A., Voelter, W., et al. (2002). Enzymatic activity associated with class IIHDACs is dependent on a multiprotein complex containing HDAC3 and SMRT/N-CoR. *Molecular Cell*, *9*(1), 45–57.

Flores, M., Hsiao, T. H., Chiu, Y. C., Chuang, E. Y., Huang, Y., & Chen, Y. (2013). Gene regulation, modulation, and their applications in gene expression data analysis. *Advances in Bioinformatics*, *2013*, 360–678.

Flores, M. & Huang, Y. F. (2012). A new algorithm for predicting competing endogenous RNAs. In: *2012 Ieee International Workshop on Genomic Signal Processing and Statistics (Gensips)*, pp. 118–121.

Flynt, A. S., Greimann, J. C., Chung, W. J., Lima, C. D., & Lai, E. C. (2010). MicroRNA biogenesis via splicing and exosome-mediated trimming in *Drosophila*. *Molecular Cell*, *38*(6), 900–907.

Fossat, P., Dobremez, E., Bouali-Benazzouz, R., Favereaux, A., Bertrand, S. S., Kilk, K., et al. (2010). Knockdown of L calcium channel subtypes: Differential effects in neuropathic pain. *Journal of Neuroscience*, *30*(3), 1073–1085.

Fox, S. B., Fawcett, J., Jackson, D. G., Collins, I., Gatter, K. C., Harris, A. L., et al. (1994). Normal human tissues, in addition to some tumors, express multiple different Cd44 isoforms. *Cancer Research*, *54*(16), 4539–4546.

Franco-Zorrilla, J. M., Valli, A., Todesco, M., Mateos, I., Puga, M. I., Rubio-Somoza, I., et al. (2007). Target mimicry provides a new mechanism for regulation of microRNA activity. *Nature Genetics*, *39*(8), 1033–1037.

Frezzetti, D., Reale, C., Cali, G., Nitsch, L., Fagman, H., Nilsson, O., et al. (2011). The microRNA-processing enzyme Dicer is essential for thyroid function. *PLoS One*, *6*(11).

Friedman, R. C., Farh, K. K. H., Burge, C. B., & Bartel, D. P. (2009). Most mammalian mRNAs are conserved targets of microRNAs. *Genome Research*, *19*(1), 92–105.

Gabory, A., Ripoche, M. A., Yoshimizu, T., & Dandolo, L. (2006). The H19 gene: Regulation and function of a non-coding RNA. *Cytogenetic and Genome Research*, *113*(1–4), 188–193.

Gantier, M. P., McCoy, C. E., Rusinova, I., Saulep, D., Wang, D., Xu, D., et al. (2011). Analysis of microRNA turnover in mammalian cells following Dicer1 ablation. *Nucleic Acids Research*, *39*(13), 5692–5703.

Garcia, D. M., Baek, D., Shin, C., Bell, G. W., Grimson, A., & Bartel, D. P. (2011). Weak seed-pairing stability and high target-site abundance decrease the proficiency of lsy-6 and other microRNAs. *Nature Structural & Molecular Biology*, *18*(10), 1139–U1175.

Gregory, R. I., Yan, K. P., Amuthan, G., Chendrimada, T., Doratotaj, B., Cooch, N., et al. (2004). The Microprocessor complex mediates the genesis of microRNAs. *Nature*, *432*(7014), 235–240.

Griffiths-Jones, S., Saini, H. K., van Dongen, S., & Enright, A. J. (2008). miRBase: Tools for microRNA genomics. *Nucleic Acids Research*, *36*, D154–D158.

Grimson, A., Farh, K. K. H., Johnsto n, W. K., Garrett-Engele, P., Lim, L. P., & Bartel, D. P. (2007). MicroRNA targeting specificity in mammals: Determinants beyond seed pairing. *Molecular Cell, 27*(1), 91–105.

Grishok, A., Pasquinelli, A. E., Conte, D., Li, N., Parrish, S., Ha, I., et al. (2001). Genes and mechanisms related to RNA interference regulate expression of the small temporal RNAs that control *C. elegans* developmental timing. *Cell, 106*(1), 23–34.

Gu, S., Jin, L., Zhang, F. J., Sarnow, P., & Kay, M. A. (2009). Biological basis for restriction of microRNA targets to the 3′ untranslated region in mammalian mRNAs. *Nature Structural & Molecular Biology, 16*(2), 144–150.

Guttman, M., Donaghey, J., Carey, B. W., Garber, M., Grenier, J. K., Munson, G., et al. (2011). lincRNAs act in the circuitry controlling pluripotency and differentiation. *Nature, 477*(7364). 295–U260.

Guy, J., Cheval, H., Selfridge, J., & Bird, A. (2011). The role of MeCP2 in the brain. *Annual Review of Cell and Developmental Biology, Vol. 27*(27), 631–652.

Hafner, M., Landthaler, M., Burger, L., Khorshid, M., Hausser, J., Berninger, P., et al. (2010). Transcriptome-wide identification of RNA-binding protein and microRNA target sites by PAR-CLIP. *Cell, 141*(1), 129–141.

Hafner, M., Lianoglou, S., Tuschl, T., & Betel, D. (2012). Genome-wide identification of miRNA targets by PAR-CLIP. *Methods, 58*(2), 94–105.

Hammell, M., Long, D., Zhang, L., Lee, A., Carmack, C. S., Han, M., et al. (2008). mirWIP: microRNA target prediction based on microRNA-containing ribonucleoprotein-enriched transcripts. *Nature Methods, 5*(9), 813–819.

Han, K., Gennarino, V. A., Lee, Y., Pang, K., Hashimoto-Torii, K., Choufani, S., et al. (2013). Human-specific regulation of MeCP2 levels in fetal brains by microRNA miR-483-5p. *Genes & Development, 27*(5), 485–490.

Hansen, T. B., Jensen, T. I., Clausen, B. H., Bramsen, J. B., Finsen, B., Damgaard, C. K., et al. (2013). Natural RNA circles function as efficient microRNA sponges. *Nature, 495*(7441), 384–388.

Hansen, T. B., Wiklund, E. D., Bramsen, J. B., Villadsen, S. B., Statham, A. L., Clark, S. J., et al. (2011). miRNA-dependent gene silencing involving Ago2-mediated cleavage of a circular antisense RNA. *EMBO Journal, 30*(21), 4414–4422.

Harfe, B. D., McManus, M. T., Mansfield, J. H., Hornstein, E., & Tabin, C. J. (2005). The RNaseIII enzyme Dicer is required for morphogenesis but not patterning of the vertebrate limb. *Proceedings of the National Academy of Sciences of the United States of America, 102*(31), 10898–10903.

Hatfield, S. D., Shcherbata, H. R., Fischer, K. A., Nakahara, K., Carthew, R. W., & Ruohola-Baker, H. (2005). Stem cell division is regulated by the microRNA pathway. *Nature, 435*(7044), 974–978.

Helwak, A., Kudla, G., Dudnakova, T., & Tollervey, D. (2013). Mapping the human miRNA interactome by CLASH reveals frequent noncanonical binding. *Cell, 153*(3), 654–665.

Hendrickson, D. G., Hogan, D. J., Herschlag, D., Ferrell, J. E., & Brown, P. O. (2008). Systematic identification of mRNAs recruited to Argonaute 2 by specific microRNAs and corresponding changes in transcript abundance. *PLoS One, 3*(5).

Hornstein, E., & Shomron, N. (2006). Canalization of development by microRNAs. *Nature Genetics, 38*(Suppl), S20–S24.

Hu, S. J., Wilson, K. D., Ghosh, Z., Han, L., Wang, Y. M., Lan, F., et al. (2013). MicroRNA-302 increases reprogramming efficiency via repression of NR2F2. *Stem Cells, 31*(2), 259–268.

Huang, T. C., Sahasrabuddhe, N. A., Kim, M. S., Getnet, D., Yang, Y., Peterson, J. M., et al. (2012). Regulation of lipid metabolism by Dicer revealed through SILAC mice. *Journal of Proteome Research, 11*(4), 2193–2205.

Hutvagner, G., & Simard, M. J. (2008). Argonaute proteins: Key players in RNA silencing. *Nature Reviews Molecular Cell Biology, 9*(1), 22–32.

Hwang, H. W., Wentzel, E. A., & Mendell, J. T. (2007). A hexanucleotide element directs microRNA nuclear import. *Science, 315*(5808), 97–100.

Ito, K., Shinomura, T., Zako, M., Ujita, M., & Kimata, K. (1995). Multiple forms of mouse PG-M, a large chondroitin sulfate proteoglycan generated by alternative splicing. *Journal of Biological Chemistry, 270*(2), 958–965.

Jeck, W. R., Sorrentino, J. A., Wang, K., Slevin, M. K., Burd, C. E., Liu, J., et al. (2013). Circular RNAs are abundant, conserved, and associated with ALU repeats. *RNA, 19*(2), 141–157.

Jeggari, A., Marks, D. S., & Larsson, E. (2012). miRcode: A map of putative microRNA target sites in the long non-coding transcriptome. *Bioinformatics, 28*(15), 2062–2063.

Jensen, K. B., & Darnell, R. B. (2008). CLIP: Crosslinking and immunoprecipitation of in vivo RNA targets of RNA-binding proteins. *Methods in Molecular Biology, 488*, 85–98.

Jeyapalan, Z., Deng, Z. Q., Shatseva, T., Fang, L., He, C. Y., & Yang, B. B. (2011). Expression of CD44 3′-untranslated region regulates endogenous microRNA functions in tumorigenesis and angiogenesis. *Nucleic Acids Research, 39*(8), 3026–3041.

Jia, W., Chen, W., & Kang, J. (2013). The functions of microRNAs and long non-coding RNAs in embryonic and induced pluripotent stem cells. *Genomics Proteomics Bioinformatics, 11*(5), 275–283.

Jiang, L., Liu, X. Q., Chen, Z. J., Jin, Y., Heidbreder, C. E., Kolokythas, A., et al. (2010). MicroRNA-7 targets IGF1R (insulin-like growth factor 1 receptor) in tongue squamous cell carcinoma cells. *Biochemical Journal, 432*, 199–205.

John, B., Enright, A. J., Aravin, A., Tuschl, T., Sander, C., & Marks, D. S. (2004). Human MicroRNA targets. *PLoS Biology, 2*(11), 1862–1879.

Johnson, C. D., Esquela-Kerscher, A., Stefani, G., Byrom, N., Kelnar, K., Ovcharenko, D., et al. (2007). The let-7 microRNA represses cell proliferation pathways in human cells. *Cancer Research, 67*(16), 7713–7722.

Jones-Rhoades, M. W., Bartel, D. P., & Bartel, B. (2006). MicroRNAs and their regulatory roles in plants. *Annual Review of Plant Biology, 57*, 19–53.

Judson, R. L., Babiarz, J. E., Venere, M., & Blelloch, R. (2009). Embryonic stem cell-specific microRNAs promote induced pluripotency. *Nature Biotechnology, 27*(5), 459–461.

Junn, E., Lee, K. W., Jeong, B. S., Chan, T. W., Im, J. Y., & Mouradian, M. M. (2009). Repression of alpha-synuclein expression and toxicity by microRNA-7. *Proceedings of the National Academy of Sciences of the United States of America, 106*(31), 13052–13057.

Kahai, S., Lee, S. C., Lee, D. Y., Yang, J., Li, M., Wang, C. H., et al. (2009). MicroRNA miR-378 regulates nephronectin expression modulating osteoblast differentiation by targeting GalNT-7. *PLoS One, 4*(10), e7535.

Kahai, S., Lee, S. C., Seth, A., & Yang, B. B. (2010). Nephronectin promotes osteoblast differentiation via the epidermal growth factor-like repeats. *FEBS Letters, 584*(1), 233–238.

Kai, Z. S., & Pasquinelli, A. E. (2010). MicroRNA assassins: Factors that regulate the disappearance of miRNAs. *Nature Structural & Molecular Biology, 17*(1), 5–10.

Kallen, A. N., Zhou, X. B., Xu, J., Qiao, C., Ma, J., Yan, L., et al. (2013). The imprinted H19 LncRNA antagonizes Let-7 MicroRNAs. *Molecular Cell, 52*(1), 101–112.

Kaller, M., Liffers, S. T., Oeljeklaus, S., Kuhlmann, K., Roh, S., Hoffmann, R., et al. (2011). Genome-wide characterization of miR-34a induced changes in protein and mRNA expression by a combined pulsed SILAC and microarray analysis. *Molecular & Cellular Proteomics, 10*(8).

Kalyana-Sundaram, S., Kumar-Sinha, C., Shankar, S., Robinson, D. R., Wu, Y. M., Cao, X. H., et al. (2012). Expressed pseudogenes in the transcriptional landscape of human cancers. *Cell, 149*(7), 1622–1634.

Kanellopoulou, C., Muljo, S. A., Kung, A. L., Ganesan, S., Drapkin, R., Jenuwein, T., et al. (2005). Dicer-deficient mouse embryonic stem cells are defective in differentiation and centromeric silencing. *Genes & Development, 19*(4), 489–501.

Kanematsu, S., Tanimoto, K., Suzuki, Y., & Sugano, S. (2013). Screening for possible miRNA–mRNA associations in a colon cancer cell line. *Gene*.

Karreth, F. A., Tay, Y., Perna, D., Ala, U., Tan, S. M., Rust, A. G., et al. (2011). In vivo identification of tumor-suppressive PTEN ceRNAs in an oncogenic BRAF-induced mouse model of melanoma. *Cell*, *147*(2), 382–395.

Karro, J. E., Yan, Y. P., Zheng, D. Y., Zhang, Z. L., Carriero, N., Cayting, P., et al. (2007). Pseudogene.org: A comprehensive database and comparison platform for pseudogene annotation. *Nucleic Acids Research*, *35*, D55–D60.

Kashyap, V., Rezende, N. C., Scotland, K. B., Shaffer, S. M., Persson, J. L., Gudas, L. J., et al. (2009). Regulation of stem cell pluripotency and differentiation involves a mutual regulatory circuit of the Nanog, OCT4, and SOX2 pluripotency transcription factors with polycomb repressive complexes and stem cell microRNAs. *Stem Cells and Development*, *18*(7), 1093–1108.

Kataoka, Y., Takeichi, M., & Uemura, T. (2001). Developmental roles and molecular characterization of a *Drosophila* homologue of *Arabidopsis* Argonaute1, the founder of a novel gene superfamily. *Genes to Cells*, *6*(4), 313–325.

Keene, J. D., Komisarow, J. M., & Friedersdorf, M. B. (2006). RIP-Chip: The isolation and identification of mRNAs, microRNAs and protein components of ribonucleoprotein complexes from cell extracts. *Nature Protocols*, *1*(1), 302–307.

Kefas, B., Godlewski, J., Comeau, L., Li, Y. Q., Abounader, R., Hawkinson, M., et al. (2008). microRNA-7 inhibits the epidermal growth factor receptor and the Akt pathway and is down-regulated in glioblastoma. *Cancer Research*, *68*(10), 3566–3572.

Keniry, A., Oxley, D., Monnier, P., Kyba, M., Dandolo, L., Smits, G., et al. (2012). The H19 lincRNA is a developmental reservoir of miR-675 that suppresses growth and lgf1r. *Nature Cell Biology*, *14*(7), 659–665.

Khvorova, A., Reynolds, A., & Jayasena, S. D. (2003). Functional siRNAs and miRNAs exhibit strand bias. *Cell*, *115*(2), 209–216.

Kim, Y. K., Yu, J., Han, T. S., Park, S. Y., Namkoong, B., Kim, D. H., et al. (2009). Functional links between clustered microRNAs: Suppression of cell-cycle inhibitors by microRNA clusters in gastric cancer. *Nucleic Acids Research*, *37*(5), 1672–1681.

Kiriakidou, M., Nelson, P. T., Kouranov, A., Fitziev, P., Bouyioukos, C., Mourelatos, Z., et al. (2004). A combined computational-experimental approach predicts human microRNA targets. *Genes & Development*, *18*(10), 1165–1178.

Kluiver, J., Slezak-Prochazka, I., Smigielska-Czepiel, K., Halsema, N., Kroesen, B. J., & van den Berg, A. (2012). Generation of miRNA sponge constructs. *Methods*, *58*(2), 113–117.

Kohler, A., & Hurt, E. (2007). Exporting RNA from the nucleus to the cytoplasm. *Nature Reviews Molecular Cell Biology*, *8*(10), 761–773.

Kozomara, A., & Griffiths-Jones, S. (2011). miRBase: Integrating microRNA annotation and deep-sequencing data. *Nucleic Acids Research*, *39*, D152–D157.

Krek, A., Grun, D., Poy, M. N., Wolf, R., Rosenberg, L., Epstein, E. J., et al. (2005). Combinatorial microRNA target predictions. *Nature Genetics*, *37*(5), 495–500.

Krishnan, V., Bryant, H. U., & Macdougald, O. A. (2006). Regulation of bone mass by Wnt signaling. *Journal of Clinical Investigation*, *116*(5), 1202–1209.

Krol, J., Loedige, I., & Filipowicz, W. (2010). The widespread regulation of microRNA biogenesis, function and decay. *Nature Reviews Genetics*, *11*(9), 597–610.

Krol, J., Sobczak, K., Wilczynska, U., Drath, M., Jasinska, A., Kaczynska, D., et al. (2004). Structural features of microRNA (miRNA) precursors and their relevance to miRNA biogenesis and small interfering RNA/short hairpin RNA design. *Journal of Biological Chemistry*, *279*(40), 42230–42239.

Kudla, G., Granneman, S., Hahn, D., Beggs, J. D., & Tollervey, D. (2011). Cross-linking, ligation, and sequencing of hybrids reveals RNA–RNA interactions in yeast. *Proceedings of the National Academy of Sciences of the United States of America*, *108*(24), 10010–10015.

Kumar, A., Wong, A. K. L., Lizard, M. L., Moore, R. J., & Lefevre, C. (2012). miRNA_Targets: A database for miRNA target predictions in coding and non-coding regions of mRNAs. *Genomics, 100*(6), 352–356.

Kumar, M. S., Erkeland, S. J., Pester, R. E., Chen, C. Y., Ebert, M. S., Sharp, P. A., et al. (2008). Suppression of non-small cell lung tumor development by the let-7 microRNA family. *Proceedings of the National Academy of Sciences of the United States of America, 105*(10), 3903–3908.

Lee, D. Y., Jeyapalan, Z., Fang, L., Yang, J., Zhang, Y., Yee, A. Y., et al. (2010). Expression of versican 3′-untranslated region modulates endogenous microRNA functions. *PLoS One, 5*(10).

Lee, D. Y., Shatseva, T., Jeyapalan, Z., Du, W. W., Deng, Z. Q., & Yang, B. B. (2009). A 3′-untranslated region (3′-UTR) induces organ adhesion by regulating miR-199a*functions. *PLoS One, 4*(2).

Lee, R. C., Feinbaum, R. L., & Ambros, V. (1993). The *C. elegans* heterochronic gene Lin-4 encodes small RNAs with antisense complementarity to Lin-14. *Cell, 75*(5), 843–854.

Lee, S. C., Fang, L., Wang, C. H., Kahai, S., Deng, Z., & Yang, B. B. (2011). A non-coding transcript of nephronectin promotes osteoblast differentiation by modulating microRNA functions. *FEBS Letters, 585*(16), 2610–2616.

Lee, Y., Ahn, C., Han, J. J., Choi, H., Kim, J., Yim, J., et al. (2003). The nuclear RNase III Drosha initiates microRNA processing. *Nature, 425*(6956), 415–419.

Lee, Y., Kim, M., Han, J. J., Yeom, K. H., Lee, S., Baek, S. H., et al. (2004). MicroRNA genes are transcribed by RNA polymerase II. *EMBO Journal, 23*(20), 4051–4060.

Leighton, P. A., Ingram, R. S., Eggenschwiler, J., Efstratiadis, A., & Tilghman, S. M. (1995). Disruption of imprinting caused by deletion of the H19 gene region in mice. *Nature, 375*(6526), 34–39.

Lewis, B. P., Burge, C. B., & Bartel, D. P. (2005). Conserved seed pairing, often flanked by adenosines, indicates that thousands of human genes are microRNA targets. *Cell, 120*(1), 15–20.

Lewis, B. P., Shih, I. H., Jones-Rhoades, M. W., Bartel, D. P., & Burge, C. B. (2003). Prediction of mammalian microRNA targets. *Cell, 115*(7), 787–798.

Li, J., Yen, C., Liaw, D., Podsypanina, K., Bose, S., Wang, S. I., et al. (1997). PTEN, a putative protein tyrosine phosphatase gene mutated in human brain, breast, and prostate cancer. *Science, 275*(5308), 1943–1947.

Li, X., Cassidy, J. J., Reinke, C. A., Fischboeck, S., & Carthew, R. W. (2009). A microRNA imparts robustness against environmental fluctuation during development. *Cell, 137*(2), 273–282.

Liang, H. W., Zhang, J. F., Zen, K., Zhang, C. Y., & Chen, X. (2013). Nuclear microRNAs and their unconventional role in regulating non-coding RNAs. *Protein & Cell, 4*(5), 325–330.

Liang, R. Q., Li, W., Li, Y., Tan, C. Y., Li, J. X., Jin, Y. X., et al. (2005). An oligonucleotide microarray for microRNA expression analysis based on labeling RNA with quantum dot and nanogold probe. *Nucleic Acids Research, 33*(2).

Licatalosi, D. D., Mele, A., Fak, J. J., Ule, J., Kayikci, M., Chi, S. W., et al. (2008). HITS-CLIP yields genome-wide insights into brain alternative RNA processing. *Nature, 456*(7221). 464–U422.

Lim, L. P., Lau, N. C., Garrett-Engele, P., Grimson, A., Schelter, J. M., Castle, J., et al. (2005). Microarray analysis shows that some microRNAs downregulate large numbers of target mRNAs. *Nature, 433*(7027), 769–773.

Lin, S. L., Chang, D. C., Chang-Lin, S., Lin, C. H., Wu, D. T. S., Chen, D. T., et al. (2008). Mir-302 reprograms human skin cancer cells into a pluripotent ES-cell-like state. *RNA - A Publication of the RNA Society, 14*(10), 2115–2124.

Lindow, M., & Gorodkin, J. (2007). Principles and limitations of computational microRNA gene and target finding. *DNA and Cell Biology, 26*(5), 339–351.

Ling, S. K., Birnbaum, Y., Nanhwan, M. K., Thomas, B., Bajaj, M., Li, Y., et al. (2013). Dickkopf-1 (DKK1) phosphatase and tensin homolog on chromosome 10 (PTEN) crosstalk via microRNA interference in the diabetic heart. *Basic Research in Cardiology, 108*(3).

Linsley, P. S., Schelter, J., Burchard, J., Kibukawa, M., Martin, M. M., Bartz, S. R., et al. (2007). Transcripts targeted by the MicroRNA-16 family cooperatively regulate cell cycle progression. *Molecular and Cellular Biology, 27*(6), 2240–2252.

Liu, C., Mallick, B., Long, D., Rennie, W. A., Wolenc, A., Carmack, C. S., et al. (2013). CLIP-based prediction of mammalian microRNA binding sites. *Nucleic Acids Research, 41*(14), e138.

Liu, C. G., Calin, G. A., Volinia, S., & Croce, C. M. (2008a). MicroRNA expression profiling using microarrays. *Nature Protocols, 3*(4), 563–578.

Liu, J. (2008). Control of protein synthesis and mRNA degradation by microRNAs. *Current Opinion in Cell Biology, 20*(2), 214–221.

Liu, Z., Sall, A., & Yang, D. C. (2008b). MicroRNA: An emerging therapeutic target and intervention tool. *International Journal of Molecular Sciences, 9*(6), 978–999.

Loewer, S., Cabili, M. N., Guttman, M., Loh, Y. H., Thomas, K., Park, I. H., et al. (2010). Large intergenic non-coding RNA-RoR modulates reprogramming of human induced pluripotent stem cells. *Nature Genetics, 42*(12), 1113. -+.

Lossner, C., Meier, J., Warnken, U., Rogers, M. A., Lichter, P., Pscherer, A., et al. (2011). Quantitative proteomics identify novel miR-155 target proteins. *PLoS One, 6*(7), e22146.

Lund, E., Guttinger, S., Calado, A., Dahlberg, J. E., & Kutay, U. (2004). Nuclear export of microRNA precursors. *Science, 303*(5654), 95–98.

Mallanna, S. K., & Rizzino, A. (2010). Emerging roles of microRNAs in the control of embryonic stem cells and the generation of induced pluripotent stem cells. *Developmental Biology, 344*(1), 16–25.

Mantenhorst, E., Danen, E. H. J., Smith, L., Snoek, M., Lepoole, I. C., Vanmuijen, G. N. P., et al. (1995). Expression of Cd44 splice variants in human cutaneous melanoma and melanoma cell-lines is related to tumor progression and metastatic potential. *International Journal of Cancer, 64*(3), 182–188.

Marco, A., Macpherson, J. I., Ronshaugen, M., & Griffiths-Jones, S. (2012). MicroRNAs from the same precursor have different targeting properties. *Silence, 3*(1), 8.

Margolis, R. U., & Margolis, R. K. (1994). Aggrecan-versican-neurocan family proteoglycans. *Methods in Enzymology, 245*, 105–126.

Marques, A. C., Tan, J., Lee, S., Kong, L., Heger, A., & Ponting, C. P. (2012). Evidence for conserved post-transcriptional roles of unitary pseudogenes and for frequent bifunctionality of mRNAs. *Genome Biology, 13*(11), R102.

Marques, A. C., Tan, J., & Ponting, C. P. (2011). Wrangling for microRNAs provokes much crosstalk. *Genome Biology, 12*(11), 132.

Marson, A., Levine, S. S., Cole, M. F., Frampton, G. M., Brambrink, T., Johnstone, S., et al. (2008). Connecting microRNA genes to the core transcriptional regulatory circuitry of embryonic stem cells. *Cell, 134*(3), 521–533.

Martin, G., Schouest, K., Kovvuru, P., & Spillane, C. (2007). Prediction and validation of microRNA targets in animal genomes. *Journal of Biosciences, 32*(6), 1049–1052.

Matouk, I. J., Abbasi, I., Hochberg, A., Galun, E., Dweik, H., & Akkawi, M. (2009). Highly upregulated in liver cancer noncoding RNA is overexpressed in hepatic colorectal metastasis. *European Journal of Gastroenterology & Hepatology, 21*(6), 688–692.

Matouk, I. J., DeGroot, N., Mezan, S., Ayesh, S., Abu-Iail, R., Hochberg, A., et al. (2007). The H19 non-coding RNA is essential for human tumor growth. *PLoS One, 2*(9).

Maziere, P., & Enright, A. J. (2007). Prediction of microRNA targets. *Drug Discovery Today, 12*(11–12), 452–458.

McLaughlin, J., Cheng, D., Singer, O., Lukacs, R. U., Radu, C. G., Verma, I. M., et al. (2008). Sustained suppression of Bcr-Abl-driven lymphoid leukemia by microRNA mimics (vol. 104, pg 20501, 2007). *Proceedings of the National Academy of Sciences of the United States of America, 105*(5). 1774–1774.

Meister, G., Landthaler, M., Patkaniowska, A., Dorsett, Y., Teng, G., & Tuschl, T. (2004). Human Argonaute2 mediates RNA cleavage targeted by miRNAs and siRNAs. *Molecular Cell, 15*(2), 185–197.

Memczak, S., Jens, M., Elefsinioti, A., Torti, F., Krueger, J., Rybak, A., et al. (2013). Circular RNAs are a large class of animal RNAs with regulatory potency. *Nature, 495*(7441), 333–338.

Mendell, J. T. (2008). miRiad roles for the miR-17-92 cluster in development and disease. *Cell, 133*(2), 217–222.

Merchan, F., Boualem, A., Crespi, M., & Frugier, F. (2009). Plant polycistronic precursors containing non-homologous microRNAs target transcripts encoding functionally related proteins. *Genome Biology, 10*(12).

Min, H., & Yoon, S. (2010). Got target? Computational methods for microRNA target prediction and their extension. *Experimental & Molecular Medicine, 42*(4), 233–244.

Miranda, K. C., Huynh, T., Tay, Y., Ang, Y. S., Tam, W. L., Thomson, A. M., et al. (2006). A pattern-based method for the identification of microRNA binding sites and their corresponding heteroduplexes. *Cell, 126*(6), 1203–1217.

Miska, E. A., Alvarez-Saavedra, E., Abbott, A. L., Lau, N. C., Hellman, A. B., McGonagle, S. M., et al. (2007). Most *Caenorhabditis elegans* microRNAs are individually not essential for development or viability. *PLoS Genetics, 3*(12), 2395–2403.

Miska, E. A., Alvarez-Saavedra, E., Townsend, M., Yoshii, A., Sestan, N., Rakic, P., et al. (2004). Microarray analysis of microRNA expression in the developing mammalian brain. *Genome Biology, 5*(9), R68.

Mohamed, J. S., Gaughwin, P. M., Lim, B., Robson, P., & Lipovich, L. (2010). Conserved long noncoding RNAs transcriptionally regulated by Oct4 and Nanog modulate pluripotency in mouse embryonic stem cells. *RNA - A Publication of the RNA Society, 16*(2), 324–337.

Mondino, A., Khoruts, A., & Jenkins, M. K. (1996). The anatomy of T-cell activation and tolerance. *Proceedings of the National Academy of Sciences of the United States of America, 93*(6), 2245–2252.

Moss, E. G. (2007). Heterochronic genes and the nature of developmental time. *Current Biology, 17*(11), R425–R434.

Mukherji, S., Ebert, M. S., Zheng, G. X. Y., Tsang, J. S., Sharp, P. A., & van Oudenaarden, A. (2011). MicroRNAs can generate thresholds in target gene expression. *Nature Genetics, 43*(9), 854–U860.

Muljo, S. A., Ansel, K. M., Kanellopoulou, C., Livingston, D. M., Rao, A., & Rajewsky, K. (2005). Aberrant T cell differentiation in the absence of Dicer. *Journal of Experimental Medicine, 202*(2), 261–269.

Na, E. S., & Monteggia, L. M. (2011). The role of MeCP2 in CNS development and function. *Hormones and Behavior, 59*(3), 364–368.

Nelson, P. T., Baldwin, D. A., Scearce, L. M., Oberholtzer, J. C., Tobias, J. W., & Mourelatos, Z. (2004). Microarray-based, high-throughput gene expression profiling of microRNAs. *Nature Methods, 1*(2), 155–161.

Nie, Z. M., Zhou, F., Li, D., Lv, Z. B., Chen, J., Liu, Y., et al. (2013). RIP-seq of BmAgo2-associated small RNAs reveal various types of small non-coding RNAs in the silkworm, *Bombyx mori. BMC Genomics, 14.*

Niranjanakumari, S., Lasda, E., Brazas, R., & Garcia-Blanco, M. A. (2002). Reversible crosslinking combined with immunoprecipitation to study RNA–protein interactions in vivo. *Methods, 26*(2), 182–190.

O'Carroll, D., Mecklenbrauker, I., Das, P. P., Santana, A., Koenig, U., Enright, A. J., et al. (2007). A Slicer-independent role for Argonaute 2 in hematopoiesis and the microRNA pathway. *Genes & Development, 21*(16), 1999–2004.

Ohshima, K., Hattori, M., Yada, T., Gojobori, T., Sakaki, Y., & Okada, N. (2003). Whole-genome screening indicates a possible burst of formation of processed pseudogenes and Alu repeats by particular L1 subfamilies in ancestral primates. *Genome Biology, 4*(11).

Okamura, K., Chung, W. J., & Lai, E. C. (2008). The long and short of inverted repeat genes in animals – microRNAs, mirtrons and hairpin RNAs. *Cell Cycle*, 7(18), 2840–2845.

Okamura, K., Ishizuka, A., Siomi, H., & Siomi, M. C. (2004). Distinct roles for argonaute proteins in small RNA-directed RNA cleavage pathways. *Genes & Development*, 18(14), 1655–1666.

Osella, M., Bosia, C., Cora, D., & Caselle, M. (2011). The role of incoherent microRNA-mediated feedforward loops in noise buffering. *PLoS Computational Biology*, 7(3).

Otaegi, G., Pollock, A., & Sun, T. (2011). An optimized sponge for microRNA miR-9 affects spinal motor neuron development in vivo. *Frontiers in Neuroscience*, 5, 146.

Panzitt, K., Tschernatsch, M. M. O., Guelly, C., Moustafa, T., Stradner, M., Strohmaier, H. M., et al. (2007). Characterization of HULC, a novel gene with striking up-regulation in hepatocellular carcinoma, as noncoding RNA. *Gastroenterology*, 132(1), 330–342.

Paraskevopoulou, M. D., Georgakilas, G., Kostoulas, N., Reczko, M., Maragkakis, M., Dalamagas, T. M., et al. (2013). DIANA-LncBase: Experimentally verified and computationally predicted microRNA targets on long non-coding RNAs. *Nucleic Acids Research*, 41(D1), D239–D245.

Pasquinelli, A. E. (2012). Non-coding RNA microRNAs and their targets: Recognition, regulation and an emerging reciprocal relationship. *Nature Reviews Genetics*, 13(4), 271–282.

Pasquinelli, A. E., Reinhart, B. J., Slack, F., Martindale, M. Q., Kuroda, M. I., Maller, B., et al. (2000). Conservation of the sequence and temporal expression of let-7 heterochronic regulatory RNA. *Nature*, 408(6808), 86–89.

Pauli, A., Rinn, J. L., & Schier, A. F. (2011). Non-coding RNAs as regulators of embryogenesis. *Nature Reviews Genetics*, 12(2), 136–149.

Penna, E., Orso, F., Cimino, D., Tenaglia, E., Lembo, A., Quaglino, E., et al. (2011). microRNA-214 contributes to melanoma tumour progression through suppression of TFAP2C. *EMBO Journal*, 30(10), 1990–2007.

Pillai, R. S., Bhattacharyya, S. N., & Filipowicz, W. (2007). Repression of protein synthesis by miRNAs: How many mechanisms? *Trends in Cell Biology*, 17(3), 118–126.

Poliseno, L., Salmena, L., Zhang, J., Carver, B., Haveman, W. J., & Pandolfi, P. P. (2010). A coding-independent function of gene and pseudogene mRNAs regulates tumour biology. *Nature*, 465(7301), 1033–1038.

Politz, J. C. R., Zhang, F., & Pederson, T. (2006). MicroRNA-206 colocalizes with ribosome-rich regions in both the nucleolus and cytoplasm of rat myogenic cells. *Proceedings of the National Academy of Sciences of the United States of America*, 103(50), 18957–18962.

Rachmilewitz, J., Goshen, R., Ariel, I., Schneider, T., de Groot, N., & Hochberg, A. (1992). Parental imprinting of the human H19 gene. *FEBS Letters*, 309(1), 25–28.

Rajewsky, N., & Socci, N. D. (2004). Computational identification of microRNA targets. *Developmental Biology*, 267(2), 529–535.

Reinhart, B. J., Slack, F. J., Basson, M., Pasquinelli, A. E., Bettinger, J. C., Rougvie, A. E., et al. (2000). The 21-nucleotide let-7 RNA regulates developmental timing in *Caenorhabditis elegans*. *Nature*, 403(6772), 901–906.

Ricciardelli, C., Sakko, A., Ween, M., Russell, D., & Horsfall, D. (2009). The biological role and regulation of versican levels in cancer. *Cancer and Metastasis Reviews*, 28(1–2), 233–245.

Rigoutsos, I. (2009). New tricks for animal microRNAs: Targeting of amino acid coding regions at conserved and nonconserved sites. *Cancer Research*, 69(8), 3245–3248.

Robb, G. B., Brown, K. M., Khurana, J., & Rana, T. M. (2005). Specific and potent RNAi in the nucleus of human cells. *Nature Structural & Molecular Biology*, 12(2), 133–137.

Roberts, P. J., & Der, C. J. (2007). Targeting the Raf-MEK-ERK mitogen-activated protein kinase cascade for the treatment of cancer. *Oncogene*, 26(22), 3291–3310.

Rodriguez, A., Griffiths-Jones, S., Ashurst, J. L., & Bradley, A. (2004). Identification of mammalian microRNA host genes and transcription units. *Genome Research*, 14(10A), 1902–1910.

Rougvie, A. E. (2005). Intrinsic and extrinsic regulators of developmental timing: From miRNAs to nutritional cues. *Development, 132*(17), 3787–3798.

Ruegger, S., & Grosshans, H. (2012). MicroRNA turnover: When, how, and why. *Trends in Biochemical Sciences, 37*(10), 436–446.

Rutnam, Z. J., & Yang, B. B. (2012). The non-coding 3′-UTR of CD44 induces metastasis by regulating extracellular matrix functions. *Journal of Cell Science, 125*(8), 2075–2085.

Rybak, A., Fuchs, H., Smirnova, L., Brandt, C., Pohl, E. E., Nitsch, R., et al. (2008). A feedback loop comprising lin-28 and let-7 controls pre-let-7 maturation during neural stem-cell commitment. *Nature Cell Biology, 10*(8), 987–993.

Saetrom, O., Snove, O., & Saetrom, P. (2005). Weighted sequence motifs as an improved seeding step in microRNA target prediction algorithms. *RNA - A Publication of the RNA Society, 11*(7), 995–1003.

Saini, H. K., Griffiths-Jones, S., & Enright, A. J. (2007). Genomic analysis of human microRNA transcripts. *Proceedings of the National Academy of Sciences of the United States of America, 104*(45), 17719–17724.

Salmena, L., Poliseno, L., Tay, Y., Kats, L., & Pandolfi, P. P. (2011). A ceRNA hypothesis: The Rosetta Stone of a hidden RNA language? *Cell, 146*(3), 353–358.

Salzman, J., Gawad, C., Wang, P. L., Lacayo, N., & Brown, P. O. (2012). Circular RNAs are the predominant transcript isoform from hundreds of human genes in diverse cell types. *PLoS One, 7*(2).

Schwarz, D. S., Hutvagner, G., Du, T., Xu, Z., Aronin, N., & Zamore, P. D. (2003). Asymmetry in the assembly of the RNAi enzyme complex. *Cell, 115*(2), 199–208.

Seitz, H. (2009). Redefining microRNA targets. *Current Biology, 19*(10), 870–873.

Selbach, M., Schwanhausser, B., Thierfelder, N., Fang, Z., Khanin, R., & Rajewsky, N. (2008). Widespread changes in protein synthesis induced by microRNAs. *Nature, 455*(7209), 58–63.

Sethupathy, P., Megraw, M., & Hatzigeorgiou, A. G. (2006). A guide through present computational approaches for the identification of mammalian microRNA targets. *Nature Methods, 3*(11), 881–886.

Sevignani, C., Calin, G. A., Siracusa, L. D., & Croce, C. M. (2006). Mammalian microRNAs: A small world for fine-tuning gene expression. *Mammalian Genome, 17*(3), 189–202.

Shahbazian, M. D., Antalffy, B., Armstrong, D. L., & Zoghbi, H. Y. (2002). Insight into Rett syndrome: MeCP2 levels display tissue- and cell-specific differences and correlate with neuronal maturation. *Human Molecular Genetics, 11*(2), 115–124.

Silber, J., Lim, D. A., Petritsch, C., Persson, A. I., Maunakea, A. K., Yu, M., et al. (2008). miR-124 and miR-137 inhibit proliferation of glioblastoma multiforme cells and induce differentiation of brain tumor stem cells. *BMC Medicine, 6.*

Sotiropoulou, G., Pampalakis, G., Lianidou, E., & Mourelatos, Z. (2009). Emerging roles of microRNAs as molecular switches in the integrated circuit of the cancer cell. *RNA - A Publication of the RNA Society, 15*(8), 1443–1461.

Stark, A., Brennecke, J., Bushati, N., Russell, R. B., & Cohen, S. M. (2005). Animal microRNAs confer robustness to gene expression and have a significant impact on 3′ UTR evolution. *Cell, 123*(6), 1133–1146.

Stark, A., Brennecke, J., Russell, R. B., & Cohen, S. M. (2003). Identification of Drosophila microRNA targets. *PLoS Biology, 1*(3), 397–409.

Su, X. Q., Xing, J. D., Wang, Z. Z., Chen, L., Cui, M., & Jiang, B. H. (2013). microRNAs and ceRNAs: RNA networks in pathogenesis of cancer. *Chinese Journal of Cancer Research, 25*(2), 235–239.

Sumazin, P., Yang, X., Chiu, H. S., Chung, W. J., Iyer, A., Llobet-Navas, D., et al. (2011). An extensive microRNA-mediated network of RNA–RNA interactions regulates established oncogenic pathways in glioblastoma. *Cell, 147*(2), 370–381.

Suzuki, H., Zuo, Y. H., Wang, J. H., Zhang, M. Q., Malhotra, A., & Mayeda, A. (2006). Characterization of RNase R-digested cellular RNA source that consists of lariat and circular RNAs from pre-mRNA splicing. *Nucleic Acids Research, 34*(8).

Swaminathan, G., Martin-Garcia, J., & Navas-Martin, S. (2013). RNA viruses and microRNAs: Challenging discoveries for the 21st century. *Physiological Genomics.*

Tan, L. P., Seinen, E., Duns, G., de Jong, D., Sibon, O. C. M., Poppema, S., et al. (2009). A high throughput experimental approach to identify miRNA targets in human cells. *Nucleic Acids Research, 37*(20).

Tang, R., Li, L. M., Zhu, D. H., Hou, D. X., Cao, T., Gu, H. W., et al. (2012). Mouse miRNA-709 directly regulates miRNA-15a/16-1 biogenesis at the posttranscriptional level in the nucleus: Evidence for a microRNA hierarchy system. *Cell Research, 22*(3), 504–515.

Tay, Y., Kats, L., Salmena, L., Weiss, D., Tan, S. M., Ala, U., et al. (2011). Coding-independent regulation of the tumor suppressor PTEN by competing endogenous mRNAs. *Cell, 147*(2), 344–357.

Thomson, D. W., Bracken, C. P., & Goodall, G. J. (2011). Experimental strategies for microRNA target identification. *Nucleic Acids Research, 39*(16), 6845–6853.

Thomson, J. M., Parker, J., Perou, C. M., & Hammond, S. M. (2004). A custom microarray platform for analysis of microRNA gene expression. *Nature Methods, 1*(1), 47–53.

Thummel, C. S. (2001). Molecular mechanisms of developmental timing in C-elegans and Drosophila. *Developmental Cell, 1*(4), 453–465.

Torrents, D., Suyama, M., Zdobnov, E., & Bork, P. (2003). A genome-wide survey of human pseudogenes. *Genome Research, 13*(12), 2559–2567.

Ule, J., Jensen, K. B., Ruggiu, M., Mele, A., Ule, A., & Darnell, R. B. (2003). CLIP identifies Nova-regulated RNA networks in the brain. *Science, 302*(5648), 1212–1215.

Ulitsky, I., & Bartel, D. P. (2013). lincRNAs: Genomics, evolution, and mechanisms. *Cell, 154*(1), 26–46.

Valadi, H., Ekstrom, K., Bossios, A., Sjostrand, M., Lee, J. J., & Lotvall, J. O. (2007). Exosome-mediated transfer of mRNAs and microRNAs is a novel mechanism of genetic exchange between cells. *Nature Cell Biology, 9*(6), 654–U672.

van Niel, G., Porto-Carreiro, I., Simoes, S., & Raposo, G. (2006). Exosomes: A common pathway for a specialized function. *Journal of Biochemistry, 140*(1), 13–21.

van Rooij, E., Sutherland, L. B., Qi, X. X., Richardson, J. A., Hill, J., & Olson, E. N. (2007). Control of stress-dependent cardiac growth and gene expression by a microRNA. *Science, 316*(5824), 575–579.

Wakiyama, M., Takimoto, K., Ohara, O., & Yokoyama, S. (2007). Let-7 microRNA-mediated mRNA deadenylation and translational repression in a mammalian cell-free system. *Genes & Development, 21*(19). 2509–2509.

Wang, J. Y., Liu, X. F., Wu, H. C., Ni, P. H., Gu, Z. D., Qiao, Y. X., et al. (2010a). CREB up-regulates long non-coding RNA, HULC expression through interaction with microRNA-372 in liver cancer. *Nucleic Acids Research, 38*(16), 5366–5383.

Wang, W. X., Wilfred, B. R., Hu, Y. L., Stromberg, A. J., & Nelson, P. T. (2010b). Anti-Argonaute RIP-Chip shows that miRNA transfections alter global patterns of mRNA recruitment to microribonucleoprotein complexes. *RNA - A Publication of the RNA Society, 16*(2), 394–404.

Wang, W. X., Wilfred, B. R., Xie, K., Jennings, M. H., Hu, Y. L., Stromberg, A. J., et al. (2010c). Individual microRNAs (miRNAs) display distinct mRNA targeting "rules". *RNA Biology, 7*(3), 373–380.

Wang, Y., Xu, Z., Jiang, J., Xu, C., Kang, J., Xiao, L., et al. (2013). Endogenous miRNA sponge lincRNA-RoR regulates Oct4, Nanog, and Sox2 in human embryonic stem cell self-renewal. *Developmental Cell, 25*(1), 69–80.

Wang, Z., Tollervey, J., Briese, M., Turner, D., & Ule, J. (2009). CLIP: Construction of cDNA libraries for high-throughput sequencing from RNAs cross-linked to proteins in vivo. *Methods, 48*(3), 287–293.

Watanabe, Y., Tomita, M., & Kanai, A. (2007). Computational methods for microRNA target prediction. *Methods in Enzymology, 427*, 65–86.

Weber, G. F., Ashkar, S., Glimcher, M. J., & Cantor, H. (1996). Receptor–ligand interaction between CD44 and osteopontin (Eta-1). *Science, 271*(5248), 509–512.

Weinmann, L., Hock, J., Ivacevic, T., Ohrt, T., Mutze, J., Schwille, P., et al. (2009). Importin 8 is a gene silencing factor that targets Argonaute proteins to distinct mRNAs. *Cell, 136*(3), 496–507.

Wienholds, E., Koudijs, M. J., van Eeden, F. J. M., Cuppen, E., & Plasterk, R. H. A. (2003). The microRNA-producing enzyme Dicer1 is essential for zebrafish development. *Nature Genetics, 35*(3), 217–218.

Wightman, B., Ha, I., & Ruvkun, G. (1993). Posttranscriptional regulation of the heterochronic gene Lin-14 by Lin-4 mediates temporal pattern-formation in *C. elegans. Cell, 75*(5), 855–862.

Witkos, T. M., Koscianska, E., & Krzyzosiak, W. J. (2011). Practical aspects of microRNA target prediction. *Current Molecular Medicine, 11*(2), 93–109.

Wu, L. G., Fan, J. H., & Belasco, J. G. (2006). MicroRNAs direct rapid deadenylation of mRNA. *Proceedings of the National Academy of Sciences of the United States of America, 103*(11), 4034–4039.

Wyman, S. K., Knouf, E. C., Parkin, R. K., Fritz, B. R., Lin, D. W., Dennis, L. M., et al. (2011). Post-transcriptional generation of miRNA variants by multiple nucleotidyl transferases contributes to miRNA transcriptome complexity. *Genome Research, 21*(9), 1450–1461.

Xiao, J. N., Yang, B. F., Lin, H. X., Lu, Y. J., Luo, X. B., & Wang, Z. G. (2007). Novel approaches for gene-specific interference manipulating actions of via microRNAS: Examination on the pacemaker channel genes HCN2 and HCN4 (Retracted article. See vol. 227, pg. 877, 2012). *Journal of Cellular Physiology, 212*(2), 285–292.

Xie, X. H., Lu, J., Kulbokas, E. J., Golub, T. R., Mootha, V., Lindblad-Toh, K., et al. (2005). Systematic discovery of regulatory motifs in human promoters and 3′ UTRs by comparison of several mammals. *Nature, 434*(7031), 338–345.

Xu, N., Papagiannakopoulos, T., Pan, G. J., Thomson, J. A., & Kosik, K. S. (2009). MicroRNA-145 regulates OCT4, SOX2, and KLF4 and represses pluripotency in human embryonic stem cells. *Cell, 137*(4), 647–658.

Yan, G. R., Xu, S. H., Tan, Z. L., Liu, L., & He, Q. Y. (2011). Global identification of miR-373-regulated genes in breast cancer by quantitative proteomics. *Proteomics, 11*(5), 912–920.

Yang, J. H., Li, J. H., Shao, P., Zhou, H., Chen, Y. Q., & Qu, L. H. (2011). starBase: A database for exploring microRNA–mRNA interaction maps from Argonaute CLIP-seq and degradome-seq data. *Nucleic Acids Research, 39*, D202–D209.

Yang, Y., Chaerkady, R., Kandasamy, K., Huang, T. C., Selvan, L. D. N., Dwivedi, S. B., et al. (2010). Identifying targets of miR-143 using a SILAC-based proteomic approach. *Molecular Biosystems, 6*(10), 1873–1882.

Yoon, H. G., Chan, D. W., Huang, Z. Q., Li, J. W., Fondell, J. D., Qin, J., et al. (2003). Purification and functional characterization of the human N-CoR complex: The roles of HDAC3, TBL1 and TBLR1. *EMBO Journal, 22*(6), 1336–1346.

Yoshimizu, T., Miroglio, A., Ripoche, M. A., Gabory, A., Vernucci, M., Riccio, A., et al. (2008). The H19 locus acts in vivo as a tumor suppressor. *Proceedings of the National Academy of Sciences of the United States of America, 105*(34), 12417–12422.

Zhang, A., Zhou, N. J., Huang, J. G., Liu, Q., Fukuda, K., Ma, D., et al. (2013). The human long non-coding RNA-RoR is a p53 repressor in response to DNA damage. *Cell Research, 23*(3), 340–350.

Zhang, L., Zhang, B. Y., Valdez, J. M., Wang, F., Ittmann, M., & Xin, L. (2010a). Dicer ablation impairs prostate stem cell activity and causes prostate atrophy. *Stem Cells, 28*(7), 1260–1269.

Zhang, Y., & Tycko, B. (1992). Monoallelic expression of the human H19 gene. *Nature Genetics, 1*(1), 40–44.

Zhang, Y., & Verbeek, F. J. (2010). Comparison and integration of target prediction algorithms for microRNA studies. *Journal of Integrative Bioinformatics, 7*(3).

Zhang, Z. D., Frankish, A., Hunt, T., Harrow, J., & Gerstein, M. (2010b). Identification and analysis of unitary pseudogenes: Historic and contemporary gene losses in humans and other primates. *Genome Biology*, *11*(3), R26.

Zhang, Z. L., Harrison, P. M., Liu, Y., & Gerstein, M. (2003). Millions of years of evolution preserved: A comprehensive catalog of the processed pseudogenes in the human genome. *Genome Research*, *13*(12), 2541–2558.

Zheng, D. Y., Zhang, Z. L., Harrison, P. M., Karro, J., Carriero, N., & Gerstein, M. (2005). Integrated pseudogene annotation for human chromosome 22: Evidence for transcription. *Journal of Molecular Biology*, *349*(1), 27–45.

Zheng, G. X. Y., Ravi, A., Calabrese, J. M., Medeiros, L. A., Kirak, O., Dennis, L. M., et al. (2011). A latent pro-survival function for the Mir-290-295 cluster in mouse embryonic stem cells. *PLoS Genetics*, 7(5).

Zhu, J. C., Sanborn, J. Z., Diekhans, M., Lowe, C. B., Pringle, T. H., & Haussler, D. (2007). Comparative genomics search for losses of long-established genes on the human lineage. *PLoS Computational Biology*, *3*(12), 2498–2509.

Zhu, Q. B., Sun, W. Y., Okano, K., Chen, Y., Zhang, N., Maeda, T., et al. (2011). Sponge transgenic mouse model reveals important roles for the microRNA-183 (miR-183)/96/182 cluster in postmitotic photoreceptors of the retina. *Journal of Biological Chemistry*, *286*(36), 31749–31760.

Zisoulis, D. G., Lovci, M. T., Wilbert, M. L., Hutt, K. R., Liang, T. Y., Pasquinelli, A. E., et al. (2010). Comprehensive discovery of endogenous Argonaute binding sites in *Caenorhabditis elegans*. *Nature Structural & Molecular Biology*, *17*(2), 173–U176.

CHAPTER FOUR

# Essential Letters in the Fungal Alphabet: ABC and MFS Transporters and Their Roles in Survival and Pathogenicity

## Michael H. Perlin[1], Jared Andrews and Su San Toh
Department of Biology, Program on Disease Evolution, University of Louisville, Louisville, KY, USA
[1]Corresponding author: e-mail address: anthersmut@louisville.edu

## Contents

*Advances in Genetics*, Volume 85
ISSN 0065-2660
http://dx.doi.org/10.1016/B978-0-12-800271-1.00004-4

## Abstract

Fungi depend heavily on their ability to exploit resources that may become available to them in their myriad of possible lifestyles. Whether this requires simple uptake of sugars as saprobes or competition for host-derived carbohydrates or peptides, fungi must rely on transporters that effectively allow the fungus to accumulate such nutrients from their environments. In other cases, fungi secrete compounds that facilitate their interactions with potential hosts and/or neutralize their competition. Finally, fungi that find themselves on the receiving end of insults, from hosts, competitors, or the overall environment are better served if they can get rid of such toxins or xenobiotics. In this chapter, we update studies on the most ubiquitous transporters, the ABC and MFS superfamilies. In addition, we discuss the importance of subsets of these proteins with particular relevance to plant pathogenic fungi. The availability of ever-increasing numbers of sequenced fungal genomes, combined with high-throughput methods for transcriptome analysis, provides insights previously inaccessible prior to the -omics era. As examples of such broader perspectives, we point to revelations about exploitive use of sugar transporters by plant pathogens, expansion of trichothecene efflux pumps in fungi that do not produce these mycotoxins, and the discovery of a fungal-specific oligopeptide transporter class that, so far, is overrepresented in the plant pathogenic fungi.

# 1. INTRODUCTION

## 1.1 Membrane-Bound Transporters

A *transporter* is an integral membrane protein that facilitates movement of macromolecules, ions, or small molecules across a biological membrane. Such proteins exist within and span the membrane across which they transport substances. The proteins may assist in the movement of substances by facilitated diffusion or active transport. The transporters can be further subdivided according to the ways in which they function. For purposes of this discussion, we will focus on primary and secondary active transporters. These proteins use chemical energy (typically from adenosine triphosphate (ATP) hydrolysis) or proton motive force, respectively, to drive transport of their target molecules. This is in contrast to transporters that act as channels or pores, which often operate by facilitated diffusion, i.e., passive transport aided by the protein. In such cases, they form openings in the membrane through which select molecules may pass. Primary and secondary transporters are also distinct from the proteins that utilize electrical potential as the driving force for transport. These latter transporters include transport proteins of the mitochondrial membrane, glucose transporters, and transporters for neurotransmitters (Lodish et al., 2000).

### 1.1.1 Primary Active Transporters

One group of transporters may be considered as primary active transporters, i.e., transporters that use chemical energy from ATP hydrolysis to facilitate transport of their target molecules. Among these, the ATP-binding cassette (ABC) superfamily permeases (transmembrane classification, 3.A.1) usually are multicomponent transporters, capable of transporting across membranes, both small molecules and macromolecules, in response to ATP hydrolysis. They transport a broad range of unrelated compounds: polysaccharides, drugs, sugars, heavy metals, oligopeptides, amino acids, and inorganic ions. They consist of transmembrane domains (TMDs, which confer specificity) and structurally conserved nucleotide-binding domains (NBDs), containing highly conserved amino acid motifs (Figure 4.1A). Regardless of the particular protein function, this latter domain is the common structural feature of all ABC proteins, performing the ATP hydrolysis that, in turn, provides the energy required for the protein activity. The structural organization of ABC transporters is more complex than that of the soluble ABC proteins and includes, for eukaryotes, examples of "full-length" and "half" transporters. The full-length transporters contain four domains, two NBDs and two TMDs, arranged in a single polypeptide. In contrast, the half transporters simply contain a TMD fused to an NBD. Many ABC proteins carry additional domains (see below). Except for the ABC-C subfamily, most subfamilies of eukaryotic ABC transporters represent both architectures, and are widely distributed among fungi. In contrast, half-size ABC-A transporters have so far only been found in the ancient fungal lineage, the chytridiomycetes (Kovalchuk and Driessen, 2010).

### 1.1.2 Secondary (Co)Transporters

Proteins involved in cotransport include several classes of integral membrane proteins. *Symporters* transport two or more ions together in the same direction, in contrast to *antiporters* which transport the ions in the opposite direction. Additionally, *uniport transport* provides transport of one ion in only one direction. The molecules to be transported usually are able to move against a concentration gradient by virtue of ion movement down the electrochemical gradient. Several molecules of each type may be transported

**(A)**

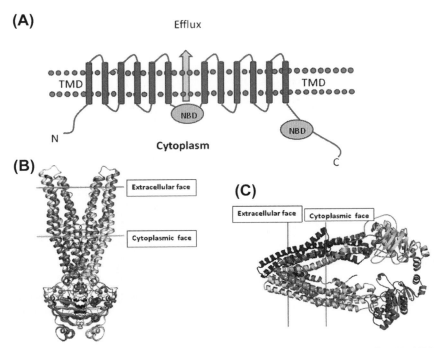

Figure 4.1 ABC Transporters. (A) Schematic of "normal" arrangement of TMD, NBD. (B) Multidrug ABC transporter SAV1866, closed state; http://pfam.sanger.ac.uk/family/PF00005. (C) P-glycoprotein ((permeability glycoprotein, abbreviated as **P-gp** or **Pgp**) also known as **multidrug resistance protein 1** (MDR1) or **ATP-binding cassette subfamily B member 1** (ABCB1)). Mouse MDR3 protein crystal stucture. http://en.wikipedia.org/wiki/File:MDR3_3g5u.png. (For color version of this figure, the reader is referred to the online version of this book.)

across the membrane via these mechanisms, and their facilitated diffusion is thereby coupled with active transport.

*The Major Facilitator Superfamily (MFS) transporters* are secondary carriers transporting small molecules in solution in response to chemiosmotic ion gradients or proton motive force (Pao, Paulsen, & Saier, 1998). Originally, MFS transporters were thought to only play a role in sugar uptake; later it was shown that additional small molecules are substrates for such proteins. At present, the list of MFS transporters includes, importantly, drug efflux systems, secretion of endogenously produced toxins (e.g., *trichothecenes*, aflatoxin, cercosporin, HC toxin) as well as metabolites of the Krebs cycle, organophosphate/phosphate exchangers, and bacterial aromatic permeases. As of this study, the MFS contains 74 families, with each usually responsible for a specific substrate type (Reddy, Shlykov, Castillo, Sun, & Saier, 2012).

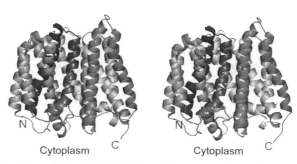

**Figure 4.2** Structure of the MFS Multidrug Transporter EmrD from *E. coli* (Yin, He, Szewczyk, Nguyen, & Chang, 2006; with permission). (For color version of this figure, the reader is referred to the online version of this book.)

The evolution of this superfamily has recently been extensively revisited (Reddy et al., 2012; http://www.tcdb.org/search/result.php?tc=2.A.1). Earlier studies suggested that a tandem intragenic duplication occurred early in the prokaryote lineage, generating a 2-transmembrane spanner (TMS) topology, which then triplicated to form the primordial 6-TMS unit (Pao et al., 1998). Possession of two 6-TMS units within a single polypeptide, through duplication, now appears to be the hallmark of all currently recognized MFS proteins (with, for example, over 180,000 sequences representing just one of the major MFS families, **PF07690**, http://pfam.sanger.ac.uk/family/PF07690; Figure 4.2), although some members contain 14 or 24 transmembrane α-helical spanners (Reddy et al., 2012). In the case of MFS permeases bearing 24 TMSs, at least some appear to have resulted from a protein fusion of two homologous MFS proteins, leading to two 12-TMS domains, with distinct function (Wood, Alizadeh, Richardson, Ferguson, & Moir, 2002).

The currently expanding wealth of genomic sequence data, transcriptomics, metabolomics, and substantial biochemical and molecular genetics investigations has revealed the occurrence of dozens of families of primary and secondary transporters. The two families found in all classifications of living organisms so far are the ABC superfamily and the MFS group, described above. Although well over 100 families of transporters have now been recognized and classified, nearly half of the transporters encoded by microorganisms are represented by the ABC superfamily and MFS transporters (Pao et al., 1998). They also occur quite often in multicellular organisms. The roles of such proteins have been examined in a variety of organisms, including fungi, plants, and animals; clearly these are a group of fundamental and particularly important proteins. For this study, we update analysis of members of the ABC and MFS superfamilies for pathogenic fungi, with a particular emphasis on plant pathogens, while drawing on recent comparative genomic and transcriptomic, as well as mutant analyses. We also provide updated information on amino acid transporters in such fungi.

## 1.2 Fungi and Their Value as Study Systems

Why do we focus on fungi? What unique aspects of fungi make them a worthy area of investigation in the description and characterization of membrane transporters? First of all, the fungi are a large and diverse kingdom within the eukaryotes, which includes the single celled yeasts, aquatic chytrids, filamentous molds, as well as the mushrooms. Though the true diversity of this group is not yet known, there are an estimated 1.5–5 million species of fungi, of which only about 5% have been formally classified (http://tolweb.org/fungi). The kingdom is further subdivided into 1 subkingdom, 7 phyla, and 10 subphyla (Ravichandra, 2013). One major difference between fungi and other eukaryotes is that fungal cells have cell walls that contain chitin, unlike the cell walls of plants and some protists, which contain cellulose, and unlike the cell walls of bacteria. Fungi form a monophyletic group, the *Eumycota* ("true fungi" or *Eumycetes*), that share a common ancestor. Although structurally similar to the oomycetes (water molds) and myxomycetes (slime molds), the fungi are a distinct group of evolutionarily related organisms. Genetic studies have shown that fungi are more closely related to animals than to plants (Shalchian-Tabrici et al., 2008).

Fungi directly benefit humans in food and industrial processes. Aside from this, at the basic science level, they are important for comparative purposes, not only because of their genetic diversity and morphological diversity but also due to the variety of their lifestyles and the biological niches in which they are found. These include cryptic lifestyles in soil, on dead matter, and as symbionts of plants, animals, or other fungi. As with most symbioses, long-term interactions between fungi and their host species can be parasitic, commensal, or mutualistic (i.e., where both partners in a mutualism benefit). Fungi perform an essential role in the decomposition of organic matter and have fundamental roles in nutrient cycling and exchange. Fungi are also used as biocontrol agents and can limit or protect against weeds, plant diseases, and insect pests. Many species produce mycotoxins, such as alkaloids and polyketides, which are toxic to plants or animals, including humans. Fungi can break down manufactured materials, and become significant pathogens of humans and other animals. Losses of crops due to fungal diseases (e.g., yellow and brown wheat rust, rice blast disease, *Fusarium* wilt, and powdery mildew of barley) or food spoilage can have strong negative consequences for human food supplies, the food industry, and local economies. For example, the wheat stem rust strain UG99 is highly virulent against most wheat cultivars and, if spread beyond its currently identified presence (several countries in Africa and the Middle East), could threaten wheat production and food security worldwide (Periyannan et al., 2013).

There has recently been some discussion as to whether certain organisms should be primarily considered as pathogens (Casadevall, Fang, & Pirofski, 2011), or whether pathogenicity/virulence should instead be considered as an emergent property. Clearly, there are some fungi for which pathogenicity is purely a matter of opportunism; many of these fungi are part of the normal flora of their associated hosts and only cause disease when the host environment is altered in such a way as to encourage the behavior on the part of the fungus that damages the host. For other fungi, there is little evidence that supports the notion of any other lifestyle in nature outside of the host and the pathway to disease. Among the pathogens, the group can be divided into those that act primarily or exclusively on animals, those whose hosts are plants, and those that parasitize other fungi.

The plant pathogenic fungi may be further subdivided into biotrophs, necrotrophs, and hemibiotrophs. Biotrophs are fungi that derive nutrients from the living tissues of their host plants. For example, *Hyaloperonospora arabidopsidis* is a species from the family Peronosporaceae. It is an obligate biotrophic parasite and the causal agent of the downy mildew of the plant model organism *Arabidopsis thaliana* (Manzoor et al., 2013). In contrast, necrotrophs kill living cells of the host and then feed on the dead matter. Hemibiotrophic fungi parasitize living tissues for a period and continue their life cycle on dead tissues. Often fungi operate with what might be termed a biphasic lifestyle. Such fungi can produce a large repertoire of secreted proteins that may serve as effector molecules. Examples of such fungi include not only strict biotrophs, such as *Ustilago maydis*, the pathogen of maize, but also the hemibiotrophic rice blast fungus, *Magnaporthe oryzae*. Further support for this idea of biphasic life cycles comes from evidence that in hemibiotrophic *Colletotrichum graminicola* plant defense is initiated during the biotrophic phase, after which the fungus switches to the necrotrophic phase (Vargas et al., 2012). Interestingly, *Piriformospora indica*, a saprobe when colonizing dead plant material, becomes a biotroph when colonizing live roots (Zuccaro et al., 2011). Further, its transcriptome mirrors that of the symbiont, *Laccaria bicolor*, when growing in living tissue, while gene expression is more similar to *Coprinus cinereus* when grown saprophytically

(Zuccaro et al., 2011). Another example is *Moniliophthora perniciosa*, the tropical fungus that causes witches' broom disease in cacao. As a hemibiotrophic fungus, it initially colonizes the living host tissues (biotrophic phase), and later grows over the dead plant (necrotrophic phase). Little is known about the mechanisms that promote these distinct fungal phases or mediate the transition between them. One candidate is the alternative oxidase (*Mp-aox*) in *M. perniciosa* (Thomazella et al., 2012). Its expression was analyzed throughout the fungal life cycle. Regulated expression of the oxidase may affect the available pool of ATP that, in turn, affects which transporters can be active.

### 1.2.1 Which Transporters might Play Different Roles in Fungi?

Together, the ABC and MFS superfamilies account for approximately half of all the identified genes encoding transporters in fungal genomes. The relative abundance of one type (i.e., ABC vs MFS) in a given species appears to follow the evolutionary trajectories of the particular fungal division/phylum. The oomycetes and ancient Chytridiomycota tend to have a significantly greater enrichment for ABC superfamily proteins, whereas the zygomycetes, ascomycetes, and basidiomycetes overall have fewer ABC representatives and a corresponding expansion of MFS transporters (Coleman and Mylonakis, 2009, Table 2). Despite their importance in virulence, there is no apparent correlation between the quantity of these transporters in fungal genomes and the pathogenicity of the fungus. For instance, *Saccharomyces cerevisiae*, the model saprobe, and its closely related cousin, *Candida albicans*, the opportunistic human pathogen, have the same number of genes encoding MFS proteins (85), and nearly the same number of genes encoding ABC proteins (24 vs 21, respectively) (Coleman and Mylonakis, 2009). In this chapter, we will examine the roles of different transporters in fungi, with particular emphasis on comparing the transporters more likely to be utilized by fungi characterized as pathogens. Within this group, we will focus most on plant pathogens, but will use animal pathogens and nonpathogenic species for purposes of comparison. Further, we plan to subdivide the analysis, where possible, to also explore whether there are transporters more often utilized by biotrophs rather than necrotrophs.

## 2. ABC TRANSPORTER SUPERFAMILY

Originally, the fungal ABC transporters were studied using either *S. cerevisiae* or *Schizosaccharomyces pombe*. Since then our understanding of the diversity and functions of this superfamily of transporters has expanded greatly as the genomes of fungi have become increasingly available. As the first eukaryote whose genome was sequenced, *S. cerevisiae* provided researchers with a peek at the complete list of ABC proteins for an organism (Decottignies & Goffeau, 1997). A wide array of physiological functions have now been determined for yeast ABC transporters, including fatty acid metabolism, protection against toxic compounds, and even secretion of mating pheromone. Moreover, there are ABC proteins found to be involved in mRNA translation and ribosome biogenesis, i.e., not just in membrane transport.

Despite this encouraging advance in information, few ABC proteins from other fungal species have been functionally characterized. The model filamentous fungus, *Emericella nidulans* (anamorph *Aspergillus nidulans*) (do Nascimento, Goldman, & Goldman, 2002) is an exception. Mainly, characterization has been limited to those ABC transporters reported to be involved in multidrug resistance (MDR) in pathogens of medical or agricultural importance (e.g., *Cryptococcus neoformans* (Sanguinetti et al., 2006), *C. albicans* (Gaur et al., 2005), *Aspergillus fumigatus* (Slaven et al., 2002), and the plant pathogen *Magnaporthe grisea* (Sun, Suresh, Deng, & Naqvi, 2006; Urban, Bhargava, & Hamer, 1999). Fortunately, the availability of genomic sequences for a growing number of fungi provided the opportunity for phylogenetic analyses that may shed additional light on the evolution and function(s) of these proteins (Kovalchuk and Driessen, 2010; Kovalchuk, Lee, & Asiegbu, 2013).

There are seven sub families of ABC transporters (A–G), as well as other group(s) of ABC proteins that do not fit into the other subfamilies, but retain the other characteristic components of the ABC superfamily. The ABC-A group may not be essential for viability in most fungi. This conclusion stems from the observation that they are notably absent from most of the fungal genomes so far analyzed (Kovalchuk and Driessen, 2010). Of course, this does not rule out their importance under certain conditions. Indeed, the only fungal ABC-A transporter which was functionally characterized so far, the *M. grisea* Abc4 protein (MGG_00937), is required for pathogenesis and appressoria formation (Gupta and Chatoo, 2008). In addition, all the ascomycetes containing two full-length ABC-A–encoding genes (*M. grisea*, *Botryotinia fuckeliana*, and *Phaeosphaeria nodorum*) turn out to be plant pathogens. Based on comparisons with mammalian homologs, there may be a role for such proteins in lipid transport and metabolism (Wenzel, Piehler, & Kaminski, 2007). Among basidiomycetes, full-length ABC-A transporters were identified in most of the species of Agaricomycotina and in *U. maydis*. Most of the analyzed species carry a single

ABC-A gene, but two are found in a few basidiomycete species (Kovalchuk et al., 2013). Both the highest number and diversity of ABC-A proteins is observed in the ancient chytrid lineages, while they are mostly lacking in the more recent branches of the fungal tree. Apparently, in this instance (genome) size doesn't matter: these genes are absent both from the compact genomes of yeast-like species, but also from much larger genomes (e.g., *Rhizopus oryzae* and *Puccinia graminis*).

In the discussion that follows, we first emphasize those subfamilies whose members contribute to multiple drug resistance phenotypes. A model for the mode of action of these types of ABC transporters has been proposed (Niimi, 2010), whereby drugs that have diffused across the membrane into the cell interact with the drug-binding sites of the transporters. ATP binding at the NBD dimer causes a conformational change in the TMDs, shifting the face of the drug-binding regions toward the outside of the cell. Then, ATP hydrolysis generates further conformational changes to complete the transport process; release of ADP and $P_i$ allows the return of the ABC transporter to the original state, ready to transport additional drug out of the cell.

In some cases, fungi may have a large repertoire of transporters from multiple categories. In the absence of high-throughput gene disruption protocols, establishing the direct roles for each transporter can thus require labor-intensive and time-consuming efforts. For example, the environmental, opportunistic human pathogen, *C. neoformans* var. *grubii*, appears to have an enriched capacity for transport, given its predicted 159 MFS and 54 ABC transporters (Cannon & Lamping, 2009). Yet, so far, only two ABC transporters have been directly linked to drug resistance for this species. The first, CneAfr1p, is a member of the pleiotropic drug resistance (PDR) family (see below), showing homology to *A. nidulans* AtrB. Overexpression of the *C. neoformans* protein leads to azole resistance, increased virulence in mouse models, and improved survival during in vitro macrophage infection. The second transporter, CneMdr1p, of the MDR-type (see below) is not associated with azole resistance in *C. neoformans*, at least not insofar as expression levels are concerned. However, overexpression of this protein (or CneAfr1p) in *S. cerevisiae* did provide resistance to azoles; interestingly, though the two *C. neoformans* proteins tested in this heterologous system have different membrane topologies, the resistance profiles of the overexpressied *S. cerevisiae* strains were quite similar (Cannon & Lamping, 2009). These examples illustrate the difficulty of understanding the full import of specific types of ABC (or other transporters) in fungi. In the sections that follow, the classes of ABC transporters are further discussed in greater detail.

The ABC proteins can be found with a wide array of "architectures" (http://pfam.sanger.ac.uk). By this we mean that different subsets of the ABC proteins contain different inventories and arrangements of Pfam domains. Analysis of these architectures for patterns among the fungi that possess them might provide insights as to their possible roles in particular fungal lifestyles. As seen in Table 4.1, some of the architectures are associated with ABC proteins found in pathogenic fungi. In most cases, the majority of such sequences in pathogens are for those that infect animals (including insects and nematodes). On the other hand, in some cases, a particular architecture is dominated by proteins that are employed by the oomycetes (i.e., *Phytophtora* sp.).

## 2.1 Pleiotropic Drug Resistance = ABC-G

This subfamily contains among its members, those associated with PDR (3.A.1.205). In the nomenclature of the Transporter Classification Database (http://www.tcdb.org/), the last number added to the subfamily designation corresponds to the substrate or range of substrates (e.g., 3.A.1.205.1 refers to the PDR or ABC-G transporters, such as Cdr1p or Cdr2p from *C. albicans*, associated with azole resistance (Cannon & Lamping, 2009; Lamping et al., 2010). As the target for such drugs is the cytochrome P450 (Yoshida, 1988), the PDR transporters thus provide a way to expel the drug prior to reaching its target. The characteristic feature distinguishing ABC-G transporters from other subfamilies is their reverse topology, i.e., the NBD precedes the TMD (opposite to what is shown in Figure 4.1A). Full-length ABC-G transporters are often referred to as PDR transporters. Although both full-length and half-size members of this subfamily are known, full-length ABC-G proteins are apparently absent from animal genomes. Recently, Kovalchuk et al. (2013; also, see Kovalchuk and Dreissen, 2010) conducted a thorough phylogenetic analysis of the ABC-G transporters. The half-size ABC-G proteins were first characterized as the *Drosophila melanogaster* white–brown complex transporters required for eye pigment formation, and serve as the characteristic members (Dreesen, Johnson, & Henikoff, 1988).

Among the basidiomycetes, the ABC-G group appears to be the most common; often, they are found in plant pathogens, but are also present in some animal pathogens (e.g., 8 in *C. neoformans*, and 2 in *M. globossa*). By way of comparison, the *S. cerevisiae* genome appears to encode 10 ABC-G transporters. Of these, five are involved in export of various chemically unrelated hydrophobic molecules, and thus, these PDR proteins contribute to resistance. ABC-G transporters described from other fungal species are an

**Table 4.1** Architectures of ABC Transporter Proteins of Fungi

| Architecture[a] | Family | Total #Sequences | #Plant Pathogens | #Animal Pathogens | Predominant Species |
|---|---|---|---|---|---|
| ABC_membrane, ABC_tran, ABC_membrane, ABC_tran | ABC-B | 4608 | | | |
| ABC_tran, ABC_tran, Chromo | ABC-F | 140 | | | |
| ABC_trans_N, ABC_tran, ABC2_membrane, PDR_CDR, ABC_tran, ABC2_membrane | ABC-G | 575 | 141 | 203 | P: *Botrytis, Nectria, Fusarium* A: *Candida, Aspergillus* |
| ABC_tran, ABC2_membrane, ABC_tran, ABC2_membrane | ABC-G | 237 | 11 | 37 | 94 Oomycete/*Phytophtora* sp.; *C. neoformans, C. glabrata, A. oryzae, A. terreus* |
| ABC_tran, ABC2_membrane, PDR_CDR, ABC_tran, ABC2_membrane | ABC-G | 211 | 44 | 76 | |
| ABC_trans_N, ABC_tran, ABC2_membrane, PDR_CDR, ABC_tran, ABC2_membrane, PDR_CDR | ABC-G | 103 | 20 | 39 | |
| ABC_tran, ABC2_membrane, ABC_tran, ABC2_membrane, PDR_CDR | ABC-G | 43 | 43 | 0 | Exclusively *Phytophtora* sp. |
| PDR_CDR | ABC-G | 23 | 22 | 1 | Mostly *Phytophtora* sp., but also *Botrytis* and *Sclerotinia* |
| ABC2_membrane, PDR_CDR, ABC_tran, ABC2_membrane | ABC-G | 16 | 7 | 3 | |

ABC_trans_N = PF14510; ABC_tran = PF00005; ABC2_membrane = PF01061; PDR_CDR = PF06422, Chromo = PF00385.
[a]Architectures predicted as predicted by www.sanger.ac.uk.

important component of drug resistance/efflux (Coleman and Mylonakis, 2009). However, PDR transporters might also be important in the translocation of various lipid molecules, as shown for *C. albicans* transporters (Sipos and Kuchler, 2006).

The PDR transporters appear to be the least conserved among fungal ABC proteins, suggesting their rapid evolution after the divergence of the main fungal lineages. Kovalchuk and Driessen (2010) identified five clusters of PDR transporters. All of them, except for Group IV, are restricted to the genomes of higher fungi (ascomycetes and basidiomycetes). However, as the phylogenies from these analyses appeared somewhat messy with the inclusion of additional basidiomycete isolates, Kovalchuk et al. (2013) redefined the ABC-G transporters to be divided into three principal groups: 1) PDR transporters, 2) proteins related to *S. cerevisiae* Yol075cp, and 3) proteins related to *S. cerevisiae* protein Adp1p. This last group is unique among fungal ABC transporters, in that its architecture includes a large *N*-terminal domain (PF07974) with epidermal growth factor-like repeats that are likely found outside the cell (Kovalchuk et al., 2013). While this domain match is not statistically significant in analyses done by NCBI or the Pfam website (http://pfam.sanger.ac.uk), it, along with the other match (Laminin_EGF-like, PF00053), suggests that this region found in other fungal ABC-G proteins would be ready to engage in protein–protein interactions outside the cell.

The largest PDR group in the phylogenetic analysis (Kovalchuk et al., 2013) contains more than 70 members. With the exception of *S. pombe*, all ascomycetes examined contained representatives of this group, while they are almost entirely missing in basidiomycetes (it is represented only by eight transporters from five species). There appears to have been an expansion of these genes in the genomes of *Penicillium chrysogenum* and *Aspergillus* species. This group includes well-characterized multidrug transporters (e.g., *S. cerevisiae* Pdr5p, Pdr10p, and Pdr15p proteins and *C. albicans* Cdr1p, Cdr2p, Cdr3p, and Cdr4p proteins). Indeed, azole resistance in clinical isolates of *C. albicans* is often associated with overexpression of both the normal target of the drugs (Erg11p) and the ABC pumps Cdr1p and Cdr2p. Often such changes in gene expression occur through aneuploidy as a result of exposure to antimicrobials or growth on different carbon sources (Cannon & Lamping, 2009). Additional factors influencing such overexpression include mutations in the promoter sites of the genes and/or alterations in the expression or function of transcription factors that normally regulate their expression.

The group of proteins related to *S. cerevisiae* Yol075cp is unusual for a variety of reasons. It is the lone group containing both full-length and half-size transporters; both types appear to be closely related. For *S. cerevisiae*, only the uncharacterized transporter, Yol075cp itself, is found in this group. Of the 28 basidiomycete species analyzed, 15 contained members belonging to this group, with most species containing a single representative. Four of the species harbor both full-length and half-size proteins, while seven species (*C. neoformans*, *Tremella mesenterica*, *Phlebia brevispora*, *Fomitopsis pinicola*, *Gloeophyllum trabeum*, *Stereum hirsutum*, and *Schizophyllum commune*) possess only full-length proteins; over half of these species are pathogens on some type of host, e.g., *T. mesenterica* is a pathogen on *S. hirsutum*, which in turn is a plant pathogen. Four species (*Coprinopsis cinerea*, *Fomitiporia mediterranea*, *U. maydis*, and *Melampsora laricis-populina*) carry only half-size transporters, and interestingly, three of these are plant pathogens.

### 2.1.1 Expression Analysis and Mutants

Expression of the PDR fungal transporters is often induced by a number of phytoalexins and other toxic compounds, despite the fact that only a few of the transporters have been shown to confer tolerance to any known phytoalexins (Pedras & Minic, 2012). Similarly, although *ABC1* of the rice blast fungus, *M. grisea*, is also strongly induced by the rice phytoalexin sakurane-tin, as well as by various fungicides (Urban et al., 1999), disruption mutants do not display increased sensitivity to these compounds. Interestingly, the mutants are blocked in disease progression shortly after penetrating rice.

The causative agent of gray mold and noble rot of grapes, *Botrytis cinerea* (teleomorph: *B. fuckeliana*), has a reputation for easily acquiring resistance to fungicides (Weber and Entrop, 2011). The roles of ABC and MFS transporters in resistance to sterol demethylation inhibitor fungicides has been investigated by comparing gene expression in wild-type and resistant isolates; in addition, disruption and overexpression mutants provided another perspective in evaluating the importance of the various transporters in susceptibility. The ABC transporters BcatrB in *B. cinerea* and GpABC1 in *Gibberella pulicaris* provide tolerance to the phytoalexins, resveratrol and rishitin, respectively (Coleman, White, Rodriguez-Carres, & Vanetten, 2011). Mutants in these genes have reduced pathogenicity and loss of virulence on the respective hosts for these plant pathogens (Fleissner, Sopalla, & Weltring, 2002). BcatrB of *B. cinerea* is also able to provide resistance to the fungicides fenpiclonil and fludioxonil in addition to resveratrol, and appears to be a major transporter of these fungicides, although the MFS transporter, Bcmfs1 (see below), may also play a minor role in transport of the demethylation inhibition fungicides (Schoonbeek, Del Sorbo, & De Waard, 2001). Moreover, while the wheat pathogen, *Mycosphaerella graminicola*, is known to possess a number of ABC transporters (Zwiers, Stergiopoulos, Gielkens, Goodall, & De Waard, 2003), only one of the five characterized genes, *MgAtr5*, seemed to be required for reduced sensitivity to resorcinol, a putative wheat defense compound, and to resveratrol. Yet, each of the five genes (*MgAtr1–5*), when expressed in the heterologous *S. cerevisiae* background, could confer protection against a variety of toxins, fungicides, and a mycotoxin.

Showing the importance of efflux transporters in the development of resistance to fungicides, Reimann and Deising (2005) found in *Pyrenophora tritici-repentis*, the fungus responsible for wheat tan spot, that both adaptation to higher levels of azole fungicides and the capability to infect plants so treated depended on expression of the efflux genes and localization of their products to the plasma membrane. Importantly, treatment of the adapted isolates with a potent hydroxyflavone derivative inhibitor of energy-dependent efflux transporters returned the fungi to their native sensitivity to the fungicides.

As suggested above (Zwiers et al., 2003), an important approach in functional analysis of efflux transporters has been heterologous expression in *S. cerevisiae* (Lamping and Cannon, 2010; Niimi et al., 2005; Niimi, 2010). For example, the importance of several domains of the *Candida glabrata* PDR transporter, CgCdr1p, was examined using an *S. cerevisiae* expression system (Lamping and Cannon, 2010). The studies allowed examination of the role of specific amino acid residues in the NBD domains, as well as the possible phosphorylation states of residues in and nearby the NBDs; the latter appeared important for glucose-dependent efflux of rhodamine 6G efflux in such recombinant strains (Wada et al., 2005).

### *2.1.2 Regulation of Efflux Pumps at the Transcriptional Level*

As hinted in the above and ongoing discussion later, one additional component in the emergence of resistance to diverse antimicrobials is changes in the levels of expression of efflux transporters. Such changes are not due to changes in the transporter genes themselves, but in those aspects that regulate their transcription: e.g., changes in their promoters, or gain-of-function mutations in transcription factors that control their expression. Control of such efflux pumps involved in PDR has been best studied in *S. cerevisiae*, where mutants of the Gal4-like transcription factors, ScPdr1p and ScPdr3p, result in activation of over 20 genes, most of which are efflux transporters of the ABC and MFS superfamilies (Cannon & Lamping, 2009). Genes controlled by these proteins contain palindromic octamer consensus binding sites known as PDR elements. Similarly, in *C. albicans* (and related *Candida* pathogenic species), there is a zinc finger transcription factor, CaTac1p, that controls expression of the ABC transporters CaCdr1p and CaCdr2p; CaTac1p gain-of-function mutations are associated with high-level drug resistance (Cannon & Lamping, 2009). Similar modes of regulation for PDR genes in other pathogenic fungi may also occur (Cannon & Lamping, 2009).

## 2.2 Multidrug Resistance = ABC-B

ABC-B proteins include both full and half-size transporters. This large group is widely distributed among eukaryotes, and they are functionally diverse. The most well-known members of this group are mammalian multidrug transporters, such as P-glycoprotein (Pgp) that is involved in drug resistance in cancer cells (Figure 4.1C). In fungi, members of this group are involved in pheromone export, MDR, mitochondrial peptides, and heavy metal resistance, among other functions. A large number (4608) of ABC-B transporters have the following architecture: **ABC_membrane, ABC_tran, ABC_membrane, ABC_tran** (Table 4.1).

Fungal full-size ABC-B transporters have been studied with the *S. cerevisiae* protein, Ste6p (YKL209c), the best characterized member of this group, representing the first ABC transporter identified in yeast (Kuchler, Sterne, & Thorner, 1989). Ste6p (and its homologs) is haploid specific, and is required for the export of a-factor, the farnesylated peptide pheromone in this and related fungi. The role of Ste6p homologs in the secretion of prenylated lipopeptides is evolutionarily highly conserved. Even in *Drosophila*, Mdr49, the Ste6p counterpart, is also required for the export of geranylated germ cell attractant (Ricardo and Lehmann, 2009). It should not be surprising that the corresponding proteins of basidiomycetes also have a role in the export of mating pheromones. Pheromone transporters were identified in all analyzed species of basidiomycetes, with most bearing a single copy of the gene (Kovalchuk et al., 2013).

Other ABC-B group members may play roles in pathogenesis. The *A. fumigatus* AbcB (Afu3g03430) protein was shown to be involved in the excretion of siderophore peptide breakdown products (Kragl et al., 2007). Since a number of fungal ABC transporters are closely associated with non-ribosomal peptide synthetase (NRPS) clusters (von Dohren, 2009) and, might have a role in the secretion of peptide-like secondary metabolites, the fact that AbcB is used for excretion of siderophore peptide breakdown products makes for a particularly tantalizing connection. The *M. grisea* Abc3 transporter (MGG_13762) is involved in the oxidative stress response, and is required for host penetration (Sun et al., 2006). Other fungal full-length ABC-B proteins are involved in MDR, as well: e.g., *S. pombe* Pmd1, *A. fumigatus* Mdr1, or *A. nidulans* AtrD (Andrade, Del Sorbo, Van Nistelrooy, & Waard, 2000; Kovalchuk et al., 2013). But there is more! One possible function of some members of this group is in efflux of mycotoxins (e.g., in β-lactam secretion) produced by the fungus. In fact, mutants of the transporter AtrD in *A. nidulans* produce less penicillin than the wild type, consistent with its role in efflux of this compound (Andrade et al., 2000).

Proteins showing similarity with the *S. pombe* Pmd1 (SPCC663.03) transporter form another cluster found in the genomes of most of the analyzed species (Kovalchuk and Driessen, 2010; Kovalchuk et al., 2013), with basidiomycetes typically having 1–4 copies. All fungal ABC-B proteins with a known function in MDR belong to this group, but the exact function of the majority of these proteins is unknown. Members of this group were found in the analyzed species of Agaricomycotina and Ustilaginomycotina but are missing from the genomes of the plant pathogenic rust fungi, *P. graminis* and *M. laricis-populina* (Kovalchuk et al., 2013). However, several proteins from the genomes of the pathogens, *U. maydis* (UM05114, UM05096) and *Malasezzia globosa* (MGL_1584, MGL_2084), appeared to be more related to the ascomycete transporters. In particular, UM05114 and MGL_1584 are closely related to the transporters of ascomycetes associated with the gene cluster for the biosynthesis of the hydroxamate-type siderophore, fusarinine (Kovalchuk and Driessen, 2010). Coincidentally, most of Agaricomycotina lack the enzymes necessary for synthesis of hydroxamate-type siderophores (Haas, Eisendle, & Turgeon, 2008), highlighting the fundamental differences in iron uptake between the ascomycetes and the basidiomycetes (Kovalchuk et al., 2013).

The importance in pathogenicity of two ABC transporters from *M. oryzae* was discovered via insertional mutagenesis (Gupta and Chattoo, 2008; Sun et al., 2006). Abc3 is required for host penetration and survival during oxidative stress (Sun et al., 2006). Similarly, the *A. fumigatus* AfuMDR4 was identified as a constitutively overexpressed or induced protein in resistant mutants exposed to the triazole drug, itraconazole (do Nascimento et al., 2003). Interestingly, the predicted ABC transporter (MVLG_04305) of *Microbotryum violaceum*, the anther smut on wildflowers (e.g., *Silene latifolia*), was found to be downregulated in infected flowers (S. S. Toh, unpublished). Moreover, *Trichophyton rubrum* (the dermatophyte pathogen) shows the upregulation of various transporter proteins as a response to keratin

(Maranhão, Paião, Fachin, & Martinez-Rossi, 2009). ABC transporters, copper ATPase, an MFS, and a permease were all identified from a subtraction suppression hybridization library. The ABC transporter, TruMDR2, was found to be required for full virulence on human nails.

## 2.3 Multidrug Resistance Associated Protein = ABC-C

The subfamily ABC-C is the largest group of full-length transporters in most of the analyzed basidiomycetes, ranging from five in the pathogens, *M. globosa* and *M. laricis-populina*, to 26 in the white rot fungus, *P. brevispora* (Kovalchuk and Driessen, 2010; Kovalchuk et al., 2013). Fungal ABC-C transporters can be subdivided into seven groups. One of the ABC-C groups is specific for *Pezizomycotina* and basidiomycetes, and there is a single corresponding gene in the genomes of ascomycetes and *C. neoformans*, and two genes in other basidiomycetes. Three of the analyzed species (*Coccidioides immitis*, *Chaetomium globosum*, and *M. globosa*; *all of which are pathogens of animals*) lack members of this group. There is a single ABC-C transporter, MLT1, required for full virulence of *C. albicans*. *MLT1* mutants are impaired in their ability to invade liver and pancreas in a mouse peritonitis model and cause less liver damage. Interestingly, due to the similarity of this vacuolar transporter to the bile pigment transporter, BPT1, of *S. cerevisiae*, it is tempting to suggest that MLT1 similarly protects the fungus against bilirubin and glutathione conjugates in the vacuole (Coleman and Mylonakis, 2009).

None of the proteins belonging to cluster VII of the ABC-C transporters has been characterized so far except for the *S. cerevisiae* Yor1p (YGR281W) itself, which serves as the prototypical member. This protein facilitates export of many different organic anions including oligomycin and, in addition, phospholipids. A concomitant increase in sensitivity to a variety of drugs and xenobiotics is associated with loss of Yor1 activity. Neither *R. oryzae* nor *S. pombe* appears to have Yor1 homologs, while most other ascomycetes contain a single Yor1-like transporter, except the two plant pathogens, *Gibberella zeae* and *M. grisea*, each with two homologs. Members of this group were found in all basidiomycetes except *M. laricis-populina*. Their copy number ranges from a single gene in *P. graminis* and *M. globosa* to nine genes in *Auricularia delicata*. The genomes of the basidiomycetes *U. maydis* and *C. neoformans* contain two Yor1 homologs, compared with five members in *C. cinerea*. In the two chytrid species, *Batrachochytrium dendrobatidis* and *Spizellomyces punctatus*, there appears to be an expansion of these relative to the ascomycete and basidiomycete species studied (Kovalchuk et al., 2013).

Recently, Kim et al. (2013) utilized a genome search of *M. oryzae*, the rice blast fungus. They were thereby able to provide an inventory of predicted ABC transporters in this fungus, along with phylogenetic analysis that subdivided these into 11 subfamilies. Among the 11 ABC-C subfamily members, transcripts were elevated under abiotic stresses, but also, importantly, during infection stages. Functional analysis of three of the genes (*MoABC5*, *MoABC6*, and *MoABC7*) demonstrated their roles in virulence, abiotic stress tolerance, and conidiation.

### 2.3.1 Additional ABC Families

ABC-D transporters were only absent from *Encephalitozoon cuniculi* and *S. pombe*. Their absence from *S. pombe* is an unexpected finding since a set of PEX genes required for peroxisome biogenesis was identified in the genome of *S. pombe* (Kiel, Veenhuis, & van der Klei, 2006). On the other hand, the absence of this group in *E. cuniculi* makes sense since this species lacks peroxisomes (Katinka et al., 2001). The presence of this set of ABC-D transporters in other fungi is a conserved feature.

Both the ABC-E and ABC-F proteins lack TMDs and consist solely of two NBDs. Their function within the cell is not related to transport. Nevertheless, members of these groups are clearly important since three of them (Rli1, Arb1, and Yef3) are essential for viability of *S. cerevisiae* cells (Schuller, Bauer, & Kuchler, 2003). The subfamily ABC-E is represented by a single gene in all analyzed basidiomycetes (Kovalchuk et al., 2013). ABC-F proteins are involved in ribosome/translation function. There are at least 140 sequences with the following architecture: **ABC_trans x 2, Chromo** (Table 4.1). Four groups (I, II, IV, and V) are represented by a single gene in nearly all analyzed basidiomycetes (Kovalchuk et al., 2013). The exceptions include two important plant pathogens, *Heterobasidion annosum* and *P. graminis*, that each carries two copies of ABC-F(V) genes.

In addition to the members of the well-defined subfamilies ABC-A–G, fungal genomes also contain several genes encoding ABC proteins that cannot be classified into any of the known groups; they lack TMDs, but their NBDs clearly show that they belong to the superfamily of ABC proteins. At least one group contains members of a transcriptional regulatory complex controlling mRNA initiation, elongation, and degradation (Liu et al., 2001).

In summing up this section, let us emphasize that the number and diversity of ABC proteins is huge and well dispersed among the fungi. On the other hand, the minimal set of such proteins required for a free-living organism may be represented by that of *S. pombe*, as it is among the smallest of the sequenced eukaryotic genomes (http://www.tcdb.org/). It lacks ABC-A proteins and even more remarkably, it is the only free-living species identified to date that lacks peroxisomal ABC-D transporters, although peroxisomes are present in this organism. Only three *S. cerevisiae* ABC genes (Rli1, Arb1, and Yef3) are required for cell viability under normal growth conditions (Schuller et al., 2003), and none of these are transporters. At the same time, there are no counterparts of two *S. pombe* transporters, Pmd1 and Hmt1, in the genomes of *Saccharomycetales*, indicating that life might be possible with even fewer ABC transporters.

The number of members of particular ABC subfamilies in the genomes of basidiomycetes tends to be lower than in those of ascomycetes, and *P. graminis* has the lowest number of ABC proteins per 1 Mb of genome among analyzed species (Kovalchuk et al., 2013). The Ascomycota show a trend of increase in number of ABC proteins, especially within the sub-phylum *Pezizomycotina* (with *Aspergillus* species and *G. zeae* as prominent examples), while *S. pombe* and members of the order *Saccharomycetales* contain a significantly reduced set of ABC proteins. The rust fungi appear to have fewer ABC genes than the Agaricomycotina: *P. graminis* and *M. laricis-populina* lack ABC-A, full-length ABC-B (MDR group), group IV of ABC-C, and group II of ABC-G, among others. On the other hand, they possess members of the group I ABC-C and group III ABC-G proteins, which are missing from the Agaricomycotina. Within the Ustilaginomycotina, *M. globosa* has the lowest number of ABC proteins of the basidiomycetes and lacks some representatives that are nearly ubiquitous. Only *M. globosa*, *U. maydis*, and *Sporosorim reilianum* possess ABC-B siderophores, represented by the ferrichrome-like siderophores of *U. maydis*. Finally, the group V ABC-C and group III ABC-G proteins are found in *U. maydis*, but are rare among the other basidiomycetes.

Another approach toward understanding the function of ABC transporters is the use of specific efflux inhibitors (often in conjunction with the antifungal agent that normally serves as a substrate). For instance, *C. glabrata* azole resistance was often found to be attributed to a gain-of-function mutation in the CgPDR1 transcriptional activator or to changes in mitochondrial function. Increased azole resistance is also often associated with increased virulence in mouse models of infection. However, milbemycin (Figure 4.3) oxim derivatives serve as inhibitors of ABC transporters and, when given in conjunction with fluconazole, they acted synergistically to remove azole-resistant isolates (Silva et al., 2013). Surprisingly, at somewhat higher dosages, the milbemycin alone displayed antifungal activity, possibly due to generation of reactive oxygen species. Consistent with this observation were transcription data indicating increased expression of genes associated with stress response, including those involved with oxydoreductive processes and protein turnover (Silva et al., 2013).

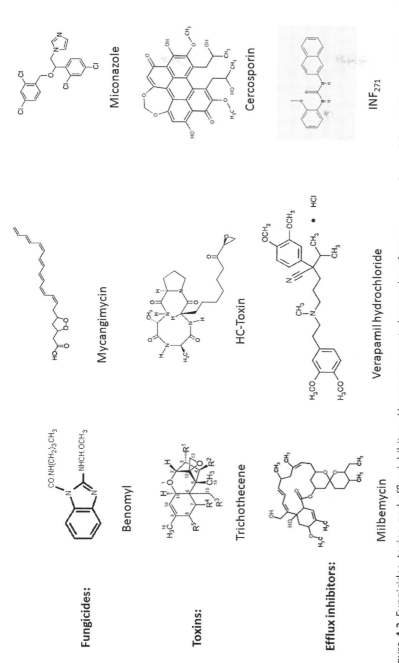

**Figure 4.3** Fungicides, toxins, and efflux inhibitors. Here are presented examples of some commonly used fungicides, mycotoxins, and inhibitors of efflux transporters. *Benomyl* is an inhibitor of microtubule function, with greater toxicity for fungal tubulins; *mycangimycin* is a polyene peroxide, originally identified for its selective inhibition of the southern pine beetle's fungal antagonist, *Ophiostoma minus* (Oh, Scott, Currie, & Clardy, 2009); *miconazole* is an imidazole antifungal agent. *Trichothecenes* are produced primarily by *Fusarium*, *Trichoderma*, and *Trichothecia* sp. and are potent inhibitors of proteins synthesis. They can have toxic effects on humans, other animals, and other fungi; *HC-toxin* is a cyclic tetrapeptide that inhibits histone deacetylases, particularly in host plant tissues (e.g., maize); *cercosporin* is a non-host-specific polyketide toxin. *Milbemycin* is a macrolide antibiotic that inhibits ABC transporters; *verapamil* is a calcium channel blocker; *INF271* is an inhibitor of multidrug transporters. (For color version of this figure, the reader is referred to the online version of this book.)

Drug efflux transporters have also been implicated in the biosynthesis of and resistance to toxins of a fungal origin. Whether the toxin is used in the process of plant pathogenesis (Daub, Herrero, & Chung, 2005, 2013) or as a toxin against competitors (*Trichoderma*, see below), mutations that reduce or abolish toxin efflux have been shown to have a severe impact on biosynthesis of the toxin, as well as additional increases in susceptibility to the toxin for the producer fungus itself. Moreover, such mutations have been associated with reduced pathogenicity (Daub et al., 2005, 2013).

Some additional insights as to roles of ABC proteins may be gleaned from expression data comparing different stages of the fungal lifecycle, with particular emphasis on pathogenic vs saprobic stages. In Table 4.2 are listed differential expression data from some of these analyses. *Hyaloperonospora arabidopsidis* is the causative agent of powdery mildew on *Arabidopsis* and is an obligate biotroph. Examination of the biotrophic phase during infection found six ABC genes upregulated (fungidb.org). In addition, a comparison of the human pathogen, *Coccidioides posadasii C735 delta SOWgp*, when grown as a saprobe compared to the infectious spherule, revealed five upregulated and six downregulated ABC genes.

## 2.4 Peptide Transporters Grouped into the ABC Superfamily

Mdl1, a member of the ABC superfamily, was identified as an intracellular peptide transporter localized in the inner membrane of yeast mitochondria (Young, Leonhard, Tatsuta, Trowsdale, & Langer, 2001). The protein was required for mitochondrial export of peptides ranging from 600 to 2100 Da generated by proteolysis of inner membrane proteins by the m-AAA protease in the mitochondrial matrix. Proteolysis by the i-AAA protease in the intermembrane space led to the release of similar sized peptides independent of Mdl1. This demonstrated the existence of two pathways that may allow communication between mitochondria and their cellular environment.

# 3. MFS SUPERFAMILY
## 3.1 Efflux Pumps
### 3.1.1 Toxin Secretion

One aspect of toxin secretion by fungi includes secondary metabolites that affect host species (HC toxin of *Cochliobolus heterostrophus*; Cercosporin; Figure 4.3). Such toxins are often produced by the products of gene clusters that encode a polyketide synthase or an NRPS or a fusion of the two. The array of metabolites produced include mycotoxins (e.g., penicillins

**Table 4.2** Differential Gene Expression of Fungal Pathogens

| Species | Host | Biotroph/ Necrotroph/ Hemibiotroph | Genes | Protein Family | Direction | Condition |
|---------|------|-----------------------------------|-------|----------------|-----------|-----------|
| *Hyaloperonospora arabidopsidis*[a] | *Arabidopsis thaliana* | Obligate biotroph | Hyaar_805654 | ABC transporter | Up | Biotrophy |
| | | | Hyaar_806395 | ABC transporter | | |
| | | | Hyaar_808962 | ABC transporter subunit (ISS) | | |
| | | | Hyaar_809334 | ABC multidrug transporter mdr2 | | |
| | | | Hyaar_813902 | ABC transporter related | | |
| | | | Hyaar_814417 | ABC transporter, ATP-binding protein | | |
| *Coccidiodes posadasii C735 delta SOWgp*[a] | Humans | | | ABC transporters | 5 genes up 6 genes down | Saprobe vs infectious spherule |
| *Microbotryum violaceum* | *Silene latifolia* | Obligate biotroph | MVLG_04305 | ABC transporter | Down | MI vs water, rich |
| *Hyaloperonospora arabidopsidis*[a] | *Arabidopsis thaliana* | Obligate biotroph | Hyaar_800306 | MFS family transporter | Up | Biotrophy |
| *Coccidiodes posadasii C735 delta SOWgp* | Humans | | | MFS family transporter | 28 genes up 22 genes down | Saprobe vs infectious spherule |

*Continued*

**Table 4.2** Differential Gene Expression of Fungal Pathogens—cont'd

| Species | Host | Biotroph/ Necrotroph/ Hemibiotroph | Genes | Protein Family | Direction | Condition |
|---|---|---|---|---|---|---|
| *Microbotryum violaceum* | *Silene latifolia* | Obligate biotroph | MVLG_00332 | MFS (PF07690) | Down | MI[b] vs water |
| | | | MVLG_00758 | MFS (PF07690) | Up | MI vs rich |
| | | | MVLG_01208 | MFS (PF07690) | Down | MI vs rich, W |
| | | | MVLG_01677 | MFS (PF07690) | Down | MI vs water |
| | | | MVLG_01678 | MFS (PF07690) | Down | MI vs water |
| | | | MVLG_03041 | MFS (PF07690) | Up | MI vs water |
| | | | MVLG_03397 | MFS (PF07690) | Up | MI vs water, R |
| | | | MVLG_04722 | MFS (PF07690) | Up | MI vs water, R |
| | | | MVLG_04984 | MFS (PF07690) | Down | MI vs water |
| | | | MVLG_05343 | MFS (PF07690) | Up | Mating type |
| | | | MVLG_05383 | MFS (PF07690) | Up | MI vs water |
| | | | MVLG_06024 | MFS (PF07690) | Up | MI vs water |
| | | | MVLG_06078 | MFS (PF07690) | Up | MI vs water |
| | | | MVLG_06792 | MFS (PF07690) | Up | MI vs water |
| *Microbotryum violaceum* | *Silene latifolia* | Obligate biotroph | MVLG_00072 | MFS, sugar PF00083; Tri12 | Down | MI vs water |
| | | | MVLG_04082 | MFS, sugar PF00083; Tri12 | Down | MI vs water |
| | | | MVLG_04942 | MFS, sugar PF00083; Tri12 | Down | MI vs water |
| *Coccidiodes posadasii C735 delta SOWgp*[a] | Humans | | | MFS family sugar transporter | 6 genes up 6 genes down | Saprobe vs infectious spherule |

| Organism | Host | Lifestyle | Gene | Family | Up/Down | Comparison |
|---|---|---|---|---|---|---|
| Microbotryum violaceum | Silene latifolia | Obligate biotroph | MVLG_00507 | Sugar PF00083 | Down | MI vs R, W |
| | | | MVLG_01903 | Sugar PF00083 | Down | MI vs rich |
| | | | MVLG_02719 | Sugar PF00083 | Down | MI vs water |
| | | | MVLG_04126 | Sugar PF00083 | Down | MI vs water |
| | | | MVLG_04712 | Sugar PF00083 | Up | MI vs water |
| | | | MVLG_05093 | Sugar PF00083 | Up | MI vs water |
| | | | MVLG_05179 | Sugar PF00083 | Up | MI vs water |
| | | | MVLG_06941 | Sugar PF00083 | Down | MI vs water, R |
| | | | MVLG_07006 | Sugar PF00083 | Up | MI vs water, R |
| Coccidioides posadasii C735 delta SOWgp[a] | Humans | | | OPT family transporter | 1 gene down | Saprobe vs infectious spherule |
| Microbotryum violaceum | Silene latifolia | Obligate biotroph | MVLG_02728 | PF03169.10::OPT | Down | MI vs water |
| | | | MVLG_02729 | PF03169.10::OPT | Down | MI vs water |
| | | | MVLG_03106 | PF03169.10::OPT | Up | MI vs water |
| | | | MVLG_04056 | PF03169.10::OPT | Down | MI vs water |
| | | | MVLG_04057 | PF03169.10::OPT | Down | MI vs water |
| | | | MVLG_05200 | PF03169.10::OPT | Down | MI vs water |

[a]Data obtained from www.fungidb.org.
[b]Conditions for M. violaceum expression: MI, late infection of male flowers; Water or W, growth of haploid cells on water agar; Rich or R, growth of haploid cells on yeast peptone dextrose (YPD) medium.

(also see above under ABC-D), aflatoxin), host-specific and general toxins (e.g., HC-toxin, T-toxin, cercosporin), as well as sideroporcs. These latter compounds can also serve as virulence factors for a variety of pathogens (Coleman and Mylonakis, 2009). The photoactivated perylenequinone toxins (e.g., cercosporin, Figure 4.3) are a particularly interesting and important class of secreted compounds required for pathogenesis of several genera of plant pathogenic fungi (Amnuaykanjanasin and Daub, 2009; Daub et al., 2005, 2013). All those identified to date are produced by the Ascomycetes (Daub et al., 2013). With few exceptions (i.e., some lichen-producing fungi), the producers are plant pathogens and their mode of action is that of photosensitizers that generate reactive oxygen species in visible and UV-A light. Damage to the host tissues often facilitates the lifestyles of these pathogens as hemibiotrophs, i.e., in the necrotrophic phase of infection. As the activated photosensitizer can damage DNA, lipids, and proteins, depending on localization, it behooves the producers of such compounds to export them safely from their own cells or risk damage to themselves. If efflux is blocked, production of the toxin by the fungus may be diminished (Choquer et al., 2007) and pathogenicity concomitantly reduced. Similarly, *Trichoderma* sp. that use efflux of toxins to inhibit competitor fungi utilize MFS proteins as part of this process. The first such transporter, Thmfs1, was recently identified and characterized for *Trichoderma harzianum* (Liu et al., 2012). Its importance in both antifungal activity against other species and fungicide tolerance for *T. harzianum* was highlighted by overexpression and disruption mutants. While expression of Thmfs1 was induced by the toxin trichodermin (and the Tri5 trichodiene synthase was strongly repressed), disruption of the *ThMFS1* gene did not dramatically reduce trichodermin production (Liu et al., 2012)

### 3.1.2 Drug Resistance

A large group of MFS proteins are associated with resistance to antimicrobials and toxic chemicals (Coleman and Mylonakis, 2009; Gulshan and Moye-Fowley, 2007; Pao et al., 1998; Reddy et al., 2012). One of the largest of the families contains **PF07690** (represented by 262 fungal species, with 16,971 sequences in the Pfam database (http://pfam.sanger.ac.uk/); in addition, for example, there are 134 representatives with this domain in *M. violaceum* (S. S. Toh, unpublished)).

Examples of these transporters are easily provided by several families from *C. albicans* and *S. cerevisiae* (Gaur et al., 2008). The Drug: H + Antiporter-1 (DHA1) Family, like the sugar porter family (TC # 2.A.1.1), is widely distributed and includes both drug-specific and MDR efflux pumps. With

22 members in *Candida*, this is one of the largest families. The *C. albicans* MDR1 was originally seen to confer resistance to benomyl and methotrexate, but later found to also provide resistance to cycloheximide, benztriazoles, 4-nitroquinolone-*N*-oxide, and sulfometuron methyl (Pasrija, Banerjee, & Prasad, 2007). Disruption of the *MDR1* gene reduced the virulence of *C. albicans* (Becker, Henry, Jiang, & Koltin, 1995). Apparently, TMD5 is particularly important for MDR1 function as it harbors a motif critical for Drug: H+ translocation (Coleman and Mylonakis, 2009). Apart from *MDR1*, which is known as a clinically relevant efflux pump protein, none of the other characterized members have been directly linked to MDR of *C. albicans*.

A second Drug: H + Antiporter-2 (DHA2) family contains proteins with 14 predicted transmembrane-spanning segments; 9 members are found in *C. albicans*. Although no *C. albicans* member has yet been characterized, we can get some sense of the role of these in resistance from *S. cerevisiae* DHA2 proteins, *SGE1* and *ATR1*. *ATR1* confers resistance to aminotriazole and 4-nitroquinolone-*N*-oxide although *ATR1* is only inducible by aminotriazole (Gompel-Klein and Brendel, 1990). *SGE1* confers resistance to crystal violet (Ehrenhofer-Murray, Wurgler, & Sengstag, 1994) and ethidium bromide (Goffeau et al., 1997).

Another group of MFS proteins is represented in the plant pathogen *M. graminoicola*, whose MgMfs1 has 14 predicted membrane spanners (DHA14). The fungus uses this protein as a protectant against both natural toxins, like cercosporin, and fungicides (Roohparvar, De Waard, Kema, & Zwiers, 2007). Heterologous expression in *S. cerevisiae* provided decreased susceptibility to these compounds. On the other hand, disruption of the gene in *M. graminicola* only increased sensitivity to strobilurin fungicides and to cercosporin, while virulence on wheat seedlings was unaffected in such mutants.

There are also several other Pfam domains found in MFS efflux proteins. These include **PF06813** (Nodulin-like; 61 fungal species, 66 sequences), mostly found in plants, but in fungi, typically there is one sequence represented per species. They are found often in animal pathogens (of mammals, skin-related pathogens are common, including Arthrodermataceae (*Microspora* and *Trichophyton*) and *Malassezia*; additional human pathogens bearing proteins with this domain include *Candida*, *Aspergillus*, *Penicillium marneffei*, *Coccidiodes*, and *Histoplasma*; and pathogens of insects/worms), as well as in chytrids, zygomycetes, and in some plant pathogens. Another group contains **PF11700** (Vacuole effluxer Atg2 like; 129 fungal species, 250 sequences), associated with efflux of leucine and other amino acids as a result of autophagy; no roles in pathogenesis have so far been indicated.

Finally, there are proteins containing **PF01554** (**the multi antimicrobial extrusion (MATE)** family (138 fungal species, 327 sequences). This is a family of drug/sodium or proton antiporters predicted to have 12 α-helical transmembrane regions. It is found in basidiomycetes, particularly among the pathogens. Some examples are found in rusts, smut fungi, as well as the human pathogens, *C. neoformans* and *M. globosa*. There is one representative found in *M. violaceum*: MVLG_06157 (TIGR00797) that matches yeast YHJ2 (2.A.66.1.5), ethionine resistance protein, ERC1. The yeast proteins are larger (up to about 700 residues) and exhibit about 12 TMSs.

The first functional characterization of an MATE-family protein for a filamentous fungus was carried out in *M. oryzae* as part of a larger analysis of catabolite repression for this fungal pathogen (Fernandez et al., 2012). Mdt1, an MATE-family pump in this organism, regulates glucose assimilation downstream of the sugar sensor, Tps1. However, Mdt1 is not only required for nutrient utilization but also for sporulation and pathogenicity. This is the first report of such roles for an MATE protein. The authors hypothesize that the adverse effects of *MDT1* disruption could be due to premature expression of cell wall-degrading enzymes.

Expression analysis to examine role(s) of various MFS proteins in infection is shown in Table 4.2. For *H. arabidopsidis*, during biotrophic expression in *A. thaliana*, only a single MFS family transporter was identified. In contrast, comparison of *C. posadasii C735 delta SOWgp* saprobe vs infectious spherule revealed 28 MFS genes upregulated and 22 MFS genes downregulated in the saprobic stage. Similarly, examination of the late phase of infection of *S. latifolia* flowers with spore development by the smut fungus, *M. violaceum*, revealed that five MFS genes were downregulated, while seven MFS genes were upregulated (S. S. Toh, unpublished).

Sutherland, Viljoen, Myburg, and Van den Berg (2013) chose another approach to identify MFS multidrug efflux pumps in *Fusarium oxysporum* that might play a role in pathogenicity and virulence. Comparative transcriptome analysis of pathogenic vs nonpathogenic races of the fungus on bananas demonstrated that transcript abundance for the transcript-derived fragment, TDF9, was fivefold higher in the pathogens.

Additionally, MFS mutants from a variety of pathogenic fungi often display increased sensitivity to naturally toxic compounds. For instance, the *B. cinerea* Bcmfs1 is apparently required to protect the fungus against camptothecin, produced by the potential host plant *Camptotheca acuminata*, as well as cercosporin, the mycotoxin, produced by a plant pathogenic rival, *Cercospora kikuchii* (Hayashi, Schoonbeek, & De Waard, 2002). Several MFS (and ABC) transporters required for full pathogenicity have been identified in forward genetic screens that employ *Agrobacterium tumefaciens*-based insertional mutagenesis (ATMT; Maruthachalam, Klosterman, Kang, Hayes, & Subbarao, 2011; Urban et al., 1999). For instance, an MFS transporter in *Verticillium dahliae*, the pathogen of lettuce, was shown to be essential for pathogenicity (Maruthachalam et al., 2011).

## 3.2 Trichothecene Efflux Pump (2.A.1.3.47, from TCDB)

As mentioned earlier, trichothecenes are one of the classes of mycotoxins that can be substrates for the MFS transporters. They are sesquiterpenoid toxins associated with the fusarium head blight disease, and include such compounds as T-2 toxin, nivalenol (NIV), and deoxynivalenol (DON) in various forms. The toxin inhibits eukaryotic protein synthesis by disrupting the peptidyltransferase. These toxins are produced by *Trichoderma*, *Fusarium*, *Trichothecium*, *Stachybotrys*, *Cephalosporium*, *Verticimonosporium*, and *Myrothecium*. The cluster associated with biosynthesis of a particular toxin in this class consists of several genes, best characterized in *Fusarium* sp. (Figure 4.4). The orthologous cluster for trichothecene biosynthesis in *Trichoderma* has been elucidated and found to have significant divergence from that of *Fusarium* (Cardoza et al., 2011). Within this biosynthetic cluster (with *Fusarium* as model) are at least eight conserved genes, including *TRI4* and *TRI11*, which belong to the cytochrome P450 superfamily. In addition, Tri6 is a zinc finger global transcriptional regulatory protein, whereas Tri10 is a positive transcriptional regulator (Nasmith et al., 2011; Peplow, Tag, Garifullina, & Beremand, 2003; Tag et al., 2001). Three additional genes are unique to the trichothene biosynthetic cluster: *TRI3* encodes a 15-O-acetyltransferase, *TRI5* encodes the

1 kb

**Figure 4.4** Typical trichothecene gene cluster, as found in *Fusarium graminearum*. The genes shown here include *TRI3* (encoding 15-O-acetyltransferase), *TRI4* and *TRI11* (encoding P450 oxygenases), *TRI5* (encoding trichodiene synthase), *TRI6* (encoding a transcription factor), *TRI8* (encoding a trichothecene C-3 esterase, bearing a secretory lipase domain (**PF03583**)), *TRI9* (for a predicted protein of unknown function), *TRI10* (regulatory function), and *TRI12* (encoding the trichothecene efflux pump). In addition, *TRI101*, encodes a trichothecene 3-O-acetyltransferase, whose function appears to be in competition with that of Tri8 (McCormick & Alexander, 2002).

trichodiene synthase, (Tijerino et al., 2011) while the product of *TRI9* is a protein of unknown function. In *Fusarium sporotrichioides*, there are additional genes, *TRI8* (secretory lipase, PF03583) and *TRI7* (membrane bound O-acyl transferase family, PF13813), located before the cluster shown in Figure 4.4, that enable the biosynthesis of the T-2 toxin (Brown et al., 2001). TRI101 is a 3-O-acetyltransferase (PF02458), whose gene is located outside the cluster, proposed to act as a self-protection mechanism during biosynthesis (McCormick and Alexander, 2002; Wuchiyama et al., 2000) since presence of a free C-3 hydroxyl group is a key component of *Fusarium* trichothecene phytotoxicity; Tri8 is the esterase that removes the C-3 protecting group, and thus, Tri8 may be considered a toxicity factor. Tri13, a cytochrome P450 monooxygenase, in conjunction with Tri7, determines the production of NIV rather than DON in *G. zeae* (anamorph *Fusarium graminearum*) and is also located within the gene cluster (Lee, Han, Kim, Yun, & Lee, 2002). Finally, Tri12 is found in fungi with the biosynthetic cluster, and this protein, a **trichothecene efflux pump (PF06609.6)**, is one subclass of MFS, with 14 TMS. This protein was described in its association with the trichothecene cluster of *Fusarium* (Ward, Bielawski, Kistler, Sullivan, & O'Donnell, 2002). Characterization of this protein in *F. sporotrichioides* suggested that the pump may play a role in self-protection against the mycotoxin produced by the fungus (Alexander, McCormick, & Hohn, 1999). Recently, with *F. graminearum*, it was found that the Tri12 protein also influences virulence in wheat (Menke, Dong, & Kistler, 2012).

With the accumulation of genomic sequence data, the Tri12 domain (PF06609) appears to be more common than one would predict. Currently, in the PFAM database (Punta et al., 2012; http://pfam.sanger.ac.uk) are curated 86 fungal species with this domain, of which all are Dikarya. Of these, 83 are ascomycetes, while we have additionally analyzed a total of 16 basidiomycete species (S. S. Toh, unpublished); within this group, species were found to harbor between 2 and 19 genes bearing the Tri12 domain. As seen in Table 4.3, the species bearing PF07690 are strongly represented by plant pathogenic species and also by animal pathogens (including as hosts, humans (18 species), insects (3 species), and nematodes (1 species)). Surprisingly, many of the fungi with expansions of these pumps are in species that do not have the inventory for biosynthesis of these mycotoxins. Thus, one may wonder what roles such efflux pumps play in the normal lifestyles of such fungi. Are they exposed to these toxins in their encounters in nature and are they susceptible? When are such pumps normally expressed or repressed? Which fungi have these?

**Table 4.3** Tri12 Presence in Taxa of Fungal Pathogens

| Taxonomic Designation | Total# Sequences | Animal Pathogen Species | Plant Pathogen Species | Symbiont Species |
|---|---|---|---|---|
| Ascomycetes | 211 | 19 | 27 | |
| *Arthrobotrys oligospora* | 1 | 1 | 0 | |
| *Exophiala dermatiditis* | 5 | 1 | 0 | |
| *Trichomaceae* | 42 | 6 (at least) | 0 | |
| *Onygenales* | 13 | 4 | 0 | |
| *Sclerotiniaceae* | 7 | 0 | 2 | |
| *Mycosphaerella graminicola* | 7 | 0 | 1 | |
| *Pleosporineae* | 19 | 0 | 3 | |
| *Nectriaceae* | 19 | 0 | 3 | |
| *Cordyceps militaris* | 2 | 1 | 0 | |
| *Trichoderma arundinaceum* | 1 | 0 | 0 | |
| *Trichoderma brevicompactum* | 1 | 0 | 0 | |
| Mitosporic *Hypocreales* (includes *Fusarium* sp.) | 32 | 0 | 14 | |
| Mitosporic *Clavicipitaceae* | 6 | 2 | 0 | |
| *Verticillium albo-atrum* | 10 | 0 | 1 | |
| *Verticillium dahliae* | 6 | 0 | 1 | |
| *Colletotrichum graminicola* | 2 | 0 | 1 | |
| *Colletotrichum higginsianum* | 6 | 0 | 1 | |
| *Candida* sp. | 5 | 5 | 0 | |

*Continued*

**Table 4.3** Tri12 Presence in Taxa of Fungal Pathogens—cont'd

| Taxonomic Designation | Total# Sequences | Animal Pathogen Species | Plant Pathogen Species | Symbiont Species |
|---|---|---|---|---|
| Basidiomycetes | 146 | 2 | 6 | 2 |
| Coprinopsis cinerea | 5 | 0 | 0 | |
| C. neoformans | 15 | 1 | 0 | 1 |
| Laccaria bicolor | 8 | 0 | 0 | |
| Malasezzia globosa | 2 | 1 | 0 | |
| Melamspora laricis-populina | 6 | 0 | 1 | |
| Microbotryum violaceum | 18 | 0 | 1 | |
| Mixia osmundae | 6 | 0 | 1 | |
| Phanerochaete chrysosporium | 19 | 0 | 0 | |
| Piriformospora indica | 15 | 0 | 0 | 1 |
| Postia placenta | 2 | 0 | 0 | |
| P. graminis | 3 | 0 | 1 | |
| Rhodotorula glutinis | 5 | 0 | 0 | |
| Schizophyllum commune | 19 | 0 | 0 | |
| Sporisorium reilianum | 8 | 0 | 1 | |
| Sporobolomyces roseus | 16 | 0 | 0 | |
| Ustilago maydis | 9 | 0 | 1 | |

Data are from the Pfam database (www.sanger.ac.uk) for Ascomycetes and from the Broad Institute (www.broadinstitute.org/) for Basidiomycetes.

To better understand the association of *TRI12* with the rest of the *TRI* cluster, we looked for the presence of other *TRI* domains in our dataset. For several of the genes (*TRI4*, *TRI6*, *TRI10*, and *TRI11*), it is difficult to ascertain convincingly whether true orthologs are present since the domains for these genes are quite general and are present... in hundreds of fungal species. On the other hand, *S. commune* has a single protein (confirmed by expression data: http://genome.jgi–psf.org/Schco3/Schco3.home.html) that carries the *TRI3* domain (PF07428), while *C. cinerea*, *Phanerochaete chrysosporium*, *L. bicolor*, *P. indica*, and *Postia placenta* have between one and nine putative genes containing the *TRI5* domain (PF06330). *Piriformospora indica* has one protein that was labeled as being related to TRI13–cytochrome P450. The *TRI9* domain (PF08195) was found only in the trichothecene producer species (i.e., *Fusarium* sp.). Not all of them have transcript evidence. It can be speculated that these enzymes perform similar biosynthetic processes but are not necessarily specific for trichothecene biosynthesis. It is possible that similar enzymes have not been characterized for their function in fungi other than those that produce trichothecenes.

It thus appears that the complete, or even partial, trichothecene biosynthetic pathway is no longer found in most other fungi. Nevertheless, the TRI12 domain has persisted and perhaps expanded. In fact, out of the seven proteins with this domain that have been found to be differentially expressed in *M. violaceum*, five are upregulated and one is downregulated in conditions when haploid cells are preparing for mating. On the other hand, only one of the seven is up- and one is downregulated late during infection of *S. latifolia* by the fungus. Three genes with this predicted domain are not differentially expressed under the conditions examined (S.S. Toh, unpublished). The variation in expression seems to suggest that the transporters have a wide substrate range and that the different transporters are required under different conditions. It may be worthwhile noting that the trichothecene producers (*Fusarium* sp.) carry between two to four predicted TRI12 proteins, whereas among the basidiomycetes, all of those we examined, except *P. placenta* and *M. globosa*, carry more than five genes with the predicted TRI12 domain. Of possible relevance to this overall question is the fact that *Trichoderma* species are used as biocontrol agents against phytopathogenic species of fungi. Some of these metabolites are phytotoxic or toxic to animals also. So, fungi that have the pump but not the biosynthesis machinery may be protecting themselves from the toxins produced by *Trichoderma* or *Fusarium* species (Malmierca et al., 2013). Perhaps those fungi with the expanded set of proteins bearing

PF07690 are in close proximity with producers. From this perspective, it is tempting to hypothesize not only an expansion in numbers of proteins bearing this domain but also the evolution of a broader substrate range to include additional, related toxins. An additional interesting side note is that in the presence of the arbuscular mycorrhizal fungi (AMF), *Glomus irregulare*, the *Fusarium sambucinum* genes *TRI5* and *TRI6* were upregulated, while *TRI4*, *TRI13*, and *TRI101* were downregulated. Thus, AMF species such as *G. irregulare* have evolved the ability to modulate mycotoxin gene expression by a plant fungal pathogen (Ismail, McCormick, & Hijri, 2011).

### 3.3 Sugar Transporters

Within the MFS, the sugar transporters are the largest family (Leandro, Fonseca, & Gonçalves, 2009; Saier, Tran, & Barabote, 2006; http://www.tcdb.org). Regardless of their substrate specificity or transport mechanisms, the sugar porters possess a five residue motif (*RXGRR*, where X is any amino acid). Recent work has provided mechanistic insights for the roles of sugar transporters in the pathogenicity of plant pathogenic fungi (Lingner, Münch, Deising, & Sauer, 2011; Wahl, Wippel, Goos, Kämper, & Sauer, 2010). Additionally, for example, newly available information from genomic and transcriptomic data (S. S. Toh, unpublished) from *M. violaceum* (Table 4.2) reveals that five sugar transporters (PF00083) are downregulated in late-stage infection, while four are upregulated. Work with yeasts has shown that some members of this family are unable to transport sugars; instead, they seem to serve as receptors, influencing gene expression (Özcan et al., 1998; Brown et al., 2006). For example, the *S. cerevisiae* Snf3p and Rgtp proteins have longer C-terminal tails than their parologous Hxt (hexose transporter) counterparts, Hxt1p–Hxt17p and Gal2p. It has been suggested that the longer tail is involved in their glucose sensing functions, especially since they do not transport glucose (Özcan et al., 1998). Additionally, some hexose transporters (Hxts) are under specific transcriptional controls. The transcriptional regulators Pdr1 and Pdr3 control expression of *HXT9* and *HXT11*; interestingly, Pdr1 and Pdr3 also regulate ABC transporters whose overexpression leads to efflux of several unrelated drugs (Section 2.1). There is a direct correlation between expression levels of *HXT9* and/or *HXT11* and susceptibility to certain drugs, either via interaction with ABC transporters or through direct action on the drugs as substrates (Nourani, Wesolowski-Louvel, Delaveau, Jacq, & Delahodde, 1997).

Deletion of either gene leads to resistance to cyclohexamide, 4-nitro-quinoline-N-oxide, and sulphomethuron methyl, while overexpression of *HXT11* increases susceptibility to the drugs.

Fungi that develop symbiotic mutualisms with their hosts must have transport systems that facilitate this relationship. For example, AMF, except for the initial spore germination and hyphal penetration of host root phases, are obligate biotrophs (Helber et al., 2011). The penetration of the root cortex must occur without elicitation of host defense response. Once inside, the plant and fungus establish a mutualism whereby, within arbuscules produced in the inner cortical cells, phosphate and other nutrients are provided to the host. In exchange, the plant provides carbohydrates to the fungal structures in the root cortex. In fact, since glucose is the likely form of carbohydrate taken up by the fungus from the apoplast, it is of note that increased apoplastic acid invertase and sucrose synthase activities are found in mycorrhizal roots (Schaarschmidt, Roitsch, & Hause, 2006). Recently, in AMF of the *Glomus* sp., Helber et al. (2011) identified a high-affinity monosaccharide transporter (*MST2*), found to function at several root locations. This is a broad spectrum transporter that may take up not only glucose but also plant cell wall-associated sugars. Of particular note is that expression of *MST2* correlates with that of the mycorrhyza-specific phosphate transporter, *PT4*, and that expression of *MST2* is required for proper mycorrhiza and arbuscule formation (Helber et al., 2011). Saprotrophic fungi derive their nutrients from dead and decaying tissue. Interestingly, an invertase (NCU04256) for *Neurospora crassa*, a saprotrophic fungus, was found to have hemicellulase activity (Tian et al., 2009). NCU04256 was also found to be upregulated when the fungus was grown in Avicel, a cellulose product, when compared to growth in minimal media (Tian et al., 2009). It is somewhat surprising that saprophytic fungi seem to lack invertases more often than plant pathogens, yet some are capable of helping to degrade plant cell walls.

It has been reported recently (Wahl et al., 2010) and variously reviewed (Doidy et al., 2012; Parrent, James, Vasaitis, & Taylor, 2009) that utilization of plant-produced sugars is of critical importance for both plant pathogenic fungi and for symbionts. As sucrose is the major transported sugar from photosynthetic source (leaves) to sink organs, fungi in intimate association with host plants must either have ways to digest sucrose to simpler sugars (e.g., glucose) for uptake or have other ways of dealing with this sugar. *Ustilago maydis*, the pathogen of maize, was the first plant intimate found to possess a sucrose transporter, UmSRT1 (Wahl et al., 2010). As this high–affinity transporter appears to have greater affinity for its substrate than the corresponding host ZmSUT1, the fungus cleverly appears to be able to directly take up this sugar from the plant apoplast, thereby avoiding detection by plant defense systems that are alerted by glucose utilization. It is emerging that similar strategies may be utilized by other fungi in symbiotic relationships with plants (e.g., AMF species, such as *Glomus interdices* (Doidy et al., 2012)).

As seen in Table 4.2, a number of genes bearing the Pfam domain for sugar transporters in MFS (PF00083) are differentially regulated during infection or in the infectious state. In *M. violaceum*, eight proteins with this predicted domain were downregulated and four were upregulated late in infection. Similarly, for *C. posadasii*, six genes were upregulated and six genes were downregulated during the infectious stage of the fungus.

For those fungi that do not employ the strategy of direct uptake of sucrose, mentioned above, their need to first begin metabolism of sucrose externally, before uptake of its breakdown products, means that they are potentially at risk when the host is monitoring production of such telltale clues to the presence of an interloper (Parrent et al., 2009). As such, it is interesting to survey the inventories of various fungi for the presence of enzymes that can metabolize sucrose. Three major classes of enzyme have the ability to metabolize sucrose into simpler sugars: *invertases* (of the glycosyl hydrolase (GH) 32 family, **PF00251, PF08244**), *sucrases* (**PF06999**), and, to a lesser extent, members of the GH31 (**PF01055**) families. Parrent et al. (2009) did an extensive survey for the GH32 family and found that mutualist fungi generally lack this family of enzymes. In our own survey (Table 4.4) of available databases, 59% of animal pathogens lacked these domains (*A. capsulatus, A. dermatitidis, C. albicans, C. glabrata, C. globosum, C. lusitaniae, C. immitis, C. posadasii, P. brasiliensis, P. marneffei, T. rubrum, M. globosa, B. dendrobatidis, A. locustae, E. cuniculi,* and *R. oryzae*), while 7% of biotrophic plant pathogenic fungi that we included lack GH32, including *M. violaceum*. However, though all the nectrophic species contained such proteins, 9% of the hemibiotrophic species lacked GH32 (e.g., *Colletotrichum higginsianum*). By way of contrast, 46% of solely saprotrophic fungi and 28% of saprotrophic/pathogenic fungi that we included lack GH32 (Table 4.4).

In the case of sucrase (**PF06999**), proportionately twice as many biotrophic plant pathogens (e.g., *M. laricis-populina, P. graminis tritici, Puccinia triticina,* and *Puccinia striiformis*) lack representatives with this domain when

**Table 4.4** Inventory of Sugar Transporter Pfam Domains

| Species | sugar_tr (PF00083) | suc_fer-like (PF06999) | glyco_hydro_32n (PF00251) | glyco_hydro_32c (PF08244) |
| --- | --- | --- | --- | --- |
| *Ajellomyces capsulatus* | 19 | 3 | 0 | 0 |
| *Ajellomyces dermatitidis* | 88 | 3 | 0 | 0 |
| *Alternaria brassicicola* | 74 | 0 | 2 | 0 |
| *Ashbya gossypii* | 13 | 2 | 1 | 1 |
| *Aspergillus clavatus* | 140 | 3 | 1 | 0 |
| *Aspergillus flavus* | 260 | 4 | 3 | 2 |
| *Aspergillus fumigatus* | 171 | 3 | 5 | 2 |
| *Aspergillus nidulans* | 212 | 3 | 3 | 2 |
| *Aspergillus terreus* | 249 | 3 | 6 | 4 |
| *Botryotinia fuckeliana* | 150 | 2 | 1 | 0 |
| *Candida albicans* | 27 | 2 | 0 | 0 |
| *Candida glabrata* | 19 | 2 | 0 | 0 |
| *Candida guilliermondii* | 43 | 2 | 1 | 0 |
| *Chaetomium globosum* | 47 | 0 | 0 | 0 |
| *Clavispora (Candida) lusitaniae* | 28 | 2 | 0 | 0 |
| *Coccidioides immitis* | 29 | 3 | 0 | 0 |
| *Coccidioides posadasii* | 29 | 3 | 0 | 0 |
| *Colletotrichum graminicola* | 83 | 2 | 3 | 0 |
| *Colletotrichum higginsianum* | 103 | 2 | 0 | 0 |
| *Fusarium oxysporum f. sp. Lycopersici* | 394 | 2 | 12 | 6 |
| *Fusarium verticillioides* | 328 | 2 | 11 | 5 |
| *Gibberella zeae* | 192 | 2 | 7 | 3 |
| *Leptosphaeria maculans* | 62 | 2 | 2 | 1 |
| *Magnaporthe oryzae (grisea)* | 69 | 2 | 4 | 3 |

*Continued*

**Table 4.4** Inventory of Sugar Transporter Pfam Domains—cont'd

| Species | sugar_tr (PF00083) | suc_fer-like (PF06999) | glyco_hydro_32n (PF00251) | glyco_hydro_32c (PF08244) |
|---|---|---|---|---|
| *Mycosphaerella fijiensis* | 91 | 2 | 5 | 2 |
| *Mycosphaerella graminicola* | 93 | 0 | 4 | 0 |
| *Nectria haematococca* | 207 | 3 | 6 | 5 |
| *Neosartorya fischeri* | 197 | 3 | 7 | 3 |
| *Neurospora crassa* | 81 | 2 | 3 | 1 |
| *Paracoccidioides brasiliensis* | 23 | 3 | 0 | 0 |
| *Penicillium marneffei* | 80 | 3 | 0 | 0 |
| *Podospora anserine* | 37 | 2 | 0 | 0 |
| *Pyrenophora tritici-repentis* | 67 | 2 | 3 | 3 |
| *Saccharomyces cerevisiae* | 28 | 2 | 1 | 1 |
| *Schizosaccharomyces pombe* | 40 | 0 | 2 | 1 |
| *Sclerotinia sclerotiorum* | 63 | 2 | 1 | 0 |
| *Stagonospora nodorum* | 176 | 2 | 6 | 2 |
| *Trichoderma atroviride* | 74 | 2 | 1 | 1 |
| *Trichoderma reesei* | 76 | 0 | 0 | 0 |
| *Trichoderma virens* | 86 | 2 | 1 | 0 |
| *Trichophyton rubrum* | 25 | 3 | 0 | 0 |
| *Verticillium dahliae* | 83 | 1 | 3 | 3 |
| *Coprinopsis cinereus* | 21 | 2 | 0 | 0 |
| *Cryptococcus gattii* | 43 | 2 | 2 | 2 |
| *Cryptococcus neoformans var. neoformans* | 57 | 2 | 1 | 1 |
| *Laccaria bicolor* | 25 | 3 | 0 | 0 |
| *Malassezia globosa* | 8 | 1 | 0 | 0 |
| *Melampsora laricis-populina* | 29 | 0 | 2 | 0 |

| | | | | |
|---|---|---|---|---|
| Microbotryum violaceum | 64 | 2 | 0 | 0 |
| Mixia osmundae | 13 | 2 | 1 | 1 |
| Phanerochaete chrysosporium | 47 | 0 | 0 | 0 |
| Postia placenta | 9 | 0 | 0 | 0 |
| Puccinia graminis tritici | 39 | 0 | 3 | 2 |
| Puccinia triticina | 47 | 0 | 3 | 2 |
| Puccinia striiformis | 45 | 2 | 3 | 2 |
| Rhodotorula graminis | 30 | 3 | 1 | 0 |
| Schizophyllum commune | 41 | 0 | 2 | 0 |
| Sporobolomyces roseus | 43 | 1 | 0 | 0 |
| Ustilago maydis | 24 | 1 | 2 | 2 |
| Batrachochytrium dendrobatidis | 2 | 0 | 0 | 0 |
| Antonospora locustae | 3 | 0 | 0 | 0 |
| Encephalitozoon cuniculi | 2 | 0 | 0 | 0 |
| Phycomyces blakesleeanus | 28 | 4 | 1 | 1 |
| Rhizopus oryzae (delemar) | 36 | 3 | 0 | 0 |
| Mucor circinelloides | 116 | 4 | 0 | 0 |
| Phytophthora ramorum | 74 | 2 | 3 | 3 |
| Phytophthora sojae | 68 | 1 | 2 | 3 |
| Phytophthora infestans | 52 | 1 | 3 | 2 |

compared with animal pathogens. On the other hand, proteins with this domain were also absent in 20% of necrotrophic plant pathogenic fungi and 9% of hemibiotrophic plant pathogenic fungi (e.g., *M. graminicola*).

## 3.4 OPTs

Transport of oligopeptides is a major component of a variety of cellular functions, including mating and pheromone sensing, sexual induction, and nitrogen storage and utilization. Peptide transporters are grouped as follows: 1) ABC superfamily (PepTs; TC 3.A.1.5); 2) peptide transporter (PTR) or proton-dependent oligopeptide transporter (POT; family TC 2.A.17); 3) peptide acetyl-CoA transporters of the MFS family (TC 2.A.1.25); 4) oligopeptide transporter (OPT) family; 5) fungal oligopeptide transporter (FOT) family. For purposes of this discussion, we will focus on the POT, OPT, and FOT families, with special emphasis, if possible, on their roles in the pathogenic program.

### 3.4.1 POT Family

POT (**PF00854**, 141 fungal species, 437 sequences) is specific for transport of dipeptides or tripeptides (Dunkel et al., 2013; Gomolplitinant and Saier, 2011). POT (2.A.17; http://www.tcdb.org/superfamily.php) is in the MFS superfamily. The evidence for this assertion stems from the membrane topology of its members (transmembrane α-helical spanners) and the limited amino acid sequence similarity of some members to members of the MFS (TC#2.A.1), as well as the recent crystallization of a representative from *Streptococcus thermophilus* (Solcan, Kwok, Fowler, & Cameron, 2012). Crystallization studies suggest facilitated opening of an intracellular gate that controls access to a central peptide-binding site; this opening appears due to a hinge-like movement within the C-terminal half of the transporter. Though this and previously crystallized POT transporters are from prokaryotes (Newstead et al., 2011), their similarity to eukaryotic POT members suggests that these analyses may provide insights to the fungal representatives as well. The POT proteins are about 450–600 amino acyl residues in length; though the family includes both eukaryotic and prokaryotic representatives, the eukaryotic proteins, in general, are longer than the bacterial proteins (Solcan et al., 2012).

The importance of individual POT transporters in fungal pathogenicity has not been tested directly although the specificity of some POTs for distinct types of small peptides has been examined via heterologous expression in *S. cerevisiae* (Dunkel et al., 2013).

### 3.4.2 OPT Family

For the "true" OPT members (**PF03169** (144 fungal species, 1008 sequences)), there is energy-dependent transfer of 2–6 amino acid peptides across membranes, where they are internalized and rapidly hydrolyzed by peptidases for use as amino acids, nitrogen, or carbon. So far, the only functionally characterized eukaryotic members of the OPT family (TC 2.A.67) are specific for oligopeptides and segregate phylogenetically from the other members of this family, the iron siderophores. Usually they have 16 (or occasionally 17; Gomolplitinant and Saier, 2011) transmembrane α-helices; they have no sequence homology to ABC or PTR transporters.

Dunkel et al. (2013) tried to examine the different roles for POT (also known as PTR; dipepetide/tripeptide) and OPTs in *C. albicans*. They found that a septuple mutant lacking both PTRs and five major OPTs (of eight total), was still pathogenic in a mouse infection model; thus, in the host niche, they are not needed. Even so, of interest is the apparent expansion of OPTs in *C. albicans* relative to the nonpathogenic *S. cerevisiae*, which only possesses two members of this family (Wiles, Cai, Naider, & Becker, 2006). One of the *C. albicans* OPTs, CaOPT1p, not a member of ABC superfamily or POT family (Wiles et al., 2006), is a tetra/pentapeptide transporter (**PF03169**, 144 fungal species; 1008 sequences). It is a highly flexible transporter, conferring the ability to transport all tripeptides, as well as a dipeptide, something not previously reported for an OPT (Dunkel et al., 2013). The OPT1,2,3Δ triple mutants have a severe growth defect, and these *C. albicans* transporters differ in their respective preferences for peptide length and sequence (Gomolplitinant and Saier, 2011). The OPTs display a wide distribution among plant and animal pathogens. Interestingly, the pathogen on Leguminosae, *Colletotrichum gloeosporioides* f. sp. *aeschynomene*, was found to have an OPT, CgOpt1, that was induced by exposure of the fungus to the plant hormone, indole-3-acetic acid (Chague, Maor, & Sharon, 2009). CgOpt1 was required for full pathogenesis and *coptg1*-silenced mutants produced fewer spores and less pigmentation than wild type. Similarly, other examples of differential expression of OPTs during the pathogenic phases of fungi include a lone *C. posadassi* gene downregulated in the saprophytic phase (fungidb.org; Table 4.2). Similarly, OPTs (here identified by PF03169) were mostly downregulated in *M. violaceum* late in the infection of its *S. latifolia* host, with only one OPT gene upregulated during this phase of its lifecycle (S. S. Toh, unpublished).

### 3.4.3 FOT Family

Recently a new family of FOT was discovered through "functional meta-transcriptomics" (Damon et al., 2011). The FOT family was identified in transcriptome analysis of soil-extracted eukaryotic-specific cDNA libraries; the soil had come from a 30-year-old spruce (*Picea abies*) forest stand. Two of the identified FOT proteins, when expressed in an *S. cerevisiae* mutant (i.e., deleted for *PTR2* and *DAL5*, its two known dipeptide transporter genes), were found to have a broad range of specificity with 60–80 different dipeptides/tripeptides serving as substrates for transport. Additional heterologous expression experiments in *Xenopus* oocytes supported a mechanism involving proton-coupled cotransport of dipeptides, for at least one transporter identified from the environmental cDNA expression library. In phylogenetic analyses, the six isolates grouped with related fungal orthologs in a distinct, statistically supported, fungal-specific clade when comparing these sequences with those of other fungal amino acid/auxin permeases. The distribution of FOTs among the fungi examined was spotty. Fungal species lacking FOT among the basidiomycetes were *L. bicolor* and *C. cinerea*; for the ascomycetes, no FOTs were identified in *N. crassa*, *Tuber melanosporum*, *A. nidulans*, and *M. grisea*. Within the same genus, they may appear or not: *A. fumigatus* had an FOT, while *A. nidulans* did not. By and large, the pathogenic species have FOT proteins, whereas the saprophytic and mutualist species generally lack them. Of note is that there was a respectable representation of fungal pathogens in the FOT clade (Damon et al., 2011). Moreover, when we used the environmental FOTs in Blastx analyses against the nr database of NCBI, many of the top hits were plant pathogenic fungi (not shown). Similar analysis using the *M. violaceum* database (www.broadinstitute.org) identified MVLG_03859 (hypothetical protein, upregulated during late infection) as the top hit; when this, in turn, was used in Blastp against the nr database, again the top hits were predominantly plant pathogenic fungal species. Phylogenetic analysis found the majority of environmental isolates reported by Damon et al. (2011) to be a sister clade to FOT representatives from *B. fuckeliana* and *Sclerotinia sclerotiorum* (Figure 4.5). Interestingly, the proteins that match the FOTs in BLAST searches contain the PF01490 (amino acid transporter) domain, which is far more pervasive in terms of identified fungal sequences (1339) and fungal species (151) than is evident from the subset characterized as FOTs. The approach of Damon et al. (2011) thus illustrates a creative alternative, functional environmental genomics, which may be used to uncover novel biological functions often missed in traditional genome sequencing of individual organisms.

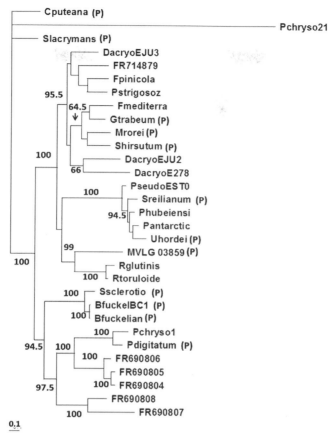

**Figure 4.5** PhyML tree of possible FOT-like representatives found from Blastx. FR714879 (*EnvFOT-A*, Damon et al., 2011) was used in Blastx search of the *Microbotryum violaceum* genome database at the Broad Institute (www.broadinstitute.org) to identify MVLG_03859 as the top hit (with an e-value of 0.0). Then MVLG_03859 was used in Blastp of the nr database at NCBI (www.ncbi.nlm.nih.gov) to find the top 20-or-so hits. These, along with the remaining five FOTs and *P. chrysogenum 21* from Damon et al. (2011) as outgroup were used in PhyML analysis (http://www.trex.uqam.ca/) with 200 Bootstrap replications. Nodes with at least 60% confidence are indicated. Plant pathogenic species are indicated (P). Cputeana, *Coniopohora puteana*; Pchryso, *Penicillium chrysogenum*; Dacryo, *Dacryopinax* sp.; Fpinicola, *Fomitopsis pinicola*; Pstrigosoz, *Puntularia strigosozonata*; Gtrabeum, *Gloeophyllum trabeum*; Mrorei, *Moniliophthora rorei*; Shirsutum, *Stereum hirsutum*; Pseudo, *Pseudozyma* sp.; Sreilianum, *Sporosorim reilianum*; Phubeiensis, *Pseudozyma hubeiensis*; Pantarctica, *Pseudozyma antarctica*; Uhordei, *Ustilago hordei*; Rglutinis, *Rhodotorula glutinis*; Rtoruloide, *Rhodosporidium toruloides*; Ssclerotio, *S. sclerotiorum*; Bfuckel or Bfuckelian, *Botryotinia fuckeliana*; Pdigitatum, *Penicillium digitatum*; FR714879, FR690804, FR690805, FR690806, FR690807, FR690808, FOTs from Damon et al. (2011).

## 4. AMINO ACID TRANSPORTERS

Amino acid transporters are found in fungi, plants, and animals (Wipf et al., 2002). The amino acid–polyamine–choline superfamily includes the amino acid permease family (**PF00324**, for which there are 162 fungal species, 2309 sequences listed). *Saccharomyces cerevisiae* has 24 members, each containing 12TMS domains; functional characterization of all the *S. cerevisiae* members has been achieved. Typically, this family is not highly specific for amino acid substrate. Instead several related or disparate amino acids may be transported. Of note, the Ssy1p acts as an extracellular amino acid sensor and transducer, thereby controlling amino acid uptake (Wipf et al., 2002). Moreover, in *C. albicans*, this SPS sensor pathway was required for virulence in a *D. melanogaster* model (Davis et al., 2011). A similar sensor role for some ammonium transporters (AMTs) has been identified for fungal representatives. This subset of AMTs are usually high–affinity transporters that additionally can "sense" ammonium availability and transduce this signal so as to effect a dimorphic switch, often associated with virulence (Biswas & Morschhäuser, 2005; Lengeler et al., 2000; Rutherford, Lin, Nielsen, & Heitman, 2008; Smith, Garcia-Pedrajas, Gold, & Perlin, 2003). As such, these "tranceptors" (Kriel, Haesendonckx, Rubio-Texeira, Van Zeebroeck, & Thevelein, 2011) have been found to affect pathogenicity and virulence in *C. albicans* (Biswas & Morschhäuser, 2005). Also, the *M. violaceum* homolog of these tranceptors is MVLG_00253, found to be upregulated late in infection (PF00909.16::Ammonium_transp).

## 5. CONCLUSIONS

In this chapter, we have highlighted ABC and MFS transporters with an eye toward delineating categories of transporters preferentially found in subsets of the fungal pathogenic species. While these are broad superfamilies of proteins, we can make some general statements. ABC-G (PDR) transporters are prevalent among pathogenic species, but may be restricted as to type/architecture. ABC-C is the largest group of full-length transporters among the basidiomycetes so far; yet, one subgroup of the ABC-C transporters is missing completely in several of the animal pathogens. One of the largest group of MFS transporters (containing PF07690) appears highly enriched among pathogenic fungal species, as is the MATE family, with examples found in the rusts, fungi, and some human pathogens.

Another subcategory of MFS, the TRI12 efflux pump (PF06609.6), is limited exclusively to fungi, and many of the species that have expansions of this domain are not producers of this mycotoxin. Finally, the newly identified FOT group of oligopeptide transporters contains a predominance of plant pathogenic species, though one caveat is that none of these genes have been functionally characterized for any of the species that bear phylogenetic similarity to the original genes isolated from cDNA libraries of soils.

## 5.1  A Combined Approach Toward Understanding the Roles of ABC and MFS Transporters in Drug Efflux?

If we want to come away with a full understanding of efflux mechanisms that promote survival of fungi, especially pathogenic fungi, we will need an approach that can identify the relative contributions of ABC and MFS transporters to the resistance phenotype(s) under investigation. Such an approach was employed recently by Prates, Kato, Ribeiro, Tegos, & Hamblin, 2011 to examine resistance to methylene blue–mediated photodynamic inactivation of *C. albicans*. Overexpression of ABC transporters reduced accumulation of the drug more than did overexpression of the MFS transporter, while selective use of either the ABC inhibitor verapamil or the MFS inhibitor $INF_{271}$ was able to show that both influx and efflux of the drug were facilitated by the MFS systems in *C. albicans*. An elegant approach was also applied toward elucidating the basis for MDR in *B. cinerea* (Kretschmer et al., 2009). Annual monitoring of field isolates in fungicide-treated vineyards identified three distinct classes of MDR *B. cinerea* that were resistant. All MDR strains overexpressed efflux transporter genes and consequently exhibited increased efflux of fungicides. One group of isolates (MDR1) contained mutations in Mrr1, a transcription factor that regulates expression of the ABC transporter, AtrB. A second group of mutants (MDR2) carried a rearranged promoter of the MFS transporter gene, *mfsM2*. Overexpression and gene disruption mutants allowed investigation of the individual roles of *mrr1*, *atrB*, and *mfsM2*. Combinations of *mrr1* and *mfsM2* alleles that could be selected for by fungicide treatment in field experiments resulted in a third MDR class (MDR3) that exhibited broad-spectrum resistance. Such studies represent an excellent example of epidemiology in an agricultural setting.

## ACKNOWLEDGMENTS

This work was supported, in part, by Grant #0947963 from the National Science Foundation to MHP. We are also grateful to the publicly available databases from which genomic (www.broadinstitute.org; http://mips.helmholtz-muenchen.de/genre/proj/ustilago; Grigoriev et al., 2001, 2012; Markowitz et al., 2012a, 2012b), transcriptomic (www.fungidb.org), Pfam (http://pfam.sanger.ac.uk), and transmembrane (http://www.tcdb.org/) data were accessible for analysis.

# REFERENCES

Alexander, N. J., McCormick, S. P., & Hohn, T. M. (1999). TRI12, a trichothecene efflux pump from *Fusarium sporotrichioides*: Gene isolation and expression in yeast. *Molecular Genetics and Genomics, 261*, 977–984. http://dx.doi.org/10.1007/s004380051046.

Amnuaykanjanasin, A., & Daub, M. E. (2009). The ABC transporter ATR1 is necessary for efflux of the toxin cercosporin in the fungus *Cercosporin nicotianae*. *Fungal Genetics and Biology, 46*, 146–158. http://dx.doi.org/10.1016/j.fgb.2008.11.007.

Andrade, A. C., Del Sorbo, G., Van Nistelrooy, J. G., & Waard, M. A. (2000). The ABC transporter AtrB from *Aspergillus nidulans* mediates resistance to all major classes of fungicides and some natural toxic compounds. *Microbiology, 146*, 1987–1997.

Becker, J. M., Henry, L. K., Jiang, W., & Koltin, Y. (1995). Reduced virulence of *Candida albicans* mutants affected in multidrug resistance. *Infection and Immunity, 63*, 4515–4518. http://dx.doi.org/10.1111/j.1365-2958.2005.04576.x.

Biswas, K., & Morschhäuser, J. (2005). The Mep2p ammonium permease controls nitrogen starvation-induced filamentous growth in *Candida albicans*. *Molecular Microbiology, 56*, 649–669. http://dx.doi.org/10.1111/j.1365-2958.2005.04576.x.

Brown, D. W., McCormick, S. P., Alexander, N. J., Proctor, R. H., & Desjardins, A. E. (2001). A genetic and biochemical approach to study trichothecene diversity in *Fusarium sporotrichioides* and *Fusarium graminearum*. *Fungal Genetics and Biology, 32*, 121–133. http://dx.doi.org/10.1006/fgbi.2001.1256.

Brown, V., Sexton, J. A., & Johnston, M. (2006). A glucose sensor in *Candida albicans*. *Eukaryotic Cell, 5*, 1726–1737.

Cannon, R. D., & Lamping, E. (2009). Efflux-mediated antifungal drug resistance. *Clinical Microbiological Reviews, 22*, 291–321. http://dx.doi.org/10.1128/CMR.00051-08.

Cardoza, R. E., Malmierca, M. G., Hermosa, M. R., Alexander, N. J., McCormick, S. P., Proctor, R. H., et al. (2011). Identification of loci and functional characterization of trichothecene biosynthesis genes in filamentous fungi of the genus *Trichoderma*. *Applied and Environmental Microbiology, 77*, 4867–4877. http://dx.doi.org/10.1128/AEM.00595-11.

Casadevall, A., Fang, F. C., & Pirofski, L.-a (2011). Microbial virulence as an emergent property: Consequences and opportunities. *PLoS Pathogens, 7*, e1002136. http://dx.doi.org/10.1371/journal.ppat.1002136.

Chague, F., Maor, R., & Sharon, A. (2009). CgOpt1 a putative oligopeptide transporter from *Colletotrichum gloeosporioides* that is involved in responses to auxin and pathogenicity. *BMC Microbiology, 9*, 173–184. http://dx.doi.org/10.1186/1471-2180-9-173.

Choquer, M., Lee, M. H., Bau, H. J., & Chung, K. R. (2007). Deletion of a MFS transporter-like gene in *Cercospora nicotianae* reduces cercosporin toxin accumulation and fungal virulence. *FEBS Letters, 581*, 489–494.

Coleman, J. J., & Mylonakis, E. (2009). Efflux in fungi: la pièce de résistance. *PLoS Pathogens, 5*, e1000486. http://dx.doi.org/10.1371/journal.ppat.1000486. PMID:19557154.

Coleman, J. J., White, G. J., Rodriguez-Carres, M., & Vanetten, H. D. (2011). An ABC transporter and a cytochrome P450 of *Nectria haematococca* MPVI are virulence factors on pea and are the major tolerance mechanisms to the phytoalexin pisatin. *Molecular Plant Microbe Interactions, 24*, 368–376. http://dx.doi.org/10.1094/MPMI-09-10-0198.

Damon, C., Vallon, L., Zimmermann, S., Haider, M. Z., Galeote, V., Dequin, S., et al. (2011). A novel family of oligopeptide transporters identified by functional metatranscriptomics of soil eukaryotes. *ISME Journal, 5*, 1871–1880. http://dx.doi.org/10.1038/ismej.2011.67.

Daub, M. E., Herrero, S., & Chung, K.-R. (2005). Photoactivated perylenequinone toxins in fungal pathogenesis of plants. *FEMS Microbiology Letters*, *252*, 197–206. http://dx.doi.org/10.1016/j.femsle.2005.08.033.

Daub, M. E., Herrero, S., & Chung, K.-R. (2013). Reactive oxygen species in plant pathogenesis: The role of perylenequinone photosensitizers. *Antioxidant & Redox Signaling*, *19*, 270–289. http://dx.doi.org/10.1089/ars.2012.5080.

Davis, M. M., Alvarez, F. J., Ryman, K., Holm, Å. A., Ljungdahl, P. O., & Engström, Y. (2011). Wild-type *Drosophila melanogaster* as a model host to analyze nitrogen source dependent virulence of *Candida albicans*. *PLoS One*, *6*, e27434. http://dx.doi.org/10.1371/journal.pone.0027434. Epub 2011 Nov 14.

Decottignies, A., & Goffeau, A. (1997). A complete inventory of the yeast ABC proteins. *Nature Genetics*, *15*, 137–145. http://dx.doi.org/10.1038/ng0297-137.

Doidy, J., Grace, E., Kühn, C., Simon-Plas, F., Casieri, L., & Wipf, D. (2012). Sugar transporters in plants and in their interactions with fungi. *Trends in Plant Sciences*, *17*, 413–422. http://dx.doi.org/10.1016/j.tplants.2012.03.00.

Dreesen, T. D., Johnson, D. H., & Henikoff, S. (1988). The brown protein of *Drosophila melanogaster* is similar to the white protein and to components of active transport complexes. *Molecular and Cellular Biology*, *8*, 5206–5215.

Dunkel, N., Hertlein, T., Franz, R., Reuß, O., Sasse, C., Schäfer, T., et al. (2013). Roles of different peptide transporters in nutrient acquisition in *Candida albicans*. *Eukaryotic Cell*, *12*, 520–528. http://dx.doi.org/10.1128/EC.00008-13. Epub 2013 Feb 2.

Ehrenhofer-Murray, A. E., Wurgler, F. E., & Sengstag, C. (1994). The *Saccharomyces cerevisiae* *SGE1* gene product: A novel drug-resistance protein within the major facilitator superfamily. *Molecular and General Genetics*, *244*, 287–294.

Fernandez, J., Wright, J. D., Hartline, D., Quispe, C. F., Madayiputhiya, N., & Wilson, R. A. (2012). Principles of carbon catabolite repression in the rice blast fungus: Tps1, Nmr1–3, and a MATE-family pump regulate glucose metabolism during infection. *PLoS Genetics*, *8*, e1002673. http://dx.doi.org/10.1371/journal.pgen.1002673.

Fleissner, A., Sopalla, C., & Weltring, K. M. (2002). An ATP-binding cassette multidrug-resistance transporter is necessary for tolerance of *Gibberella pulicaris* to phytoalexins and virulence on potato tubers. *Molecular Plant Microbe Interactions*, *15*, 102–108. http://dx.doi.org/10.1094/MPMI. 2002.15.2.102.

Gaur, M., Choudhury, D., & Prasad, R. (2005). Complete inventory of ABC proteins in human pathogenic yeast, *Candida albicans*. *Journal of Molecular Microbiology and Biotechnology*, *9*, 3–15. http://dx.doi.org/10.1159/000088141.

Gaur, M., Puri, N., Manoharlal, R., Rai, V., Mukhopadhayay, G., Choudhury, D., & Prasad, R. (2008). MFS transportome of the human pathogenic yeast *Candida albicans*. *BMC Genomics*, *9*, 579.

Goffeau, A., Park, J., Paulsen, I. T., Jonniaux, J. L., Dinh, T., Mordant, P., et al. (1997). Multidrug-resistant transport proteins in yeast: Complete inventory and phylogenetic characterization of yeast open reading frames with the major facilitator superfamily. *Yeast*, *13*, 43–54. http://dx.doi.org/10.1002/(SICI)1097-0061(199701)13. 1<43::AID-YEA56>3.0.CO;2-J.

Gomolplitinant, M., & Saier, M. H., Jr. (2011). Evolution of the oligopeptide transporter family. *Journal of Membrane Biology*, *240*, 89–110. http://dx.doi.org/10.1007/s00232-011-9347-9. Epub 2011 Feb 24.

Gompel-Klein, P., & Brendel, M. (1990). Allelism of *SNQ1* and *ATR1*, genes of the yeast *Saccharomyces cerevisiae* required for controlling sensitivity to 4-nitroquinoline-N-oxide and aminotriazole. *Current Genetics*, *18*, 93–96.

Grigoriev, I. V., Cullen, D., Goodwin, S. B., Hibbett, D., Jeffries, T. W., Kubicek, C. P., et al. (2011). Fueling the future with fungal genomics. *Mycology*, *2*, 192–209.

Grigoriev, I. V., Nordberg, H., Shabalov, I., Aerts, A., Cantor, M., Goodstein, D., et al. (2012). The genome portal of the Department of Energy Joint Genome Institute. *Nucleic Acids Research*, *40*, D26–D32.

Gulshan, K., & Moye-Rowley, W. S. (2007). Multidrug resistance in fungi. *Eukaryotic Cell, 6,* 1933–1942. http://dx.doi.org/10.1128/EC.00254-07.

Gupta, A., & Chatoo, B. B. (2008). Functional analysis of a novel ABC transporter ABC4 from *Magnaporthe grisea. FEMS Microbiology Letters, 278,* 22–28. http://dx.doi.org/10.1111/j.1574-6968.2007.00937.x.

Haas, H., Eisendle, M., & Turgeon, B. G. (2008). Siderophores in fungal physiology and virulence. *Annual Reviews of Phytopathology, 46,* 149–187. http://dx.doi.org/10.1146/annurev.phyto.45.062806.094338.

Hayashi, K., Schoonbeek, H. J., & De Waard, M. A. (2002). Bcmfs1, a novel major facilitator superfamily transporter from *Botrytis cinerea,* provides tolerance towards the natural toxic compounds camptothecin and cercosporin and towards fungicides. *Applied and Environmental Microbiology, 68,* 4996–5004. http://dx.doi.org/10.1128/AEM.68.10.4996-5004.2002.

Helber, N., Wippel, K., Sauer, N., Schaarschmidt, S., Hause, B., & Requena, N. (2011). A versatile monosaccharide transporter that operates in the arbuscular mycorrhizal fungus *Glomus* sp. is crucial for the symbiotic relationship with plants. *Plant Cell, 23,* 3812–3823. http://dx.doi.org/10.1105/tpc.111.089813. PMID:21972259.

Ismail, Y., McCormick, S., & Hijri, M. (2011). A fungal symbiont of plant-roots modulates mycotoxin gene expression in the pathogen *Fusarium sambucinum. PLoS ONE, 6,* e17990. http://dx.doi.org/10.1371/journal.pone.0017990.

Katinka, M. D., Duprat, S., Cornillot, E., Méténier, G., Thomarat, F., Prensier, G., et al. (2001). Genome sequence and gene compaction of the eukaryote parasite *Encephalitozoon cuniculi. Nature, 414,* 450–453. http://dx.doi.org/10.1038/35106579.

Kiel, J. A., Veenhuis, M., & van der Klei, I. J. (2006). PEX genes in fungal genomes: Common, rare or redundant. *Traffic, 7,* 1291–1303. http://dx.doi.org/10.1111/j.1600-0854.2006.00479.x.

Kim, Y., Park, S.-Y., Kim, D., Choi, J., Lee, Y.-H., Lee, J.-H., et al. (2013). Genome-scale analysis of ABC transporter genes and characterization of the ABCC type transporter genes in *Magnaporthe oryzae. Genomics, 101,* 354–361. http://dx.doi.org/10.1016/j.ygeno.2013.04.003.

Kovalchuk, A., & Driesen, A. J. (2010). Phylogenetic analysis of fungal ABC transporters. *BMC Genomics, 11,* 177. http://dx.doi.org/10.1186/1471-2164-11-177.

Kovalchuk, A., Lee, Y. H., & Asiegbu, F. O. (2013). Diversity and evolution of ABC proteins in basidiomycetes. *Mycologia.* 13-001. http://dx.doi.org/10.3852/13-001.

Kragl, C., Schrettl, M., Abt, B., Sarg, B., Lindner, H. H., & Haas, H. (2007). EstB-mediated hydrolysis of the siderophore triacetylfusarinine C optimizes iron uptake of *Aspergillus fumigatus. Eukaryotic Cell, 6,* 1278–1285. http://dx.doi.org/10.1128/EC.00066-07.

Kretschmer, M., Leroch, M., Mosbach, A., Walker, A. S., Fillinger, S., Mernke, D., et al. (2009). Fungicide-driven evolution and molecular basis of mutidrug resistance in field populations of the grey mould fungus *Botrytis cinerea. PloS Pathogens, 5,* e1000696. http://dx.doi.org/10.1371/journal.ppat.1000696.

Kriel, J., Haesendonckx, S., Rubio-Texeira, M., Van Zeebroeck, G., & Thevelein, J. M. (2011). From transporter to transceptor: Signaling from transporters provokes re-evaluation of complex trafficking and regulatory controls: Endocytic internalization and intracellular trafficking of nutrient transceptors may, at least in part, be governed by their signaling function. *BioEssays, 33,* 870–879. http://dx.doi.org/10.1002/bies.201100100. Epub 2011 Sep 13.

Kuchler, K., Sterne, R. E., & Thorner, J. (1989). *Saccharomyces cerevisiae STE6* gene product: A novel pathway for protein export in eukaryotic cells. *EMBO Journal, 8,* 3973–3984.

Lamping, E., Baret, P.V., Holmes, A. R., Monk, B. C., Goffeau, A., & Cannon, R. D. (2010). Fungal PDR transporters: Phylogeny, topology, motifs, function. *Fungal Genetics and Biology, 47,* 127–142. http://dx.doi.org/10.1016/j.fgb.2009.10.007.

Lamping, E., & Cannon, R. D. (2010). Use of a yeast-based membrane protein expression technology to overexpress drug resistance efflux pumps. *Methods in Molecular Biology, 666*, 219–250. http://dx.doi.org/10.1007/978-1-60761-820-1_15.

Leandro, M. J., Fonseca, C., & Gonçalves, P. (2009). Hexose and pentose transport in ascomycetous yeasts: An overview. *FEMS Yeast Research, 9*, 511–525. http://dx.doi.org/10.1111/j.1567-1364.2009.00509.x.

Lee, T., Han, Y. K., Kim, K. H., Yun, S. H., & Lee, Y. W. (2002). *Tri13* and *Tri7* determine deoxynivalenol- and nivalenol-producing chemotypes of *Gibberella zeae*. *Applied and Environmental Microbiology, 68*, 2148–2154. http://dx.doi.org/10.1128/AEM.68.5.2148-2154.2002.

Lengeler, K. B., Davidson, R. C., D'souza, C., Harashima, T., Shen, W. C., Wang, P., et al. (2000). Signal transduction cascades regulating fungal development and virulence. *Microbiology and Molecular Biology Reviews, 64*, 746–785. http://dx.doi.org/10.1128/MMBR.64.4.746-785.2000.

Lingner, U., Münch, S., Deising, H. B., & Sauer, N. (2011). Hexose transporters of a hemibiotrophic plant pathogen: Functional variations and regulatory differences at different stages of infection. *The Journal of Biological Chemistry, 286*, 20913–20922. http://dx.doi.org/10.1074/jbc.M110.213678.

Liu, H. Y., Chiang, Y. C., Pan, J., Chen, J., Salvadore, C., Audino, D. C., Badarinarayana, V., Palaniswamy, V., Anderson, B., & Denis, C. L. (2001). Characterization of CAF4 and CAF16 reveals a functional connection between the CCR4–NOT complex and a subset of SRB proteins of the RNA polymerase II holoenzyme. *Journal of Biological Chemistry, 276*, 7541–7548.

Liu, M., Liu, J., & Wang, W. M. (2012). Isolation and functional analysis of Thmfs1, the first major facilitator superfamily transporter from the biocontrol fungus *Trichoderma harzianum*. *Biotechnology Letters, 34*, 1857–1862. http://dx.doi.org/10.1007/s10529-012-0972-x.

Lodish, H., Berk, A., Zipursky, L., Matsudaira, P., Baltimore, D., & Darnell, J. (2000). *Molecular Cell Biology* (4th ed.). New York: W. H. Freeman (Chapter 15).

Malmierca, M. G., Cardoza, R. E., Alexander, N. J., McCormick, S. P., Collado, I. G., Hermosa, R., et al. (2013). Relevance of trichothecenes in fungal physiology: Disruption of tri5 in *Trichoderma arundinaceum*. *Fungal Genetics and Biology, 53*, 22–33. http://dx.doi.org/10.1016/j.fgb.2013.02.001. Epub 2013 Feb 27.

Manzoor, H., Kelloniemi, J., Chiltz, A., Wendehenne, D., Pugin, A., Poinssot, B., et al. (2013). Involvement of the glutamate receptor AtGLR3.3 in plant defense signaling and resistance to *Hyaloperonospora arabidopsidis*. *Plant Journal, 76*, 466–480. http://dx.doi.org/10.1111/tpj.12311. Epub 2013 Sep 19.

Markowitz, V. M., Chen, I.-M. A., Palaniappan, K., Chu, K., Szeto, E., Pillay, M., et al. (2012a). IMG 4 version of the integrated microbial genomes comparative analysis system. *Nucleic Acids Research, 40*, D123–D129.

Markowitz, V. M., Chen, I.-M. A., Chu, K., Szeto, E., Palaniappan, K., Grechkin, Y., et al. (2012b). IMG/M: The integrated metagenomes data management and comparative analysis system. *Nucleic Acids Research, 40*, D123–D129. http://dx.doi.org/10.1093/nar/gkr975.

Maranhão, F. C., Paião, F. G., Fachin, A. L., & Martinez-Rossi, N. M. (2009). Membrane transporter proteins are involved in *Trichophyton rubrum* pathogenesis. *Journal of Medical Microbiology, 58*, 163–168. http://dx.doi.org/10.1099/jmm.0.002907-0.

Maruthachalam, K., Klosterman, S. J., Kang, S., Hayes, R. J., & Subbarao, K. V. (2011). Identification of pathogenicity-related genes in the vascular wilt fungus *Verticillium dahliae* by *Agrobacterium tumefaciens*-mediated T-DNA insertional mutagenesis. *Molecular Biotechnology, 49*, 209–221. http://dx.doi.org/10.1007/s12033-011-9392-8.

McCormick, S. P., & Alexander, N. J. (2002). *Fusarium Tri8* encodes a trichothecene C-3 esterase. *Applied and Environmental Microbiology, 68*, 2959–2964. http://dx.doi.org/10.1128/AEM.68.6.2959-2964.2002.

Menke, J., Dong, Y., & Kistler, H. C. (2012). *Fusarium graminearum* Tri12p influences virulence to wheat and trichothecene accumulation. *Molecular Plant Microbe Interactions*, *25*, 1408–1418. http://dx.doi.org/10.1094/MPMI-04-12-0081-R.

do Nascimento, A. M., Goldman, M. H., & Goldman, G. H. (2002). Molecular characterization of ABC transporter-encoding genes in *Aspergillus nidulans*. *Genetics and Molecular Research*, *1*, 337–349.

do Nascimento, A. M., Goldman, G. H., Park, S., Marras, S. A., Delmas, G., Oza, U., et al. (2003). Multiple resistance mechanisms among *Aspergillus fumigatus* mutants with high-level resistance to itraconazole. *Antimicrobial Agents and Chemotherapy*, *47*, 1719–1726.

Nasmith, C. G., Walkowiak, S., Wang, L., Leung, W. W., Gong, Y., Johnston, A., et al. (2011). Tri6 is a global transcription regulator in the phytopathogen *Fusarium graminearum*. *PLoS Pathogens*, *7*, e1002266. http://dx.doi.org/10.1371/journal.ppat.1002266.

Newstead, S., Drew, D., Cameron, A. D., Postis, V. L., Xia, X., Fowler, P. W., et al. (2011). Crystal structure of a prokaryotic homologue of the mammalian oligopeptide-proton symporters, PepT1 and PepT2. *EMBO Journal*, *30*, 417–426. http://dx.doi.org/10.1038/emboj.2010.309. Epub 2010 Dec 3.

Niimi, M., Wada, S., Tanabe, K., Kaneko, A., Takano, Y., Umeyama, T., et al. (2005). Functional analysis of fungal drug efflux transporters by heterologous expression in *Saccharomyces cerevisiae*. *Japanese Journal of Infectious Diseases*, *58*, 1–7.

Niimi, M. (2010). Characterization of the multi-drug efflux Systems of pathogenic fungi using functional hyperexpression in *Saccharomyces cerevisiae*. *Japanese Journal of Medical Mycology*, *51*, 79–87.

Nourani, A., Wesolowski-Louvel, M., Delaveau, T., Jacq, C., & Delahodde, A. (1997). Multiple-drug-resistance phenomenon in the yeast *Saccharomyces cerevisiae*: Involvement of two hexose transporters. *Molecular and Cellular Biology*, *17*, 5453–5460.

Oh, D-. C., Scott, J. J., Currie, C. R., & Clardy, J. (2009). Mycangimycin, a polyene peroxide from a mutualist *Streptomyces* sp.. *Organic Letters*, *11*, 633–636. http://dx.doi.org/10.1021/ol802709x.

Özcan, S., Dover, J., & Johnston, M. (1998). Glucose sensing and signaling by two glucose receptors in the yeast Saccharomyces cerevisiae. *EMBO Journal*, *17*, 2566–2573.

Pao, S. S., Paulsen, I. T., & Saier, M. H., Jr. (1998). Major facilitator superfamily. *Microbiology and Molecular Biology Reviews*, *62*, 1–34. Review. PMID:9529885.

Parrent, J. L., James, T. Y., Vasaitis, R., & Taylor, A. F. (2009). Friend or foe? Evolutionary history of glycoside hydrolase family 32 genes encoding for sucrolytic activity in fungi and its implications for plant-fungal symbioses. *BMC Evolutionary Biology*, *9*, 148. http://dx.doi.org/10.1186/1471-2148-9-148.

Pasrija, R., Banerjee, D., & Prasad, R. (2007). Structure and function analysis of CaMdr1p, a major facilitator superfamily antifungal efflux transporter protein of *Candida albicans*: Identification of amino acid residues critical for drug/H+ transport. *Eukaryotic Cell*, *6*, 443–453.

Pedras, M. S. C., & Minic, Z. (2012). Differential protein expression in response to the phytoalexin brassinin allows the identification of molecular targets in the phytopathogenic fungus *Alternaria brassicicola*. *Molecular Plant Pathology*, *13*, 483–493.

Peplow, A. W., Tag, A. G., Garifullina, G. F., & Beremand, M. N. (2003). Identification of new genes positively regulated by Tri10 and a regulatory network for trichothecene mycotoxin production. *Applied and Environmental Microbiology*, *69*, 2731–2736. http://dx.doi.org/10.1128/AEM.69.5.2731-2736.2003.

Periyannan, S., Moore, J., Ayliffe, M., Bansal, U., Wang, X., Huang, L., et al. (2013). The gene *Sr33*, an ortholog of barley *Mla* genes, encodes resistance to wheat stem rust race Ug99. *Science*, *341*, 786–788. http://dx.doi.org/10.1126/science.1239028. Epub 2013 Jun 27.

Prates, R. A., Kato, I. T., Ribeiro, M. S., Tegos, G. P., & Hamblin, M. R. (2011). Influence of multidrug efflux systems on methylene blue-mediated photodynamic inactivation of *Candida albicans*. *Journal of Antimicrobial Chemotherapy, 66*, 1525–1532. http://dx.doi.org/10.1093/jac/dkr160.

Punta, M., Coggill, P. C., Eberhardt, R. Y., Mistry, J., Tate, J., Boursnell, C., et al. (2012). The Pfam protein families database. *Nucleic Acids Research, 40*(Database issue), D290–D301. http://dx.doi.org/10.1093/nar/gkr1065. Epub 2011 Nov 29.

Ravichandra, N. G. (2013). *Fundamentals of Plant Pathology*. Delhi: Ghosh, PHI Learning Private Limited.

Reddy, V. S., Shlykov, M. A., Castillo, R., Sun, E. I., & Saier, M. H., Jr. (2012). The major facilitator superfamily (MFS). Revisted *FEBS Journal, 279*, 2022–2035. http://dx.doi.org/10.1111/j.1742-4658.2012.08588.x. Epub 2012 May 8. Erratum in: FEBS Journal 2013 Aug; 280(16): 3975.

Reimann, S., & Deising, H. B. (2005). Inhibition of efflux transporter-mediated fungicide resistance in *Pyrenophora tritici-repentis* by a derivative of natural 4′-hydroxyflavone and potentiation of fungicide activity. *Applied and Environmental Microbiology, 71*, 3269–3275. http://dx.doi.org/10.1128/AEM.71.6.3269-3275.2005.

Ricardo, S., & Lehmann, R. (2009). An ABC transporter controls export of a *Drosophila* germ cell attractant. *Science, 323*, 943–946. http://dx.doi.org/10.1126/science.1166239.

Roohparvar, R., De Waard, M. A., Kema, G. H., & Zwiers, L. H. (2007). MgMfs1, a major facilitator superfamily transporter from the fungal wheat pathogen *Mycosphaerella graminicola*, is a strong protectant against natural toxic compounds and fungicides. *Fungal Genetics and Biology, 44*, 378–388. http://dx.doi.org/10.1016/j.fgb.2006.09.007.

Rutherford, J. C., Lin, X., Nielsen, K., & Heitman, J. (2008). Amt2 permease is required to induce ammonium-responsive invasive growth and mating in *Cryptococcus neoformans*. *Eukaryotic Cell, 7*, 237–246. http://dx.doi.org/10.1128/EC.00079-07. Epub 2007 Nov 30.

Saier, M. H., Jr., Tran, C. V., & Barabote, R. D. (2006). TCDB: The Transporter Classification Database for membrane transport protein analyses and information. *Nucleic Acids Research, 34*, D181–D186. http://dx.doi.org/10.1093/nar/gkj001.

Sanguinetti, M., Posteraro, B., La Sorda, M., Torelli, R., Fiori, B., Santangelo, R., et al. (2006). Role of AFR1, an ABC transporter-encoding gene, in the in vivo response to fluconazole and virulence of *Cryptococcus neoformans*. *Infection and Immunity, 74*, 1352–1359.

Schaarschmidt, S., Roitsch, T., & Hause, B. (2006). Arbuscular mycorrhiza induces gene expression of the apoplastic invertase LIN6 in tomato (*Lycopersicon esculentum*) roots. *Journal of Experimental Botany, 57*, 4015–4023. Epub 2006 Oct 18.

Schoonbeek, H. J., Del Sorbo, G., & De Waard, M. A. (2001). The ABC transporter BcatrG affects the sensitivity of *Botrytis cinerea* to the phytoalexin resveratrol and the fungicide fenpiclonil. *Molecular Plant Microbe Interactions, 14*, 562–571.

Schuller, C., Bauer, B. E., & Kuchler, K. (2003). Inventory and evolution of fungal ABC protein genes. In I. B. Holland, S. P. C. Cole, K. Kuchler, & C. F. Higgins (Eds.), *ABC proteins: from bacteria to man*. Elsevier.

Shalchian-Tabrizi, K., Minge, M. A., Espelund, M., Orr, R., Ruden, T., Jakobsen, K. S., et al. (2008). Multigene phylogeny of choanozoa and the origin of animals. *PLoS ONE, 3*, e2098. http://dx.doi.org/10.1371/journal.pone.0002098.

Silva, L. V., Sanguinetti, M., Vandeputte, P., Torelli, R., Rochat, B., & Sanglard, D. (2013). Milbemycins: More than efflux inhibitors for fungal pathogens. *Antimicrobial Agents and Chemotherapy, 57*, 873–886. http://dx.doi.org/10.1128/AAC.02040-12. Epub 2012 Dec 3.

Sipos, G., & Kuchler, K. (2006). Fungal ATP-binding cassette (ABC) transporters in drug resistance & detoxification. *Current Drug Targets, 7*, 471–481.

Slaven, J. W., Anderson, M. J., Sanglard, D., Dixon, G. K., Bille, J., Roberts, I. S., et al. (2002). Increased expression of a novel *Aspergillus fumigatus* ABC transporter gene, *atrF*, in the presence of itraconazole in an itraconazole resistant clinical isolate. *Fungal Genetics and Biology*, *36*, 199–206. http://dx.doi.org/10.1016/S1087-1845(02)00016-6.

Smith, D. G., Garcia-Pedrajas, M. D., Gold, S. E., & Perlin, M. H. (2003). Isolation and characterization from pathogenic fungi of genes encoding ammonium permeases and their roles in dimorphism. *Molecular Microbiology*, *50*, 259–275. http://dx.doi.org/10.1046/j.1365-2958.2003.03680.x.

Solcan, N., Kwok, J., Fowler, P. W., & Cameron, A. D. (2012). Alternating access mechanism in the POT family of oligopeptide transporters. *The EMBO Journal*, *31*, 3411–3421. http://dx.doi.org/10.1038/emboj.2012.157.

Sun, C. B., Suresh, A., Deng, Y. Z., & Naqvi, N. I. (2006). A multidrug resistance transporter in *Magnaporthe* is required for host penetration and for survival during oxidative stress. *Plant Cell*, *18*, 3686–3705. http://dx.doi.org/10.1105/tpc.105.037861. Epub 2006 Dec 22.

Sutherland, R., Viljoen, A., Myburg, A. A., & Van den Berg, N. (2013). Pathogenicity associated genes in *Fusarium oxysporum* f. sp. *cubense* race 4. *South African Journal of Sciences*, *109*. http://dx.doi.org/10.1590/sajs.2013/20120023. Art. #0023, 10 pages.

Tag, A. G., Garifullina, G. F., Peplow, A. W., Ake, C., Jr., Phillips, T. D., Hohn, T. M., et al. (2001). A novel regulatory gene, Tri10, controls trichothecene toxin production and gene expression. *Applied and Environmental Microbiology*, *67*, 5294–5302. http://dx.doi.org/10.1128/AEM.67.11.5294-5302.2001.

Thomazella, D. P., Teixeira, P. J., Oliveira, H. C., Saviani, E. E., Rincones, J., Toni, I. M., et al. (2012). The hemibiotrophic cacao pathogen *Moniliophthora perniciosa* depends on a mitochondrial alternative oxidase for biotrophic development. *New Phytologist*, *194*, 1025–1034. http://dx.doi.org/10.1111/j.1469-8137.2012.04119.x. Epub 2012 Mar 23.

Tian, C., Beeson, W. T., Iavarone, A. T., Sun, J., Marletta, M. A., Cate, J. H. D., et al. (2009). Systems analysis of plant cell wall degradation by the model filamentous fungus *Neurospora crassa*. *Proceedings of the National Academy of Sciences*, *106*, 22157–22162.

Tijerino, A., Cardoza, R. E., Moraga, J., Malmierca, M. G., Vicente, F., Aleu, J., et al. (2011). Overexpression of the trichodiene synthase gene *tri5* increases trichodermin production and antimicrobial activity in *Trichoderma brevicompactum*. *Fungal Genetics and Biology*, *48*, 285–296. http://dx.doi.org/10.1016/j.fgb.2010.11.012. Epub 2010 Dec 7.

Urban, M., Bhargava, T., & Hamer, J. E. (1999). An ATP-driven efflux pump is a novel pathogenicity factor in rice blast disease. *EMBO Journal*, *18*, 512–521. http://dx.doi.org/10.1093/emboj/18.3.512.

Vargas, W. A., Martín, J. M., Rech, G. E., Rivera, L. P., Benito, E. P., Díaz-Mínguez, J. M., et al. (2012). Plant defense mechanisms are activated during biotrophic and necrotrophic development of *Colletotrichum graminicola* in maize. *Plant Physiology*, *158*, 1342–1358. http://dx.doi.org/10.1104/pp.111.190397. Epub 2012 Jan 12.

von Dohren, H. (2009). A survey of nonribosomal peptide synthetase (NRPS) genes in *Aspergillus nidulans*. *Fungal Genetics and Biology*, *46*, S45–S52. http://dx.doi.org/10.1016/j.fgb.2008.08.008.

Wada, S., Tanabe, K., Yamazaki, A., Niimi, M., Uehara, Y., Niimi, K., et al. (2005). Phosphorylation of *Candida glabrata* ATP-binding cassette transporter Cdr1p regulates drug efflux activity and ATPase stability. *Journal of Biological Chemistry*, *280*, 94–103. http://dx.doi.org/10.1074/jbc.M408252200.

Wahl, R., Wippel, K., Goos, S., Kämper, J., & Sauer, N. (2010). A novel high-affinity sucrose transporter is required for virulence of the plant pathogen *Ustilago maydis*. *PLoS Biology*, *8*, e1000303. http://dx.doi.org/10.1371/journal.pbio.1000303.

Ward, T. J., Bielawski, J. P., Kistler, H. C., Sullivan, E., & O'Donnell, K. (2002). Ancestral polymorphism and adaptive evolution in the trichothecene mycotoxin cluster of phytopathogenic *Fusarium*. *Proceedings of the National Academy of Sciences of the United States of America*, *99*, 9278–9283. http://dx.doi.org/10.1073/pnas.142307199.

Weber, R. W. S., & Entrop, A.-P. (2011). Multiple fungicide resistance in *Botrytis*: A growing problem in German soft-fruit production. In N. Thajuddin (Ed.), *Fungicides – Beneficial and Harmful Aspects*. (ISBN: 978-953-307-451-1, InTech, DOI: 10.5772/27199. Available from http://www.intechopen.com/books/fungicides-beneficial-and-harmful-aspects/multiple-fungicide-resistance-in-botrytis-a-growing-problem-in-german-soft-fruit-production.

Wenzel, J. J., Piehler, A., & Kaminski, W. E. (2007). ABC A-subclass proteins: Gatekeepers of cellular phosphor- and sphingolipid transport. *Frontiers in Biosciences*, *12*, 3177–3193. http://dx.doi.org/10.2741/2305.

Wiles, A. M., Cai, H., Naider, F., & Becker, J. M. (2006). Nutrient regulation of oligopeptide transport in *Saccharomyces cerevisiae*. *Microbiology*, *152*, 3133–3145. http://dx.doi.org/10.1099/mic.0.29055-0.

Wipf, D., Ludewig, U., Tegeder, M., Rentsch, D., Koch, W., & Frommer, W. B. (2002). Conservation of amino acid transporters in fungi, plants and animals. *Trends in Biochemical Science*, *27*, 139–147. http://dx.doi.org/10.1016/S0968-0004(01)02054-0.

Wood, N. J., Alizadeh, T., Richardson, D. J., Ferguson, S. J., & Moir, J. W. B. (2002). Two domains of a dual-function NarK protein are required for nitrate uptake, the first step of denitrification in *Paracoccus pantotrophus*. *Molecular Microbiology*, *44*, 157–170. 10.1046/j.1365-2958.2002.02859.x.

Wuchiyama, J., Kimura, M., & Yamaguchi, I. (2000). A trichothecene efflux pump encoded by Tri102 in the biosynthesis gene cluster of *Fusarium graminearum*. *Journal of Antibiotics (Tokyo)*, *53*, 196–200. http://dx.doi.org/10.1046/j.1365-2958.2002.02859.x.

Yin, Y., He, X., Szewczyk, P., Nguyen, T., & Chang, G. (2006). Structure of the multidrug transporter EmrD from *Escherichia coli*. *Science*, *312*, 741–744.

Yoshida, Y. (1988). Cytochrome P450 of fungi: Primary target for azole antifungal agents. *Current Topics in Medical Mycology*, *2*, 388–418. http://dx.doi.org/10.1007/978-1-4612-3730-3_11.

Young, L., Leonhard, K., Tatsuta, T., Trowsdale, J., & Langer, T. (2001). Role of the ABC transporter Mdl1 in peptide export from mitochondria. *Science*, *291*, 2135–2138. http://dx.doi.org/10.1126/science.1056957.

Zuccaro, A., Lahrmann, U., Güldener, U., Langen, G., Pfiffi, S., Biedenkopf, D., et al. (2011). Endophytic life strategies decoded by genome and transcriptome analyses of the mutualistic root symbiont *Piriformospora indica*. *PLoS Pathogens*, *7*, e1002290. http://dx.doi.org/10.1371/journal.ppat.1002290. Epub 2011 Oct 13.

Zwiers, L. H., Stergiopoulos, I., Gielkens, M. M., Goodall, S. D., & De Waard, M. A. (2003). ABC transporters of the wheat pathogen *Mycosphaerella graminicola* function as protectants against biotic and xenobiotic toxic compounds. *Molecular Genetics and Genomics*, *269*, 499–507. http://dx.doi.org/10.1007/s00438-003-0855-x.

# CHAPTER FIVE

# Unisexual Reproduction

**Kevin C. Roach, Marianna Feretzaki, Sheng Sun and Joseph Heitman[1]**

Department of Molecular Genetics and Microbiology, Duke University Medical Center, Durham, NC, USA
[1]Corresponding author: email address: heitm001@duke.edu

## Contents

## Abstract

Sexual reproduction is ubiquitous throughout the eukaryotic kingdom, but the capacity of pathogenic fungi to undergo sexual reproduction has been a matter of intense debate. Pathogenic fungi maintained a complement of conserved meiotic genes but the populations appeared to be clonally derived. This debate was resolved first with the discovery of an extant sexual cycle and then unisexual reproduction. Unisexual reproduction is a distinct form of homothallism that dispenses with the requirement for an opposite mating type. Pathogenic and nonpathogenic fungi previously thought to be asexual are able to undergo robust unisexual reproduction. We review here recent advances in our understanding of the genetic and molecular basis of unisexual reproduction throughout fungi and the impact of unisex on the ecology and genomic evolution of fungal species.

One of the defining characteristics of eukaryotic life is the ability to reproduce sexually. Sexual reproduction was likely an attribute of the most recent common ancestor of all extant eukaryotic lineages (Baldauf, 2003; Baldauf, Roger, Wenk-Siefert, & Doolittle, 2000; Cavalier-Smith, 2004; Dacks & Roger, 1999; Derelle et al., 2006; Fritz-Laylin et al., 2010; Patterson, 1999;

*Advances in Genetics,* Volume 85
ISSN 0065-2660
http://dx.doi.org/10.1016/B978-0-12-800271-1.00005-6

Ramesh, Malik, & Logsdon, 2005; Simpson & Roger, 2004). Less than 1% of angiosperm plants and just 0.1% of characterized animals are thought to lack the ability to undergo meiosis and indeed, since the first sexual fungus *Syzygites megalocarpus* was described, eukaryotic microorganisms have been discovered to be not just sexual or asexual but homothallic, that is self-fertile, as well as heterothallic, self-incompatible (Asker & Jerling, 1992; Blakeslee, 1904; Vrijenhoek, 1998; White, 1978; Whitton, Sears, Baack, & Otto, 2008). In fungi, heterothallic species are self-sterile and require two partners with compatible mating types to initiate the sexual cycle, while homothallic species are self-fertile and can initiate sexual reproduction between cells of the same mating type and, in some species, even from a single cell or between cells descended from a single cell.

Fungal organisms possess diverse sexual strategies for both homothallic reproduction and maintaining heterothallic incompatibility. Fungi establish cell identify using either a bipolar mating-type system, that is one locus that produces two mating types, or tetrapolar mating-type system, that is two loci that are able to produce many mating types. The basidiomycetous fungus *Ustilago maydis* has a tetrapolar mating system with the two mating-type loci, *a* and *b*, located on different chromosomes. The *a* mating locus encodes pheromones and pheromone receptors involved in mate recognition and cell fusion during mating (Bölker, Urban, & Kahmann, 1992). The *b* locus encodes bE and bW homeodomain proteins that heterodimerize to regulate the expression of genes required for filamentous growth and completion of meiosis (Kämper, Reichmann, Romeis, Bölker, & Kahmann, 1995; Kronstad & Leong, 1990; Schulz et al., 1990). There are two alleles of the *a* locus and at least 25 alleles of the *b* locus; mating requires cells to carry different alleles at both the *a* and *b* loci. As a result, meiotic segregants can mate with only 25% of the products of the cross. In contrast, the budding yeast *Saccharomyces cerevisiae* possesses a mating-type system with features common to bipolar fungi; cells that inherit the *MAT*α idiomorph are α-cells, and those that inherit *MAT***a** are **a**-cells. *MAT*α and *MAT***a** are idiomorphs as they encode nonallelic genes, α1 and α2 in *MAT*α and **a**1 in *MAT***a**. The two idiomorphs of the *MAT* locus encode critical DNA-binding proteins that regulate expression of cell identity genes throughout the genome. Mating between α- and **a**-cells results in cellular and nuclear fusion to generate an α/**a** diploid that can undergo meiosis in which any given segregant can mate with 50% of the progeny of the cross.

Mating-type systems enable self-recognition and incompatibility in heterothallic fungi; however, homothallic species overcome self-incompatibility through three general mechanisms. *Saccharomyces cerevisiae* is an example of the first homothallic mechanism as only one allele is expressed from the active *MAT* locus, while two additional silent *MAT* cassettes are present and enable mating-type switching through gene conversion of the *MAT* locus. Thus, mating-type switching enables even a clonal population to mate but only after conversion to opposite mating-type haploid cells. More recent work has revealed a second mechanism of homothallism, the presence of both mating types in a single genome, at either linked or unlinked loci. *Sordaria macrospora*, a filamentous ascomycete, contains tandem copies of the *MATA* and *MATa* loci allowing expression of both mating-type loci in its genome and self-fertility (Debuchy & Turgeon, 2006; Pöggeler, Risch, Kück, & Osiewacz, 1997). The third type of homothallism, unisex, in which the requirement for an opposite mating type is dispensed with entirely, has most recently been discovered. While the molecular mechanisms and phylogenetic extent of unisex are not yet fully elucidated, new research has revealed that unisexual fungi are an important and medically relevant subset of homothallic fungi.

## 1. HOMOTHALLISM IN ASCOMYCETES AND BASIDIOMYCETES

Of the two major fungal divisions, the relative frequency of homothallism is higher in the Ascomycetes than in the Basidiomycetes (Whitehouse, 1949). This higher frequency of homothallism in Ascomycetes could be a reflection of the differences in the self-incompatibility systems in the two groups of fungi. In Ascomycetes, self-sterility is governed by a bipolar (unifactoral) system, while for the majority of the Basidiomycetous species, self-incompatibility is controlled by a tetrapolar (bifactoral) system. Thus, if self-fertility evolved from ancestral self-sterility, theoretically this transition would require less change in the incompatibility system in Ascomycetes than in Basidiomycetes. However, it should be noted that homothallism has also been identified in Basidiomycetous species that are close relatives of self-sterile species with tetrapolar mating systems. In addition, the debate about whether the ancestral state of the mating system in Ascomycetes and Basidiomycetes is homothallism or heterothallism is yet to be settled, with incongruent and inconclusive results generated by studies of different groups of species (Galagan et al., 2005; Gioti, Mushegian, Strandberg, Stajich, & Johannesson, 2012; Inderbitzin, Harkness, Turgeon, & Berbee, 2005; Nygren et al., 2011; Rydholm, Dyer, & Lutzoni, 2007).

Remarkably, there are fungal species that have the ability to transition between heterothallism and homothallism. For example, *S. cerevisiae Ho*$^+$ strains can undergo mating-type switching and are homothallic, while naturally occurring *ho*$^-$ strains are heterothallic. Among the fungal species that show plasticity in their reproductive modes are two important human pathogenic fungi, *Candida albicans* and *Cryptococcus neoformans*. Both species have bipolar mating systems and can undergo heterothallic reproduction between cells with opposite mating types. However, if appropriate conditions are present, unisexual reproduction can also be initiated and completed involving cells of the same mating type in the absence of cells of the opposite mating type.

*Saccharomyces cerevisiae* is the classic example of homothallism using mating-type switching. It had been long debated whether *S. cerevisiae* is homothallic or heterothallic, as both modes of reproduction were observed in this species (Hicks & Herskowitz, 1977; Hicks, Strathern, & Herskowitz, 1977; Lin & Heitman, 2007; Lindegren & Lindegren, 1943a, 1943b, 1944; Winge, 1935; Winge & Laustsen, 1937; Winge & Roberts, 1949). It turned out that both arguments are correct, and it depends on which strain is under examination. *Saccharomyces cerevisiae* has an unifactorial self-incompatibility system, and the mating type is governed by the allele at the active mating-type (*MAT*) locus. In addition to the active *MAT* locus, there are also two silent *MAT* cassettes, *HML* and *HMR*, in the genome that contain two compatible *MAT* alleles. *HML* and *HMR* are normally kept silent by a modified chromatin structure that requires the action of the Sir protein complex. However, during mitosis, a special endonuclease encoded by the gene *HO* initiates a gene conversion process that replaces the existing allele at the active *MAT* locus with the opposite allele from one of the two silent *MAT* cassettes. Colonies established by an *Ho*$^+$ strain will contain cells with compatible *MAT* alleles at the active *MAT* locus. Therefore, *Ho*$^+$ strains are homothallic. On the other hand, there are naturally occurring *ho*$^-$ *S. cerevisiae* strains that are unable to undergo mating-type switching and are heterothallic (McCusker, 2006).

A similar mating-type switching system has been identified in another model yeast, the fission yeast *Schizosaccharomyces pombe*. The two mating-type switching systems in *S. cerevisiae* and *S. pombe* share similarities in that 1) there are three mating-type cassettes present on the same chromosome, with one of them being active and the other two silenced, and 2) efficient unidirectional switching through gene conversion is initiated by a DNA lesion during mitosis. However, significant differences also exist between the two systems. First, the two species differ in the structure, sequence, as well as the nature of the genes encoded within *MAT*. In addition, the DNA lesion that initiates switching differs between the two systems. While a DSB is provoked by the Ho endonuclease in *S. cerevisiae*, the switch-initiating lesion in *S. pombe* appears to require a certain unusual type of replication-induced break that involves a nick and possibly retention of ribonucleotides from the primer of an Okazaki fragment (Arcangioli & de Lahondes, 2000; Dalgaard & Klar, 1999; Kaykov & Arcangioli, 2004; Vengrova & Dalgaard, 2004, 2006). Furthermore, the underlying mechanisms for the silencing of the two silent *MAT* cassettes also appear to be different between *S. cerevisiae* and *S. pombe*. It is thus apparent that the similar mating-type switching systems have evolved independently in the two lineages.

Remarkably, it has recently been shown that another budding yeast, *Kluyveromyces lactis*, also undergoes mating-type switching through yet another mechanism that is distinct from *S. cerevisiae* and *S. pombe* (Barsoum, Martinez, & Åström, 2010). *Kluyveromyces lactis* has lost the Ho endonuclease gene and was thought to switch stochastically by mitotic gene conversion. However, recently it was discovered that a special protein, α3, is essential for switching in *K. lactis*. α3 is homologous to transposases and the amino acids conserved among transposases are required for successful mating-type switching in *K. lactis*, suggesting α3 originated through domestication of a transposase (Barsoum et al., 2010). The fact that several similar yet distinct homothallic mating systems involving mating-type switching have evolved independently in different fungal lineages suggests that there are selection pressures that favor selfing, at least under certain circumstances.

Homothallism in fungi could also be achieved through the coexistence of compatible *MAT* alleles within one individual. The compatible *MAT* alleles could be located in different nuclei (i.e. binuclei spore, so-called pseudohomothallism), within the same nucleus but separated from each other (e.g. *Aspergillus nidulans*), or within the same nucleus and fused together (e.g. *Cochliobolus* sp.). Several basidiomycetous species produce heterokaryotic spores, such as *Agrocybe semiorbicularis*, *Conocybe tenera* for. *bispora*, *Coprinus ephemerus*, and *Aleurodiscus canadensis* (Raper, 1966). These species produce two-spored basidia. Following meiosis, two nuclei migrate into each of the two spores. Germination of each individual spore can establish dikaryotic mycelia, and in most cases, form clamp connections (Raper, 1966). In these pseudohomothallic species, occasionally there are spores that produce stable mycelia that lack clamp connections. Further mating analyses using these "nonhomothallic" individuals suggest the existence of an underlying bipolar self-incompatibility system in each of these species (Raper, 1966). Thus, it is likely that homothallism by the production of heterokaryotic spores is derived from an ancestral bipolar mating system. Similar examples of pseudohomothallism based on the production of heterokaryotic spores with underlying bipolar incompatibility also occur in several Ascomycetous species, such as *Neurospora tetrasperma* and *Podospora anserina* (Ames, 1934; Dodge, 1927; Dodge, Singleton, & Rolnick, 1950; Raper, 1966).

There are also a few Basidiomycetous species that are homothallic without clamp connections, such as *Calocera cornea, Coprinus ephemeroides, Octojuga pleurotelloides, Octojuga pseudopinsitus,* and *Filobasidiella depauperata* (Raper, 1966; Rodriguez-Carres, Findley, Sun, Dietrich, & Heitman, 2010). *Filobasidiella depauperata* appears to be an obligate sexual species that is closely related to the human pathogenic *Cryptococcus* species complex. *Filobasidiella depauperata* spores are uninucleate, and each germinated spore establishes a monokaryotic mycelia that lacks clamp connections but produces basidia decorated with four long spore chains (Kwon-Chung et al., 1995; Rodriguez-Carres et al., 2010). It is likely that the obligate selfing in *F. depauperata* evolved from an ancestral bipolar self-incompatibility system. However, it is not yet clear how this transition was achieved although analyses of genes located within the *MAT* locus of *C. neoformans* suggest they have been involved in extensive chromosomal rearrangements, such as translocations and inversions (Fraser et al., 2004; Rodriguez-Carres et al., 2010).

## 2. UNISEXUAL REPRODUCTION IN PATHOGENIC FUNGI

The numerous and distinct fungal life cycles illustrate the plasticity of reproductive strategies fungi employ in nature to preserve the benefits of sexual reproduction. The transition between the traditional heterothallic sexual cycle requiring mating to take place between different mating-type cells and homothallic cycles that allow more promiscuous mating are common and occur throughout the fungal kingdom. Individual isolates of the same species are capable of choosing between a homothallic or a heterothallic lifestyle that will either ameliorate the cost of mating or fit the environmental needs of the species and allow them to expand and survive in hostile niches. The frequent transitions illustrate the balance between outcrossing and inbreeding, or between vegetative growth and increased frequency of sexual reproduction, are likely a response to specific environmental cues that favor one or the other strategy.

Pathogens such as *C. neoformans* and *C. albicans* highlight a conundrum for mycologists; both have extant sexual cycles yet they generate largely clonal populations in nature. *Cryptococcus neoformans* has an opposite α–**a** sexual cycle that induces a dimorphic transition from yeast growth to hyphae (Kwon-Chung, 1975, 1976a, 1976b). However, the predominance of the α mating type in clinical and environmental isolates led to the hypothesis that this species might have been largely asexual. Recent studies revealed that α mating-type isolates of *C. neoformans* evolved a self-fertile strategy and are able to complete a unisexual cycle in the absence of an opposite mating partner (Lin, Hull, & Heitman, 2005). Population studies show that unisexual reproduction occurs in several of the lineages of the *Cryptococcus* species and it is likely an important strategy for reproduction in nature (Bui, Lin, Malik, Heitman, & Carter, 2008; Fraser et al., 2005; Lin et al., 2007; Lin et al., 2009; Ni et al., 2013).

Similarly, *C. albicans* was considered to be a strictly asexual species for over a century. Yet the discovery of the mating-type-like (*MTL*) loci, *MTL*α and *MTL***a**, the presence of mating- and meiosis-specific genes in the genome, the isolation of α and **a** mating-competent strains, and finally the discovery that the white–opaque transition dramatically enhances mating established the presence of a parasexual cycle (Bennett & Johnson, 2003; Hull & Johnson, 1999; Hull, Raisner, & Johnson, 2000; Lockhart et al., 2002; Magee & Magee, 2000; Miller & Johnson, 2002). *Candida albicans* also undergoes unisexual reproduction in the absence of the Bar1 protease (Alby, Schaefer, & Bennett, 2009). However, despite retaining many meiotic genes, meiosis has yet to be observed in *C. albicans* (Sherwood & Bennett, 2009). Other pathogenic *Candida* species are missing key meiotic regulators, and yet undergo extant meiotic sexual cycles (Reedy, Floyd, & Heitman, 2009; Sherwood & Bennett, 2009). *Cryptococcus* and *Candida* species serve as excellent examples of reproductive plasticity that allows cryptic sexual cycles to enable outcrossing and inbreeding, generating genetic diversity or preserving genetic configurations that confer a fitness advantage in a specific environmental niche.

*Cryptococcus neoformans* is able to undergo unisexual reproduction and generate genetic diversity, even in clonal populations of pathogenic fungi (Ni et al., 2013). From a medical perspective, unisexual reproduction poses a challenge for confronting and controlling emerging new strains that have been found to be either hypervirulent or resistant to current antifungal treatments (Fraser et al., 2005; Ni et al., 2013). Similar studies on pathogenic parasites reveal that unisexual reproduction is widespread in pathogenic microbes, and can generate "superbugs" that are resistant to current treatments and responsible for local outbreaks (Heitman, 2006, 2010; Wendte et al., 2010).

## 2.1 Cryptococcus neoformans

The basidiomycetous fungus *C. neoformans* and its sibling species are the most common fungal agents of meningoencephalitis, which is fatal if untreated. The pathogenic species of *Cryptococcus* comprise four serotypes based on capsular antigens: *C. neoformans* (serotype D), *C. neoformans* var. *grubii* (serotype A), and the sister species *Cryptococcus gattii* (serotypes B and C) (Hull & Heitman, 2002). *Cryptococcus neoformans* serotypes A and D are prevalent worldwide but most commonly infect immunocompromised individuals while *C. gattii* is usually restricted to tropical and subtropical regions and is able to infect immunocompetent individuals (Ellis & Pfeiffer, 1990; Speed & Dunt, 1995). These pathogenic *Cryptococcus* species are typically found as haploid yeasts in nature and inside the host. They comprise two mating types, α and **a**, and display a well-defined sexual cycle. In response to nutrient limitation, cells of opposite mating type secrete pheromones that are sensed by the pheromone receptors and initiate the formation of conjugation tubes extending toward the pheromone source. Pheromone production triggers a cell–cell fusion event; however, the nuclei remain separate, producing a dikaryon that undergoes a dimorphic transition to filamentous hyphae. The hyphae grow and produce yeast cells, called blastospores, via budding, which harbor one of the two haploid nuclei of the hyphal compartment. At the apex of the hyphae, specialized structures (basidia) are formed where nuclear fusion occurs, followed by meiosis. Multiple rounds of mitosis and budding produce four chains of basidiospores (reviewed in Heitman, 2006, 2010; Heitman, Sun, & James, 2013; Idnurm et al., 2005).

Sexual reproduction and virulence are linked as the cycle can generate and disperse *Cryptococcus* spores in the environment. Infection is thought to be acquired by inhalation of spores, which are readily aerosolized and of an ideal size to lodge in the alveoli of the lung to cause pulmonary infection. Both spores and yeast are capable of causing fatal infections in mice, and particles small enough to be spores are found in air samples from *Cryptococcus* outbreak locations on Vancouver Island. Taken together, these observations support the hypothesis that spores are infectious propagules of *Cryptococcus* (Giles, Dagenais, Botts, Keller, & Hull, 2009; Kidd et al., 2007a, 2007b; Springer, Saini, Byrnes, Heitman, & Frothingham, 2013; Velagapudi, Hsueh, Geunes-Boyer, Wright, & Heitman, 2009).

*Cryptoccoccus* exhibits an extant sexual cycle under laboratory conditions with readily observed production of hyphae and spores. However, natural populations exhibit such a marked bias toward mating-type α that it is not clear how relevant these laboratory conditions are for environmental populations. A small number of environmental and clinical isolates are diploid hybrids of serotype A and D (**a**ADα and αAD**a** isolates) that harbor both mating types and serve as indirect evidence of mating in nature (Lengeler, Cox, & Heitman, 2001; Litvintseva, Lin, Templeton, Heitman, & Mitchell, 2007). However, the population of environmental and clinical isolates sampled thus far is almost exclusively of mating-type α, raising doubts about the frequency of α–**a** heterosexual reproduction that occurs in nature (Lengeler et al., 2001; Litvintseva et al., 2007). The first isolate of mating-type **a** was isolated from a clinical sample belonging to the highly pathogenic serotype A group (Lengeler, Wang, Cox, Perfect, & Heitman, 2000). Further genotyping of ~3000 strains showed that only 3 were *MAT***a** isolates that also exhibited low competence for mating (Keller, Viviani, Esposto, Cogliati, & Wickes, 2003; Viviani, Nikolova, Esposto, Prinz, & Cogliati, 2003). Frequencies of mating-type alleles do vary considerably between geographically isolated populations. For instance, a sub–Saharan African population of *C. neoformans* comprised ~25% mating-type **a** individuals (Litvintseva et al., 2003). *MAT***a** isolates from this population exhibit evidence of recombination, and are robustly fertile under laboratory conditions (Litvintseva et al., 2003). The presence of these *MAT***a** isolates is geographically restricted and the opportunities for α–**a** heterosexual reproduction beyond this locale seems likely to be severely limited in nature.

As the vast majority of *Cryptococcus* isolates are mating-type α, it was thought that the organism was asexual and reproduced mitotically. The absence of a sexual cycle was hypothesized to have evolved concomitantly with the emergence of the highly pathogenic isolates. The same notion was advanced and popularized for many other pathogenic eukaryotic microorganisms, including *C. albicans*, and eukaryotic parasites. This theory was fundamentally challenged by evidence of sexual recombination in natural populations that are exclusively of mating-type α (Brandt, Hutwagner, Kuykendall, & Pinner, 1995; Xu, Vilgalys, & Mitchell, 2000b). Extensive genotyping and sequencing revealed evidence for both clonal expansion and recombination in serotype A and D isolates from sub–Saharan Africa and the United States (Litvintseva, Kestenbaum, Vilgalys, & Mitchell, 2005; Litvintseva et al., 2003).

Mating-type α isolates were observed to form hyphae, basidia, and spores upon culture on mating media; however, due to the absence of an opposite mating partner, this process was considered to be strictly mitotic and asexual, and was termed monokaryotic or haploid fruiting (Wickes, Mayorga, Edman, & Edman, 1996). Monokaryotic fruiting shares many characteristic features with sexual reproduction, including the production of infectious spores, and was eventually discovered to be a sexual cycle involving cells of one mating type (Lin et al., 2005). In a process similar to heterosexual reproduction, nutrient limitation induces a dimorphic transition from yeast to hyphae in cells of one mating type. Unisexual hyphae grow to form basidia at the tips, where meiosis and multiple rounds of mitosis produce abundant spores. In response to nutrient limitation, haploid cells can generate hyphae that grow to produce the basidium where a late endoreplication event occurs and produces a transient diploid nucleus that undergoes meiosis and sporulation. Diploidization can also occur early during α–α mating, either through endoreplication or cell–cell fusion between genetically different or clonal cells. In this case, the diploid cells initiate the formation of diploid hyphae that lead to basidia where meiosis and budding occurs to produce haploid spores (Lin et al., 2005). Unisexual reproduction generates a mixture of haploid and diploid hyphae, reflecting either a late or an early diploidization event, indicating considerable plasticity in unisexual reproduction (Feretzaki & Heitman, 2013).

Unisexual reproduction is a sexual cycle that involves ploidy changes (1N→2N→1N) and hyphal development. Unisexual rates of recombination are similar to those observed in heterosexual reproduction, suggesting that both sexual programs are meiotic cycles (Lin et al., 2005). Deletion of the highly conserved meiotic factors Spo11 or Dmc1 severely impairs sporulation during unisexual and heterosexual reproduction (Feretzaki & Heitman, 2013; Lin et al., 2005). Spo11 induces double strand breaks on the homologous chromosomes that initiate recombination, while Dmc1 is responsible for repairing DNA breaks by facilitating the invasion of DNA strands leading to the formation of Holliday junctions. Both genes are dispensable for hyphal and basidia development but critical for meiosis and sporulation, providing further evidence that unisexual reproduction is an extant sexual meiotic cycle.

Unisexual reproduction was initially observed only in α mating-type cells, and thus it was thought the cycle was restricted to the α mating type. Further analysis revealed that some *MAT*a isolates also possess the ability to undergo unisexual reproduction (Hull & Heitman, 2002; Lin, Huang, Mitchell, & Heitman, 2006; Tscharke, Lazera, Chang, Wickes, & Kwon-Chung, 2003). Subsequent studies revealed that hyphal development is a quantitative trait and a scan of ~25% of the genome identified five major quantitative trait loci (QTLs) that orchestrate hyphal initiation and elongation during unisexual reproduction. The *MAT* locus is the most prominent of the five QTLs, and the *MAT*α allele is linked to increased unisexual reproduction (Lin et al., 2006). This is perhaps not surprising as numerous regulatory molecules of unisexual and heterosexual reproduction are encoded by the mating-type locus (Feretzaki & Heitman, 2013; Lin et al., 2005; Lin, Jackson, Feretzaki, Xue, & Heitman, 2010). Pheromone production and sensing gene products activate the pheromone-sensing pathway, highly conserved in fungi, which controls unisexual reproduction through the G-protein-activated MAP kinase cascade (Figure 5.1) (Hsueh, Lin, Kwon-Chung, & Heitman, 2011). The pheromone receptors are coupled to G-protein signaling subunits that transfer the signal to the downstream kinases. Upon pheromone stimulation, the G-proteins activate the signal transduction protein Ste20α/**a** that then triggers the three-tiered phosphorylation cascade of the major kinases Ste11α/**a** (MAPKKK), Ste7 (MAPKK), and Cpk1 (MAPK) (Davidson, Nichols, Cox, Perfect, & Heitman, 2003; Hsueh et al., 2011).

Few genetic differences between unisexual and heterosexual reproduction have been identified thus far. The first pathway-specific elements to be discovered were the mating-type-specific *SXI1*α and *SXI2***a** genes, which encode homeodomain transcription factors that control mating-type identify and sexual development. Sxi1α and Sxi2**a** are required for heterosexual reproduction but are dispensable for unisexual reproduction (Hull, Boily, & Heitman, 2005; Hull, Davidson, & Heitman, 2002). These differences suggest divergent unisexual and heterosexual pathways with distinct downstream effectors of cell identity or sexual development.

The pheromone-signaling pathway is a critical regulator of mating in fungi and plays an essential role in recognizing and responding to opposite mating type cells (reviewed in Jones & Bennett, 2011). The pathway is structurally and functionally conserved among fungi closely (*U. maydis*) and distantly (*S. cerevisiae* and *C. albicans*) related to *Cryptococcus* and thus the majority of these components were identified in *Cryptococcus* through homology-based approaches. However, the downstream pathway of the pheromone-signaling cascade exhibits significant rewiring since its divergence from the most recent common ancestor shared with *S. cerevisiae*. The major transcription factor target of the pheromone signaling cascade is Ste12 in *S. cerevisiae*; however, in *Cryptococcus* Ste12 is dispensable for hyphal development during both unisexual and heterosexual reproduction (Fields & Herskowitz, 1985; Hartwell, 1980; Yue et al., 1999).

**(A)**                                                    **(B)**

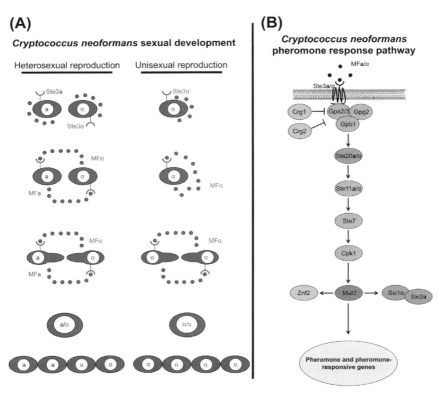

**Figure 5.1 Sexual development of _Cryptococcus neoformans_.** (A) During bisexual reproduction, cells of opposite mating-type secrete pheromones that are sensed by the pheromone receptor Ste3α/**a**. Pheromone sensing leads to cell–cell fusion producing a dikaryotic and then diploid intermediate that undergoes meiosis to generate recombinant progeny. During unisexual reproduction, nutrient-limiting conditions trigger pheromone production that enhances cell–cell fusion or endoreplication and the diploid intermediate undergoes meiosis, similar to bisexual reproduction. (B) The pheromone-signaling pathway governs sexual development in _C. neoformans_. Binding of pheromone to the pheromone receptor stimulates the pathway through a G-protein-coupled receptor complex that activates the Ste20α/**a** kinase and the three tiered MAPK phosphorelay system. The transcription factor target of the pathway is Mat2, which regulates the expression of sexual- and filamentation-specific genes. Although the signaling cascade is highly conserved among fungi, the downstream transcriptional network exhibits extensive rewiring in different species. (For color version of this figure, the reader is referred to the online version of this book.)

The major transcription factor target of pheromone sensing in _Cryptococcus_ is the High Mobility Group protein Mat2 (Lin et al., 2010). Mat2, along with other components of the pathway, is required for cell–cell fusion indicating an early role during hyphal development. Mat2 induces pheromone production and subsequently triggers pheromone-evoked responses by binding directly to a *cis*-regulatory sequence in the promoter region of the pheromone genes known as the pheromone-response element (Kruzel, Giles, & Hull, 2012). The transcriptional circuit of the pheromone cascade became more complex with the identification of Znf2, a novel zinc finger transcription factor. Znf2 is required for hyphal development, however, it is dispensable for cell–cell fusion. Surprisingly, deletion of the gene increases the efficiency of cell fusion events during heterosexual reproduction and stimulates pheromone expression during unisexual reproduction (Lin et al., 2010). All these components are essential for hyphal development during both unisexual and heterosexual reproduction. The only known exceptions are Sxi1$\alpha$ and Sxi2**a**, which are required for heterosexual but not unisexual reproduction. It is surprising that no other components of the pheromone-signaling or -response pathway have been identified as uniquely essential for either the unisexual or heterosexual pathways, especially given these pathways' role in restricting heterothallic systems to the heterosexual cycle. It may be that there are additional components that act downstream of the mating pathway to specifically regulate unisexual but not heterosexual reproduction. Further experimentation will allow the identification of these hypothesized factors and further elucidate the pathways.

The components of the mating pathway are highly conserved in *C. neoformans* and the sibling species *C. gattii*. Unisexual reproduction was first directly observed for laboratory strains of the *C. neoformans* serotype D lineage (Lin et al., 2005). More recent evidence suggests that unisexual reproduction also occurs in *C. neoformans* serotype A. Genotypic analysis of diploid serotype AD hybrid strains revealed a homozygous $\alpha/\alpha$ mating-type locus ($\alpha$AD$\alpha$) produced via mating between mating-type $\alpha$ isolates of different serotypes (Lin et al., 2007). Moreover, independent population genetic studies revealed evidence of recombination in serotype A isolates from trees in India and infected animals in Australia that were derived exclusively from mating-type $\alpha$ populations (Bui et al., 2008; Hiremath et al., 2008). Under laboratory conditions, $\alpha$AA$\alpha$ diploids generate abundant hyphae but few spores when cultured solo on mating media, indicating that unisexual reproduction occurs in this population. Robust hyphal growth but a paucity of spore production could suggest either a limited unisexual cycle or a defect of these isolates, which have been growing for extended periods in the intermediate diploid state. This diploid intermediate or unisexual product is relatively common in the population; screening ~500 environmental and clinical isolates found that ~8% were diploid, the majority of which were $\alpha$AA$\alpha$ diploids (Lin et al., 2009).

*Cryptococcus gattii*, the sibling species of *C. neoformans*, thought to be restricted to tropical and subtropical regions, is emerging as a pathogen of significant global public health importance (Byrnes, Bartlett, Perfect, & Heitman, 2011; Datta et al., 2009). Of particular interest is the link between sexual cycles with both the increasing geographic range of outbreaks and virulence (Byrnes et al., 2010; Fraser et al., 2005; Voelz et al., 2013). Although *C. gattii* undergoes heterosexual reproduction under laboratory conditions, like *C. neoformans*, the natural population is predominantly *MAT*α, providing further evidence of a correlation between unisexual reproduction and pathogenesis (Byrnes et al., 2010; Fraser et al., 2005; Fraser, Subaran, Nichols, & Heitman, 2003). Geographically isolated populations from *Eucalyptus* trees in Australia exhibit similar evidence of recombination in both α–**a** mixed populations and exclusively α mating-type populations (Saul, Krockenberger, & Carter, 2008). Moreover, the strains associated with the ongoing outbreak of meningoencephalitis on Vancouver Island and in the Pacific Northwest are exclusively mating-type α haploids that are fertile under laboratory conditions. Genotypic analysis suggests that the genotype associated with the outbreak could be the progeny of unisexual reproduction between two α mating-type parents (Fraser et al., 2003, 2005). In addition, the identification of an α/α diploid intermediate and the isolation of particles small enough to be spores present in air samples from Vancouver Island further supports that *C. gattii* undergoes unisexual reproduction in nature (Fraser et al., 2005; Kidd et al., 2007b). These studies show that unisexual reproduction occurs in nature in several independent lineages, and may facilitate the expansion of the geographic range of pathogens and contribute to the production of infectious spores.

## 2.2 *Candida albicans*

*Candida albicans* is the most common human fungal pathogen, normally associated with asymptomatic commensal colonization of the gastrointestinal tract and oral and vaginal mucosa of most of the world's population. *Candida albicans* is not commonly a burden in immunocompetent hosts but candidiasis of the oral cavity and candidemia leading to colonization of internal organs and in some cases the central nervous system are prevalent and cause serious infections in immunocompromised individuals (Klein et al., 1984). Diagnosis of candidemia is frequently too late for antifungal treatments to be effective, leading to mortality rates between 30% and 50% (Kibbler et al., 2003; Pfaller, Jones, Messer, Edmond, & Wenzel, 1998). Until recently, *C. albicans* was thought to be an obligate diploid organism with an asexual life cycle. However, the identification of the *MTL* locus and the isolation of mating-competent partners led to the discovery of a parasexual cycle (Hull & Johnson, 1999; Hull et al., 2000; Xie et al., 2013). Two diploid cells of the opposite mating type fuse to produce a transient tetraploid that undergoes stochastic chromosome loss to return to a diploid or aneuploid state (Bennett & Johnson, 2003; Legrand et al., 2004; Magee & Magee, 2000).

Sequencing of *C. albicans* clinical isolates revealed that the *MTL* locus contains highly conserved genes involved in mating, including the idiomorphic transcriptional regulators of cell identity **a**1, **a**2, α1, and α2 (Hull & Johnson, 1999). The *C. albicans MTL* locus is larger than the closely related ascomycetous fungus *S. cerevisiae MAT* locus, harboring additional genes that encode phosphatidylinositol kinases (*PIK*), oxysterol-binding proteins (*OBP*), and poly A polymerases (*PAP*). Almost 90% of environmental and clinical diploid isolates are heterozygous at the *MTL*, harboring both mating types (α/**a**) and do not mate (Miller & Johnson, 2002; Tavanti et al., 2005). Genetic manipulation to induce gene deletion or homozygosis of the chromosome containing the *MTL* locus generates mating-competent strains (*MTL*α/α, *MTL***a**/**a**, *MTL*Δ/α, and *MTL*Δ/**a**) and these strains mate successfully in infected animals and the laboratory (Hull & Johnson, 1999; Magee & Magee, 2000).

A further breakthrough in understanding mating was the discovery of a link to the phenotypic white–opaque cell-type switch. The switch from a white to an opaque cell enhances mating efficiency 1 million-fold (Miller & Johnson, 2002). The white–opaque switch was initially described more than two decades ago, where virulent white cells spontaneously switched to avirulent opaque cells (Slutsky et al., 1987). The phenotypic switch from white to opaque is driven by the transcription factor white–opaque regulator 1 (Wor1) through a positive feedback loop, during which Wor1 binds its own promoter to increase its transcription 40-fold (Huang et al., 2006; Zordan, Galgoczy, & Johnson, 2006). *WOR1* transcription is repressed by the *MTL* **a**1–α2 heterodimer in heterozygous diploid cells, which causes *MTL*α/**a** clinical isolates to repress white–opaque switching and mating (Figure 5.2) (Miller & Johnson, 2002). Other transcriptional regulators promote white–opaque switching, including Czf1 and Wor2, which act in concert with Wor1 to downregulate the expression of Efg1, a promoter of the white phenotype (Lachke, Srikantha, & Soll, 2003; Vinces, Haas, & Kumamoto, 2006; Zordan, Miller, Galgoczy, Tuch, & Johnson, 2007). In addition to genetic control, numerous environmental cues affect white–opaque switching. Opaque cells, but not white cells, colonize skin where mating occurs,

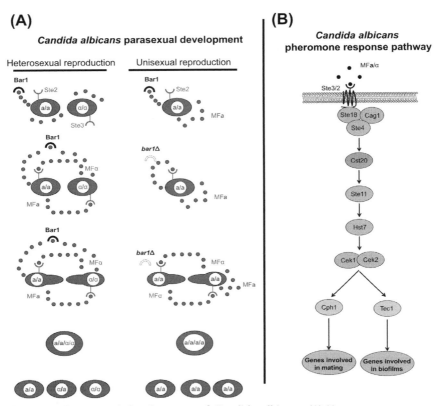

**Figure 5.2 Parasexual development of** *Candida albicans*. (A) Homozygous opaque *MTLα/α* cells secrete α pheromone that is sensed by the Ste2 pheromone receptor on the surface of *MTL***a/a** cells. During bisexual reproduction, pheromone sensing induces polarized growth toward the pheromone source and initiates cell–cell fusion. The intermediate tetraploid *MTLα/α/***a/a** cell undergoes stochastic chromosome loss that generates *MTL* homozygous and *MTL* heterozygous diploid progeny with significant rates of aneuploidy. Surprisingly, homozygous opaque *MTL***a/a** cells can secrete both **a** and α pheromones. The Bar1 protease degrades α pheromone to prevent autocrine activation of the mating pathway. In the absence of the Bar1 protease, *MTL***a/a** cells secrete α pheromone that accumulates and binds to the Ste2 pheromone receptor on the same or neighboring cell. The autocrine pheromone-signaling pathway drives unisexual reproduction that leads to cell–cell fusion and results in a tetraploid intermediate. The *MTL***a/a/a** cell undergoes stochastic chromosome loss to return to the diploid state, generating significant aneuploidy in the process. (B) The pheromone-response pathway is highly conserved in the Saccharomycotina subphylum and all the components of *C. albicans* are orthologs of the *S. cerevisiae* pheromone pathway. The pathway consists of the G-protein-coupled receptor complex and the Cst20 kinase followed by the MAPK signaling cascade. In opaque cells, the transcription factor target of the pathway is Cph1 that governs the expression of mating-specific genes, while in white cells, the Tec1 transcription factor regulates the abundance of biofilm-specific factors. (For color version of this figure, the reader is referred to the online version of this book.)

perhaps because the opaque phenotype is unstable at higher physiological temperatures (Kvaal et al., 1999; Lachke et al., 2003; Slutsky et al., 1987). In addition, despite the instability of opaque cells at high temperatures, elevated $CO_2$ and high concentration of N-acetylglucosamine promote white–opaque switching, conditions that mimic the environment of the GI tract where mating is likely to occur (Dumitru et al., 2007; Huang, Srikantha, Sahni, Yi, & Soll, 2009; Huang et al., 2010; Hull et al., 2000).

Sexual development of *C. albicans* is driven by pheromones that activate the highly conserved pheromone signaling cascade. Just as in *S. cerevisiae*, pheromones bind to the receptors (Ste2 or Ste3), which activate the coupled trimeric G-protein complex and the downstream Cst20 kinase (Magee, Legrand, Alarco, Raymond, & Magee, 2002). Subsequently, the signal triggers the sequential phosphorylation of the MAP kinase cascade components Ste11 (MAPKKK), Hst7 (MAPKK), and Cek1/Cek2 (MAPKs) (Chen, Chen, Lane, & Liu, 2002; Magee et al., 2002).

The same MAPK pathway operates in both white and opaque cells, but the response to pheromones is quite different. In opaque cells, pheromone binding leads to the formation of conjugation tubes and cell–cell fusion, while white cells respond to pheromone by increasing adhesion and forming biofilms (Daniels, Srikantha, Lockhart, Pujol, & Soll, 2006; Lockhart, Daniels, Zhao, Wessels, & Soll, 2003). The differential response to pheromones led to the hypothesis that, although opaque and white cells share the same pathway, the main transcription factor target is different. It was initially proposed that MAPK signaling activates the Cph1 transcription factor, the homolog of *S. cerevisiae* Ste12, which mediates the expression of mating-specific genes in opaque cells, while in white cells, pheromone signaling was proposed to activate Tec1, which orchestrates the expression of genes involved in adhesion and biofilm formation (Sahni et al., 2010; Yi et al., 2008). Although Tec1 is required for pheromone-induced biofilm formation, evidence has been presented that Cph1 is the sole transcriptional target of the pheromone signaling cascade in both white and opaque cells. The downstream targets of Cph1 diverge in white and opaque cells, mediating the expression of genes required for biofilm formation and mating (including pheromone secretion and sensing), respectively (Lin et al., 2013). All these components are essential for pheromone production in *C. albicans* and deletion of the genes severely impairs mating; however, their function is unknown during unisexual reproduction. Marked strains of α/α or **a/a** opaque cells cocultured with opposite mating-type cells produce marked unisexual mating products as well as heterosexual mating products, indicating that unisexual reproduction can occur in both α/α and **a/a** backgrounds in the presence of high pheromone concentrations (Alby et al., 2009). Based on these and other findings, we can infer that the same pheromone-signaling cascade likely operates in both α/α and **a/a** opaque cells during heterosexual and unisexual reproduction.

The specialized mating-competent opaque cells respond to pheromones to produce conjugation tubes that lead to cell–cell fusion and nuclear fusion to generate a uninucleate tetraploid cell (Bennett, Miller, Chua, Maxon, & Johnson, 2005). In contrast, white cells, which are unable to mate, respond to pheromone sensing by activating a specialized transcriptional pathway regulating biofilm formation (Sahni et al., 2009). If white cells are mating incompetent, it raises the question of the purpose of the white cell response to pheromones. The pheromones secreted have limited diffusion range, particularly the prenylated 14 amino acid MF**a** secreted by **a** cells. Opaque cells are rare in natural *Candida* populations; thus the majority of the population, comprised of white cells, forms biofilms in response to pheromones produced by minority opaque cells, facilitating a pheromone gradient that enables distant mating competent cells to locate opposite mating partners (Daniels et al., 2006). The initial steps of mating between opaque cells closely mirrors *S. cerevisiae* sexual reproduction and putative regulators of cell fusion may have conserved functions in *C. albicans* (Bennett, Uhl, Miller, & Johnson, 2003; Lockhart et al., 2003).

Unlike *S. cerevisiae*, which undergoes meiosis and sporulation to produce asci with four spores, *C. albicans* can mate to produce an **a/a**/α/α tetraploid but neither meiosis nor asci with spores have been observed. Although tetraploid *C. albicans* cells are stable, certain conditions induce a parasexual cycle. During parasex, ploidy is reduced through stochastic chromosome loss that returns tetraploid cells to the diploid or near-diploid state, generating considerable aneuploidy in the process (Bennett & Johnson, 2003). Although meiosis is absent, the parasexual cycle generates genetic recombination and gene conversion in the progeny (Forche et al., 2008). Moreover, *SPO11*, the meiosis-specific gene whose product is responsible for initiating recombination by inducing DNA double strand breaks, is dispensable for parasexual ploidy reduction; however, it is required for the observed recombination during concerted chromosome loss. Given that *SPO11* is a highly conserved meiotic "toolkit" gene conserved across eukaryotes, it suggests that a meiosis-like process may be operating during the parasexual cycle although it has also been suggested that Spo11 may have been reconfigured to play a novel mitotic recombination role, as it is expressed in mitotically dividing cells (Forche et al., 2008).

Candida albicans has well-established heterosexual mating between α/α and **a/a** cells followed by parasexual reduction; however, MLST analysis indicated that the populations of C. albicans are predominantly clonal, hinting that outbreeding is rare in nature (Odds et al., 2007). This raised the question of how prevalent white–opaque switching and heterosexual mating is in natural populations. However, in addition to opposite-sex mating, C. albicans utilizes an alternate mating pathway between cells of the same mating type (Alby et al., 2009). Unlike S. cerevisiae, where the cells express only one of the two pheromones, C. ablicans opaque **a/a** cells are able to express and secrete both α- and **a**-mating pheromones (Alby et al., 2009; Bennett & Johnson, 2006). Typically, the aspartyl protease Bar1 degrades the α pheromone produced by **a/a** cells and prevents autoactivation of the mating pathway in the absence of the α/α mating partner (Figure 5.2) (Schaefer, Cote, Whiteway, & Bennett, 2007). However, in the absence of this protease, **a/a** cells produce and respond to α pheromone and initiate mating through autocrine pheromone signaling. The α pheromone secreted by **a/a** cells binds the Ste2 pheromone receptor, stimulating the mating pathway and leading to pheromone responses. Cell and nuclear fusion occurs producing an **a/a/a/a** tetraploid that undergoes stochastic chromosome loss to return to the **a/a** diploid or aneuploid state, while undergoing some limited recombination (Alby et al., 2009). Cell fusion of **a/a** cells can also be induced by a minority of α/α cells that serve as donors of α pheromone in a ménage a trois mating. Inactivation of Bar1 might also occur in certain environmental niches such as the acidic environment of the vaginal mucosa, where lower pH could inhibit Bar1 and induce unisexual reproduction in vivo. A striking morphological difference between the heterosexual and unisexual cycle is that the α/α/**a/a** tetraploid undergoes an opaque–white switch due to the expression of the **a**1/α2 heterodimer (that inhibits opaque-specific transcriptional regulators) generating white sterile cells, while the **a/a/a/a** tetraploid remains opaque. The production of mating pheromones is crucial to the induction of unisexual reproduction, as in Cryptococcus species; however, the specific mechanisms of pheromone production and sensing are significantly different, emphasizing the diversity and independent origin of unisexual reproduction in the two species.

Candida albicans, long thought to be an obligate diploid, was recently discovered to be capable of producing haploid cells (Hickman et al., 2013). The obligate diploid hypothesis was proposed and historically suggested to result from balanced recessive lethal mutations that may be present throughout the genome. Haploid C. albicans cells have been observed only under selective laboratory conditions and in some chromosomes contained alleles from only one parental homolog, consistent with a limited number of haploid deleterious or lethal mutations. These haploids are proposed to arise through a concerted chromosome loss mechanism, similar to that of the tetraploid parasexual cycle (Bennett & Johnson, 2003; Hickman et al., 2013). As with diploids homozygous at the *MTL* locus, the haploid cells, which have a single copy of either the *MTL***a** or *MTL*α idiomorphs, efficiently switch from white to the opaque state and undergo heterosexual but not unisexual mating (Hickman et al., 2013). *Candida albicans* haploids are inherently unstable and undergo autodiploidization during propagation. The mechanism of autodiploidization is as yet unknown although it is not thought to involve homothallic mating, as neither haploid white or opaque cells of the same mating type were able to mate on mating media. Haploids exhibit a reduced growth rate and attenuated virulence although this fitness defect was rescued by mating but not autodiploidization, leading to the hypothesis that a burden of recessive mutations may be the cause of slow growth, and not the ploidy state of the cells.

Although recent studies revealed genetic evidence of recombination in clonal populations of *C. albicans*, the frequency of unisexual reproduction in nature remains unclear. Unisexual reproduction has been observed only in fertile opaque homozygous **a**/**a** or α/α cells, while the majority of clinical and environmental isolates are sterile white heterozygous α/**a** cells unable to switch to the opaque state in vitro. However, it is possible that certain conditions may stimulate white–opaque switching of α/**a** cells in vivo. Lower levels of hemoglobin in an in vitro model alter the expression of **a**1, **a**2, α1, and α2, which cause white α/**a** cells to behave phenotypically as **a**/**a** cells, inducing white–opaque switching and heterosexual reproduction (Pendrak, Yan, & Roberts, 2004). These white α/**a** cells are mating competent, leading to the hypothesis that unisexual reproduction can also occur under conditions in the mammalian host. When an α/**a** cell switches to **a**/**a**, Bar1 may function to allow the new **a** cell to escape the cell-cycle arrest invoked by the pheromone produced by the α progenitor, however, unisex is also repressed by the activity of Bar1 (Chan & Otte, 1982). Unisexual and heterosexual reproduction are both maintained in *C. albicans* and Bar1 may regulate the balance between inbreeding and outbreeding with each strategy generating genetic diversity in response to environmental cues in respective niches.

It is now appreciated that pheromones govern unisexual reproduction and general intercellular communication. Pheromones are implicated in both heterothallic and homothallic reproduction in a variety of fungi (Paoletti et al., 2007; Spellig, Bolker, Lottspeich, Frank, & Kahmann, 1994). Surprisingly, *C. albicans* pheromone/receptor binding, which regulates the pheromone cascade, exhibits considerable plasticity. Mutations in much of the 13 amino acid sequence of the α-mating pheromone did not affect signaling or unisexual reproduction of **a/a** cells. Moreover, *C. albicans* can respond to a variety of pheromone analogs, including pheromones from the related species *Candida dubliniensis*, *Candida parapsilosis*, and *Candida tropicalis*, which are able to bind to the receptor and induce unisexual reproduction in *C. albicans* WT opaque cells and biofilm formation in white cells (Alby & Bennett, 2011; Chen et al., 2002; Magee et al., 2002). This suggests that alternative pheromones may activate the pathway and promote homothal-lism in niches inhabited by diverse flora. *Candida albicans*, *C. dubliniensis*, *C. parapsilosis*, and *C. tropicalis* coinhabit the oral cavity of healthy indi-viduals, providing just such a diverse niche in nature (Daniels et al., 2006; Ghannoum et al., 2010; Lockhart et al., 2003; Martins et al., 2010; Melton et al., 2010). It is hypothesized that similar signals from the host or other microbiota may stimulate *C. albicans*' promiscuous pheromone receptor and signaling cascade to drive biofilm formation and unisexual reproduction in WT cells, as interspecies pheromone signaling may ameliorate the repres-sion of the Bar1 protease if these cross-species pheromones are resistant to Bar1 cleavage.

## 2.3 Neurospora

The *Neurospora* genus provides an interesting case study for the evolution of mating systems and unisex. The genus contains species with multiple mat-ing systems including heterothallism, homothallism, and pseudohomothallism (Wik, Karlsson, & Johannesson, 2008). By examining the synteny of the *mat* loci of homothallic *Neurospora* species and comparing them to the *mat* loci of closely related heterothallic species, it was discovered that the *mat* locus is conserved among heterothallic species belonging to distinct phylogenetic clades (Figure 5.3) (Gioti et al., 2012). However, the *mat* loci in the homo-thallic species have undergone distinct chromosomal rearrangements, suggest-ing that the ancestral state in *Neurospora* is heterothallism and homothallism has evolved independently in different lineages (Gioti et al., 2012). Of four transitions from heterothallism to homothallism examined in *Neurospora* spe-cies, three involve acquisition of compatible *mat* alleles into the same haploid genome. In two species, *Neurospora pannonica* and *Neurospora terricola*, unequal cross-over events or chromosomal translocations are hypothesized to have led

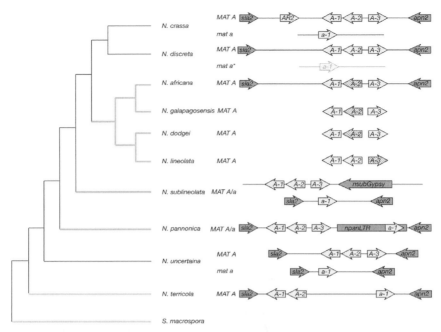

**Figure 5.3 *Neurospora* genus *mat* locus organization.** The organization of the *mat* locus of selected *Neurospora* species rooted with the homothallic outgroup species *Sordaria macrospora* (adapted from Gioti et al., 2012). Blue branches of the phylogram represent heterothallic species and red branches depict homothallic species. Gray arrows represent genes flanking the *mat* locus and their transcriptional orientation. Yellow arrows indicate mat genes, with *A-1, A-2, A-3*, and *a-1* corresponding to *mat A-1, mat A-2, mat A-3, and mat a-1*, respectively. Transposable elements are represented by purple and orange elements. The *N. discreta mat a* idiomorph has not yet been sequenced and is inferred phylogenetically. Genes truncated by stop codons are striped arrows (Wik et al., 2008). Sequencing of the coding regions and cDNA of *mat A-1, A-2*, and *A-3* revealed independent mutations that may result in pseudogenization. mat A-2, an HPG domain protein, suffered nonsense mutations at codon 39 in *N. dodgei* and codon 288 in *N. africana* and *N. galapagosensis*; however, *mat A-2* is still transcribed in *N. dodgei* and *N. galapagosensis*. The HMG transcription factor *mat A-3* is highly divergent in the homothallic species related to *N. africana; N. lineolata* has a stop codon at position 215, amino acids 100–177 are deleted in *N. africana* and *N. galapagosensis*, and *N. dodgei* has a mutation of the canonical start codon. (For interpretation of the references to color in this figure legend, the reader is referred to the online version of this book.)

to the linkage of the *mat A* and *mat a* loci on a single chromosome. In the *Neurospora sublineolata* genome, conversely, the *mat* loci are unlinked but both present in a haploid. Interestingly, the mating–type loci of homothallic species contain remnants of two novel retrotransposons, suggesting that the transition from heterothallism to homothallism might have been facilitated by these selfish genetic elements (Gioti et al., 2012).

*Neurospora* also contains a homothallic species with a single mating-type locus in the population. *Neurospora africana* retains only the *mat A* genes: a *mat a* idiomorph has not been discovered in the population and is not required for mating (Glass & Smith, 1994). The *mat A* idiomorph is conserved in sequence and arrangement between homothallic *N. africana* and heterothallic *Neurospora crassa*. It is not yet known what genetic mechanism enabled self-fertility but *N. africana* lacks any requirement for a contribution from an opposite mating-type cell, similar to other unisexual organisms. Intriguingly, *N. africana* is not unique. Three closely related species that share with *N. africana* a monophyletic origin, *Neurospora galapagosensis*, *Neurospora dodgei*, and *Neurospora lineolata*, are also homothallic and have only one mating-type idiomorph (Gioti, Stajich, & Johannesson, 2013; Glass, Metzenberg, & Raju, 1990; Nygren et al., 2011; Wik et al., 2008). This raises the possibility that the ancestor to these species was unisexual and this sexual cycle has persisted over significant evolutionary timescales.

## 3. CRYPTIC UNISEXUAL SPECIES

It is difficult to ignore the pattern of historical thought about the life cycles of fungi thought to be asexual. It was often observed that each population was asexual and clonal, often with just a single idiomorph at the mating-type locus but retaining well-conserved meiotic genes. Further investigation of population genetics revealed evidence of recombination before discovery of a cryptic sexual cycle. We hypothesize that this pattern may be relevant to additional fungal species, including pathogenic species, which are thought to be clonal and lack a recognized sexual cycle. *Trichophyton rubrum* has long been thought to be clonal, however, signatures of recombination were detected (Gräser, Kühnisch, & Presber, 1999). Only one mating-type idiomorph has been reported, but both mating and meiosis genes are well conserved, suggesting that there might be an extant unisexual cycle (Kano et al., 2013; Li, Metin, White, & Heitman, 2010; Martinez et al., 2012). *Alternaria* species are important agricultural pathogens that were first reported as clonal, however, evidence now suggests there may be a cryptic sexual or unisexual cycle (Peever et al., 1999; Simmons, 1999). Two highly conserved mating-type idiomorphs are found in *Alternaria alternata*, in addition to ample evidence of recombination both between opposite mating type populations and in populations with a single mating type (Berbee, Payne, Zhang, Roberts, & Turgeon, 2003; Stewart, Kawabe, Abdo, Arie, & Peever, 2011; Stewart et al., 2013). Following a similar trajectory, evidence of recombination in populations of *Batrachochytrium dendrobatidis*, a global zoonotic pathogen thought to be asexual, was discovered (Farrer et al., 2011, 2013; James et al., 2009; Morgan et al., 2007). While the mating-type locus has not yet been identified, meiotic genes are present and conserved in the *B. dendrobatidis* genome (Halary et al., 2011). *Ashbya gossypii* is a preduplication filamentous ascomycete closely related to *S. cerevisiae* (Dietrich et al., 2004; Simmons, 1986, 1999). The *A. gossypii* genome contains three identical unlinked copies of *MATa*, a truncated *MATα*, and a single complete *MATα* locus (Dietrich, Voegeli, Kuo, & Philippsen, 2013; Wendland & Walther, 2011). The standard lab strain contains only *MATa* loci, but pheromone sensing is not required for *A. gossypii* to sporulate, as strains deleted for the α-pheromone receptor *STE2* or the **a**-pheromone receptor *STE3* still are able to sporulate (Wendland, Dünkler, & Walther, 2011). Strains deleted for the mating transcription factor *STE12* exhibited increased sporulation, further complicating the question of what, if any sexual or unisexual cycle *A. gossypii* engages in (Wendland, Dünkler, & Walther, 2011). A detailed investigation may yield evidence of heretofore-unrecognized heterothallic or homothallic sexual cycles in these species.

## 4. EVOLUTIONARY ORIGIN OF UNISEX

Transitions between heterothallism and homothallism, and an increased potential for inbreeding, are common throughout the fungal kingdom (Debuchy & Turgeon, 2006; Gioti et al., 2012; Inderbitzin et al., 2005; Yun, Berbee, Yoder, & Turgeon, 1999). Homothallism may be hypothesized to confer long-term evolutionary costs, primarily because the selfing population will inevitably experience reduced effective recombination rates and population size, and consequently have reduced efficacy of purifying selection and become more prone to genetic drift (Charlesworth & Wright, 2001; Hill & Robertson, 1966; Otto & Lenormand, 2002; Pollak, 1987), which could lead to genomic maladaptation, including the spread of deleterious selfish genetic elements (e.g. transposable elements) as well as accelerated rates of protein evolution or decay. However, given the frequency with which species have become homothallic, we must consider that homothallism could be a neutral mutational event (Lynch, 2007). When a mutation occurs that reduces self-incompatibility, there may be no immediate negative or positive selection on that mutation. Mitotically dividing fungal populations show a spectrum of high-frequency mutations, including both adaptive and neutral changes (Lang et al., 2013). Mutations conferring self-compatibility may occur in species and a homothallic sexual cycle could then be fixed stochastically by drift or carried to fixation by linked adaptive mutations.

Against this neutral hypothesis, it must be considered that changes from a heterothallic to homothallic sexual cycle impact a species' life cycle and genome in remarkable ways. As predicted by population genetic theory, homothallic *Neurospora* species show relaxation of purifying selection in protein-coding genes, reduced efficiency in silencing transposable elements, as well as reduced codon usage bias in highly expressed genes (Gioti et al., 2013). Additionally, a study of genome sequences from 16 yeast species in the family Saccharomycetaceae found that the *MAT* locus appears to be a deletion hotspot, in that the distance between *MAT* and *HML* is gradually eroding over evolutionary time as genes near *MAT* are repeatedly deleted, truncated, and transposed (Gordon et al., 2011). The authors proposed that this evolutionary erosion of the yeast sex chromosomes is caused by accidents occurring during mating-type switching, coupled with the selection pressure to keep *MAT* and *HML* on the same chromosome. There is ample evidence supporting the hypothesis that selfing can have long-term negative effects. On the other hand, self-fertility ensures sexual reproduction, which could provide short-term and even long-term benefits to the species such as generation of spores that can better cope with harsh environments and generation of novel genotypes through meiosis (Table 5.1). The reproductive strategy being selected for in a particular species depends on interactions among many factors, including both environmental

**Table 5.1** Putative Impacts of Unisexual Reproduction

*Access to Recombination*
　　Avoid Muller's Ratchet
　　Recombination within the mating-type locus
*Increased Mating*
　　Universal mating partner
　　Unisexual mating improves fitness for heterosexual mating
　　Meiotic rejuvenation (Ünal, Kinde, & Amon, 2011)
　　Production of hyphae and spores to explore new environments
*Increased Genetic Diversity*
　　Access to the diploid state as a capacitor for evolution
　　Generation of epimutations through sex-induced silencing (SIS)
　　Meiotic *de novo* generation of phenotypic and genotypic diversity
　　Biparental mitochondrial inheritance and mitochondrial recombination
*Other Impacts*
　　Toggle ploidy to enhance survival in environments where the haploid or diploid is more fit
　　Suppression of transposons

(e.g. nutrient availability) and biological (e.g. population structure). Given all the long-term disadvantages potentially associated with selfing, the fact that many species maintain the ability to undergo homothallic reproduction and selfing suggests the conditions that favor homothallism may be widespread.

## 5. UNISEX INCREASES THE OPPORTUNITIES FOR MEIOSIS

That most eukaryotes reproduce sexually during their life cycle is puzzling when considering the twofold cost of sex (Maynard Smith, 1978; Williams, 1975). The first cost is genome dilution; each allele will be passed on to each daughter during clonal division but has just a 50% chance to be passed on during sexual reproduction. The second cost is that each daughter requires contributions from two parents while a single mother can produce a daughter mitotically. This is typically thought of as a cost of males but other mating systems have similar costs (Lloyd, 1988). Homothallic reproductive strategies such as unisex may have initially been selected, in part, to avoid the twofold cost of modern sex.

The ubiquity of sex must be the result of significant advantages to counterbalance the costs of sex (Bell, 1982; Williams, 1975). Hermann Muller recognized that asexually dividing populations faced a fundamental challenge; their genomes would gradually accumulate deleterious mutations in an irreversible manner, termed Muller's Ratchet (Figure 5.4) (Felsenstein, 1974; Haldane, 1937; Muller, 1932, 1964). Every individual in a population would eventually suffer a harmful mutation, and selection will not be sufficient to maintain fitness without the ability to recombine and purge deleterious mutations. Sexual reproduction allows organisms to use recombination to purge the genome of these deleterious mutations to avoid Muller's Ratchet. In a few species with characterized unisexual pathways, both sexual and unisexual cycles utilize the same post-cell-fusion meiotic and recombinational pathways (Bui et al., 2008; Feretzaki & Heitman, 2013; Lin et al., 2005). Therefore, unisexual reproduction, like heterosexual sexual reproduction, can be hypothesized to allow populations to avoid Muller's Ratchet but this remains to be documented experimentally. Using unisex to access recombination is especially important for pathogenic species, where a host might be colonized by a small number of cells of a single mating type.

Sex increases genetic and phenotypic diversity of populations and species through recombination (Dobzhansky, Levene, Spassky, & Spassky, 1959;

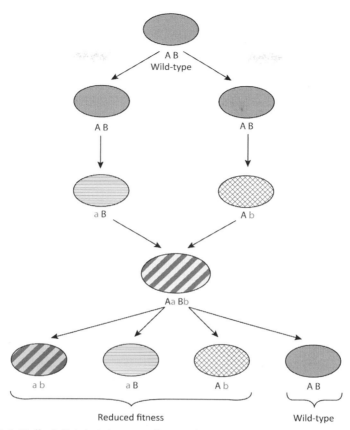

**Figure 5.4 Muller's Ratchet.** In asexually growing organisms, every cell or lineage will eventually suffer a deleterious mutation. There is no mechanism for selection to remove all the deleterious mutations ("a", "b") from the population, and therefore, fitness will irreversibly decline in mitotically dividing populations. Heterosexual and unisexual cycles give access to recombination that may restore wild-type ("AB") fitness to some lineages in the population. (For interpretation of the references to color in this figure legend, the reader is referred to the online version of this book.)

Levene, 1959; Spassky, Spassky, Levene, & Dobzhansky, 1958; Spiess, 1959). Recombination is not the only mechanism for unisex to promote fungal genetic diversity; however, diversity is also generated through aneuploidy in fungi, which can be deleterious in many eukaryotes (Pavelka et al., 2010; Selmecki, Forche, & Berman, 2010; Torres et al., 2007, 2010). Aneuploidy is a facilitator of fungal responses to changing or stressful environmental conditions (Dunham et al., 2002; Rancati et al., 2008; Yona et al., 2012). In addition, in fungal pathogens such as *C. neoformans* and *C. albicans*, aneuploid chromosomes commonly confer antifungal drug resistance (Selmecki, Dulmage, Cowen, Anderson, & Berman, 2009; Selmecki, Forche, & Berman, 2006; Sionov, Lee, Chang, & Kwon-Chung, 2010). Unisexual reproduction is able to generate de novo genetic diversity and aneuploidy (Ni et al., 2013). In *C. neoformans*, changes in pathogenic phenotypes were observed in ~7% of the progeny of unisexual reproduction and two-thirds of these variant progeny were aneuploid for at least one chromosome and aneuploidy was the cause of their variant phenotype. Understanding how unisex generates genetic and phenotypic diversity, and how this enables pathogens to adapt to new environments and medical interventions, is of particular importance in dealing with emerging and established fungal pathogens.

Transition from a heterothallic to homothallic sexual cycle changes the gene flow and selection dynamics of a population, in particular presenting new barriers to the invasion of selfish genetic elements. Transposons have been found in all organisms with sequenced genomes but there is still some debate as to the way they invade populations. Transposons, like all selfish genetic elements, are fundamentally sexually transmitted parasites of a genome. Invasion of these selfish elements relies on being transmitted to offspring at a frequency higher than expected under Mendelian segregation. Population genetic models predict that a change from outcrossing to selfing will reduce the effectiveness of this selfish strategy in diploid organisms, as increased homozygosity of these deleterious transposons will increase the strength of selection in the population until the elements are purged from the genome (Boutin, Le Rouzic, & Capy, 2012; Wright, Ness, Foxe, & Barrett, 2008; Wright & Schoen, 1999). Comparisons between selfing and outcrossing species of plants, as well as hermaphroditic and outcrossing nematodes, found that selfing lineages had significantly fewer transposable elements in their genomes (Dolgin, Charlesworth, & Cutter, 2008; Morgan, 2001). A change to a unisexual cycle may leverage increased selection to reduce the number of transposons in the genome.

The switch to homothallism is likely to lead to increased opportunities for meiosis and the significant meiotic transcriptional changes and marshaling of genomic defenses. Some fungal species recognize repeated sequences and utilize a repeat-induced point mutation (RIP) or methylation-induced premeiotically (MIP) mechanism to disable transposons that are highly active during the sexual cycle (Graia et al., 2001; Hamann, Feller, & Osiewacz, 2000; Hua-Van, Hericourt, Capy, Daboussi, & Langin, 1998; Ikeda et al., 2002; Nakayashiki, Nishimoto, Ikeda, Tosa, & Mayama, 1999; Neuveglise, Sarfati, Latge, & Paris, 1996; Selker & Stevens, 1985). RIP detects linked and unlinked sequences that are repeated two or more times in the genome and causes G–C to A–T transition mutations, efficiently degrading duplicated genetic elements, although RIP is less efficient in selfing populations (Cambareri, Jensen, Schabtach, & Selker, 1989; Selker, Cambareri, Jensen, & Haack, 1987). MIP targets similar duplicated sequence for methylation and silencing prior to meiosis (Barry, Faugeron, & Rossignol, 1993; Faugeron, Rhounim, & Rossignol, 1990; Freedman & Pukkila, 1993; Goyon & Faugeron, 1989).

In addition to silencing repeated sequences, unpaired DNA is targeted for modification and repression during meiosis. A new insertion of a transposon in a haploid will have no match on the homologous chromosome during mating. This unpaired DNA activates the RNAi response to silence all homologs of the unpaired DNA (Aramayo & Metzenberg, 1996; Janbon et al., 2010; Shiu & Metzenberg, 2002; Shiu, Raju, Zickler, & Metzenberg, 2001; Son, Min, Lee, Raju, & Lee, 2011). RNAi silencing is significantly upregulated during heterosexual cycles suppressing new transposon insertions in meiosis. In addition to suppressing transposable elements, RNAi can increase phenotypic diversity during environmental challenges. *Cryptococcus* uses RNAi to epigenetically silence repeated genes and generate phenotypic diversity. This silencing can be activated during mitotic or meiotic growth, however, the silencing is 250 times more efficient during heterosexual and unisexual reproduction cycles (Wang, Darwiche, & Heitman, 2013; Wang et al., 2010, 2012).

Outcrossing enhances the advantages of recombination through increasing the genetic diversity of populations while decreasing the risk of a diploid carrying a homozygous deleterious mutation that recombination cannot restore. Plants and animals often have molecular and developmental systems to impede diploid selfing, such as dioecy. Fungi such as *Cryptococcus* determine sexual compatibility by their haploid genotypes and typically cannot inhibit diploid selfing. Though predominately diploid, *Candida* species can grow and are competent to mate as haploids or diploids. Restriction of mating to opaque, homozygous *MTL* locus cells combined with a paucity of haploids in the natural population is likely to effectively promote outcrossing during heterosex because *MTL* α/**a** cells do not undergo meiosis to produce mating-compatible haploids. In addition to genetic or molecular mechanisms, outcrossing can be promoted through developmental strategies, such as wide dispersal of spores (Murphy & Zeyl, 2010; Wright et al., 2008). Fungal heterosexual reproduction promotes genetic diversity and outcrossing through incompatibility with cells of the same mating type and the depression of haploid selfing. One cost of self-incompatibility is a relative dearth of potential mating partners in the populations. In bipolar species with equal frequencies of *MAT* alleles, the chance of any two cells being compatible is 50% but for those species with skewed *MAT* distributions, the chances that any two cells are mating compatible is much lower. These tradeoffs make heterothallism, or obligate outcrossing, a high-risk and high-reward strategy. Some portion of the population may not be able to find a compatible mating partner but because there is no chance of selfing, the progeny of successful matings will have higher genetic diversity.

Unisexual reproduction has arisen in fungal species that are both predominantly diploid, such as *Candida* species, or haploid, such as *Cryptococcus* species. The costs posed by incompatibility with same mating-type partners are reduced when all cells are available for mating through unisex for predominantly haploid species (Iwasa & Sasaki, 1987). If an opposite mating-type partner is not available, a same mating-type partner or even a daughter can be a unisexual partner. Haploids expend considerable resources signaling, responding, and interacting with potential mating partners in the often stressful conditions that prompt mating (Xu, 2005). Unisexual haploids may be able to expend less energy finding suitable mates, and any haploid encountered will be compatible. Unisexual reproduction may have contradictory effects on the genetic diversity of species by increasing the number of successful matings with unrelated same mating-type partners while also increasing inbreeding depression by making daughters available for mating. The benefits of increased outcrossing are even more pronounced in a species like *C. neoformans*, a species with uneven mating-type ratios (Halliday et al., 1999; Kwon-Chung & Bennett, 1978; Litvintseva et al., 2005; Yan, Li, & Xu, 2002). The changes in outcrossing rate will be more modest in species that grow predominately as diploid, undergoing meiosis and returning quickly to the diploid state, as even in diploids that are strictly heterothallic spores produced by a diploid are able to mate with sisters. A unisexual cycle does increase the proportion of other siblings from a single meiotic event that a spore is compatible with from 50% to 100%. When diploids carry lethal or deleterious recessive mutations, as *C. albicans* does, this promiscuity may be even more critical to enable unfit haploids to return to the favored diploid state. After sporulation, these deleterious mutations can render a portion of the opposite mating-type haploids inviable or unsuitable for mating. Unisex allows a greater number of haploids to return to the favored diploid condition in such circumstances.

Several fungi with identified unisexual cycles grow mitotically primarily as haploids. While the full impact of ploidy on evolution and response to selection is not yet clear, a diploid cell clearly has several evolutionary advantages over the haploid (Gerstein & Otto, 2009; Haldane, 1937). First, diploids simply have twice the genetic material for mutation to act upon. Each gene is present twice and each copy is replicated during cell division providing additional substrate on which mutation can act (Orr & Otto, 1994). This doubling of mutational targets is especially important if the population facing selection is small, such as during an incipient infection (Zeyl, Vanderford, & Carter, 2003). Second, haploids face a challenge when confronting selective pressure that diploids avoid. Any genetic changes a haploid suffers are immediately expressed and potentially impact the phenotype. It may not be possible for haploids to acquire certain phenotypes or genetic changes directly. A new phenotype may require several genetic changes, one or more of which are detrimental but act epistatically or combinatorially to be advantageous. Any one of the changes may be detrimental, and subject to strong negative selection in haploids, but once several mutations occur, the changes are advantageous. It is difficult for haploids to cross these selective valleys to reach a global peak. On the other hand, diploids are able to complement initially deleterious heterozygous mutations with a dominant wild-type allele and carry those mutations for generations until the epistatic or compensatory mutation occurs. Thus, diploids can serve as a capacitor for evolution by increasing the quantity of DNA substrate for mutation to act on and buffering mutations from selection. As diploids complete a sexual cycle and return to the haploid state, mutations and new phenotypes are exposed to selection. The unisexual cycle allows haploid cells to reach the diploid state more frequently and without the need to find an opposite mating-type cell, expanding the evolutionary potential of a haploid.

While sex is nearly universal in the eukaryotic lineage, it is not clear how meiotic ability is maintained in facultative sexual organisms such as fungi. Conditions suitable for mating, low-nutrient conditions, and the presence of a compatible mating partners may not arise for many generations, resulting in relaxed selection on mating pathway genes. While a population may be able to divide mitotically indefinitely, continual mitotic growth results in a rapid decrease in fecundity (Lang, Murray, & Botstein, 2009; Xu, 2002; Xu et al., 2000a). Unisexual reproduction is an intermittent substitute when conditions are unfavorable for heterosexual mating or compatible mating-type partners are not available. Unisexual reproduction utilizes the same, or very similar, genetic pathways as heterosexual reproduction (Feretzaki & Heitman, 2013). Frequent unisexual mating continually selects for functional or increasingly functional alleles of mating pathway genes, acting as practice for heterosexual mating. Thus, unisexual mating may help increase heterosexual fecundity through maintenance of shared pathways.

## 6. UNISEX ALTERS THE LIFE CYCLE OF FUNGI

The transition from haploid to diploid or diploid to haploid brings about a large number of transcriptional, phenotypic, and morphological changes. These changes can be environmentally dependent, or specific to the lifecycle of a species. In general, under limiting growth conditions, diploid cells are larger, increasing the ratio of volume to surface area. The change in surface area changes the number of transporters and receptors the cell can utilize while a change in volume can affect the protein content and number of organelles. Changes in the ratio of these components interact with specific environmental conditions. In addition, there are species-specific morphological changes that are dependent on changes in ploidy state as well, such as the transition from yeast to hyphae during the sexual cycle in *Cryptococcus* and *U. maydis* (Snetselaar, Bölker, & Kahmann, 1996). The unisexual cycle can give access to these advantageous morphological and transcriptional changes.

Many unisexual species undergo significant morphological changes during the sexual cycle. *Cryptococcus neoformans* grows mitotically as a haploid yeast and only produces hyphae, basidia, diploid cells, and haploid spores during the unisexual or heterosexual cycles. These cell types have significant morphological changes that can be advantageous. *Cryptococcus neoformans* uses the hyphae produced by the unisexual cycle to explore the environment and forage for nutrients (Phadke, Feretzaki, & Heitman, 2013). The exploration of the environment by hyphae may also have other salutary effects such as bringing mating-ready cells into contact with relatively distant mating partners (Table 5.1). In this way, unisex mating can increase the number and frequency of outcrossing by enabling foraging for mates, akin to courtship in *S. cerevisiae* (Jackson & Hartwell, 1990a, 1990b).

Perhaps the most important morphological state that a sexual cycle gives access to is the spore. Sporulation is a response to adverse environmental conditions and is essential for dispersal to new niches. Species from the *Cryptococcus* genus are able to produce spores only via a unisexual or heterosexual cycle. A unisexual cycle may be essential for lineage survival in the case of a relative absence of opposite mating-type partners. Other species, such as *Neurospora*, are able to generate mitotic spores called conidia that are able to effectively disperse and fill many roles of a sexual spore. However, even in these species, there can be significant differences between the survival rates between mitotically generated conidia and meiotically generated spores (Trapero-Casas & Kaiser, 2007).

In addition to morphological changes, unisex changes the ploidy from haploid to diploid and allows the cell to return to the haploid state. Significant effort has been made to model under what environmental conditions a diploid or haploid state would be favored. Factors that have been considered include nutrient conditions, mutational load, and life cycle but experimental work had thus far been unable to support general conditions that favor a particular ploidy state (Adams & Hansche, 1974; Anderson, Sirjusingh, & Ricker, 2004; Crow & Kimura, 1965; Lewis, 1985; Mable, 2001; Mable & Otto, 2001; Orr & Otto, 1994; Weiss, Kukora, & Adams, 1975). Recently, an extensive study of the interactions between ploidy and environment in the *Saccharomyces* group found that an impact of ploidy on growth is common. Under some environmental conditions, ploidy had no impact on growth but for those conditions where the growth rates of different ploidy states differ, nearly equal numbers of conditions favored diploids and haploids (Zörgö et al., 2013). Advantages in an environmental condition of a particular ploidy were consistent over significant evolutionary time but there were no consistent growth advantages across environmental conditions for a ploidy state. Even in DNA damaging conditions, where an additional copy of each gene was hypothesized to be advantageous, there was no consistent growth advantage for diploids. Phleomycin, an inducer of DNA lesions, strongly favored diploids while hydroxyurea, which impedes DNA repair, favored haploids. Nutrient-limiting conditions did not clearly favor either the haploid or diploid state, with some limiting nutrients favoring haploids and others favoring diploids. Thus, each ploidy state confers advantages depending on the environment. A unisexual organism can easily access both the haploid and diploid state by mating quickly in response to conditions that favor diploids and sporulating when conditions are favorable for haploid growth. A unisexual cycle that simply toggles a cell from haploid to diploid and then back to haploid could provide benefits to the organism in a fluctuating environment with no need to admix genetic diversity, or recombine.

Cellular and organelle genomic interactions are common and fraught with implications for sexual reproduction. Multicellular organisms uniparentally inherit mitochondrial and chloroplast genomes to limit organelle heteroplasmy. While uniparental inheritance is common in fungal species, it is by no means the rule. The model yeasts *S. cerevisiae* and *S. pombe* inherit mitochondrial genomes from either parent but *Cryptococcus* species uniparentally inherit their genome from the mating-type **a** parent (Callen, 1974; Seitz-Mayr, Wolf, & Kaudewitz, 1978; Strausberg & Perlman, 1978; Xu et al., 2000a). Mitochondrial heteroplasmy causes reduced metabolic activity and accentuated stress responses in multicellular eukaryotes and is a common cause of disease (Sharpley et al., 2012) but it is not yet clear what effect mitochondrial heteroplasmy has on fungi. *Cryptococcus neoformans* uniparentally inherits the mitochondria from the mating-type **a** parent but mating-type α is the prevalent unisexual mating type (Kwon-Chung & Bennett, 1978; Yan & Xu, 2003). Unisex is likely to disrupt the uniparental inheritance of mitochondria. In *C. neoformans*, mating-type α will retain its mitochondria instead of receiving a mating-type **a**-derived mitochondria. There is no known mechanism to differentiate between two mating-type α mitochondria during unisex, potentially introducing mitochondrial heteroplasmy and enabling mitochondrial genomic recombination (Yan, Hull, Sun, Heitman, & Xu, 2007). In addition, it would allow mitochondria that had evolved an epistatic, commensal, or selfish interaction with a mating-type α genome to be transmitted through the unisexual cycle.

## 7. MEDICAL IMPACTS OF UNISEX

Despite the ubiquity of sexual reproduction in complex eukaryotes and early observance in fungal lineages, asexuality was thought to be common among pathogenic eukaryotes. The widespread clonal reproduction among pathogenic and parasitic microorganisms engaged in an evolutionary arms race with hosts was even more surprising given the "Red Queen" arms race that pathogens must engage in with their hosts (Van Valen, 1974a, 1974b). A host–pathogen arms race places special emphasis on an organism's ability to rapidly evolve and take advantage of genetic diversity (Delsing, Bleeker-Rovers, Kullberg, & Netea, 2012; Maor & Shirasu, 2005; Maynard Smith, 1978). Indeed, genetic conflict between hosts and pathogens is one of the leading evolutionary hypotheses for the ubiquity of sex in eukaryotic hosts (Ladle, Johnstone, & Judson, 1993; Otto & Michalakis, 1998; West, Lively, & Read, 1999). Selection of pathogens leads to intense specialization for common host genotypes, leading to the reduction of those hosts' fitness relative to rare host genotypes (Haldane, 1949; Williams, 1975). Sex and recombination in hosts helps to increase genetic diversity and generate new and rare genotypes (Hamilton, 1980; Hamilton, Axelrod, & Tanese, 1990). Eukaryotic pathogens, such as *C. albicans* and *C. neoformans*, can employ sexual reproduction to maintain diversity and produce genotypes that are competent to infect new hosts (Byrnes et al., 2009; Carriconde et al., 2011; Kidd et al., 2004).

It was first noted that although there was little evidence of recombination, the genomes of these pathogenic eukaryotes retained a full complement of meiotic genes (Bougnoux et al., 2008; Loftus et al., 2005; Ramesh et al., 2005; Sturm, Vargas, Westenberger, Zingales, & Campbell, 2003; Tzung et al., 2001). Only in the last decade has it been considered possible that eukaryotic microorganisms are both clonal and sexually reproducing (Heitman, 2006, 2010). The discovery of both canonical sexual and a new novel unisexual cycle in multiple pathogenic fungi, and also parasites, revealed how eukaryotic microbial pathogens are equipped to engage in the arms race with their hosts (Alby et al., 2009; Lin et al., 2005). Thus, we now appreciate that rather than being asexual, clonal, and strictly mitotic as proposed as recently as a decade ago, eukaryotic microbial pathogens are unisexual, or cryptically sexual, resolving this decades-long paradox.

Unisex plays a particularly important role in *C. neoformans* virulence as the infectious propagule is thought to be the spores produced by sexual reproduction (Giles et al., 2009; Sukroongreung, Kitiniyom, Nilakul, & Tantimavanich, 1998; Velagapudi et al., 2009). The *C. neoformans* population is >95% α dramatically limiting the opportunities for heterosexual spore generation leaving unisexual spores as the likely major infectious propagule (Halliday et al., 1999; Kwon-Chung & Bennett, 1978; Litvintseva et al., 2005). In addition, *C. neoformans* deploys unisex to generate de novo genetic variation through mutation and aneuploidy, after infection bottleneck (Ni et al., 2013; Sionov et al., 2010). These factors highlight how unisex may be integral to the virulence of this medically important fungal pathogen.

Homothallism is widespread in the fungal kingdom and the full extent of alternative homothallic mechanisms such as unisex is just beginning to be investigated. It is particularly intriguing that extant unisexual cycles were discovered in phylogenetically divergent human pathogens only relatively recently. Unisex may be a particularly advantageous means of homothallism for animal pathogens. Further research into the molecular mechanisms underlying unisex, how outcrossing versus inbreeding is regulated, the diversity and extent of unisexual species, and the environmental impacts of unisex will increase our understanding of fungal diversity in addition to its impact and evolutionary origin.

## ACKNOWLEDGMENTS

We thank Dr Eve Chow, Blake Billmyre, and Shelly Clancey for critical reading and thoughtful discussions. KCR is supported by the NIH Molecular Mycology and Pathogenesis Training Program (AI52080). This work was supported by NIH/NIAID R37 grant AI39115-16 and R01 grant AI50113-10 to J.H.

# REFERENCES

Adams, J., & Hansche, P. E. (1974). Population studies in microorganisms I. Evolution of diploidy in *Saccharomyces cerevisiae*. *Genetics, 76*(2), 327–338.

Alby, K., & Bennett, R. J. (2011). Interspecies pheromone signaling promotes biofilm formation and same-sex mating in *Candida albicans*. *Proceedings of the National Academy of Sciences of the United States of America, 108*(6), 2510–2515. http://dx.doi.org/10.1073/pnas.1017234108.

Alby, K., Schaefer, D., & Bennett, R. J. (2009). Homothallic and heterothallic mating in the opportunistic pathogen *Candida albicans*. *Nature, 460*(7257), 890–893. http://dx.doi.org/10.1038/nature08252.

Ames, L. M. (1934). Hermaphroditism involving self-sterility and cross-fertility in the ascomycete *Pleurage anserina*. *Mycologia, 26*(5), 392–414. http://dx.doi.org/10.2307/3754255.

Anderson, J. B., Sirjusingh, C., & Ricker, N. (2004). Haploidy, diploidy and evolution of antifungal drug resistance in *Saccharomyces cerevisiae*. *Genetics, 168*(4), 1915–1923. http://dx.doi.org/10.1534/genetics.104.033266.

Aramayo, R., & Metzenberg, R. L. (1996). Meiotic transvection in fungi. *Cell, 86*(1), 103–113. http://dx.doi.org/10.1016/S0092-8674(00)80081-1.

Arcangioli, B., & de Lahondes, R. (2000). Fission yeast switches mating type by a replication-recombination coupled process. *The EMBO Journal, 19*(6), 1389–1396. http://dx.doi.org/10.1093/emboj/19.6.1389.

Asker, S. E., & Jerling, L. (1992). *Apomixis in Plants*. Boca Raton, FL: CRC Press.

Baldauf, S. L. (2003). The deep roots of eukaryotes. *Science, 300*(5626), 1703–1706. http://dx.doi.org/10.1126/science.1085544.

Baldauf, S. L., Roger, A. J., Wenk-Siefert, I., & Doolittle, W. F. (2000). A kingdom-level phylogeny of eukaryotes based on combined protein data. *Science, 290*(5493), 972–977. http://dx.doi.org/10.1126/science.290.5493.972.

Barry, C., Faugeron, G., & Rossignol, J. L. (1993). Methylation induced premeiotically in *Ascobolus*: Coextension with DNA repeat lengths and effect on transcript elongation. *Proceedings of the National Academy of Sciences of the United States of America, 90*(10), 4557–4561.

Barsoum, E., Martinez, P., & Åström, S. U. (2010). α3, a transposable element that promotes host sexual reproduction. *Genes & Development, 24*(1), 33–44. http://dx.doi.org/10.1101/gad.557310.

Bell, G. (1982). *The Masterpiece of Nature. The Evolution and Genetics of Sexuality*. Los Angeles, CA: University of California Press.

Bennett, R. J., & Johnson, A. D. (2003). Completion of a parasexual cycle in *Candida albicans* by induced chromosome loss in tetraploid strains. *The EMBO Journal, 22*(10), 2505–2515.

Bennett, R. J., & Johnson, A. D. (2006). The role of nutrient regulation and the Gpa2 protein in the mating pheromone response of *C. albicans*. *Molecular Microbiology, 62*(1), 100–119. http://dx.doi.org/10.1111/j.1365-2958.2006.05367.x.

Bennett, R. J., Miller, M. G., Chua, P. R., Maxon, M. E., & Johnson, A. D. (2005). Nuclear fusion occurs during mating in *Candida albicans* and is dependent on the *KAR3* gene. *Molecular Microbiology, 55*(4), 1046–1059. http://dx.doi.org/10.1111/j.1365-2958.2005.04466.x.

Bennett, R. J., Uhl, M. A., Miller, M. G., & Johnson, A. D. (2003). Identification and characterization of a *Candida albicans* mating pheromone. *Molecular and Cellular Biology, 23*(22), 8189–8201.

Berbee, M. L., Payne, B. P., Zhang, G., Roberts, R. G., & Turgeon, B. G. (2003). Shared ITS DNA substitutions in isolates of opposite mating type reveal a recombining history for three presumed asexual species in the filamentous ascomycete genus *Alternaria*. *Mycological Research, 107*(Pt 2), 169–182.

Blakeslee, A. F. (1904). Sexual reproduction in the Mucorineae. *Proceedings of the American Academy of Arts and Sciences*, *40*(4), 205–319. http://dx.doi.org/10.2307/20021962.

Bölker, M., Urban, M., & Kahmann, R. (1992). The a mating type locus of *U. maydis* specifies cell signaling components. *Cell*, *68*(3), 441–450.

Bougnoux, M.-E., Pujol, C., Diogo, D., Bouchier, C., Soll, D. R., & d'Enfert, C. (2008). Mating is rare within as well as between clades of the human pathogen *Candida albicans*. *Fungal Genetics and Biology*, *45*(3), 221–231. http://dx.doi.org/10.1016/j.fgb.2007.10.008.

Boutin, T. S., Le Rouzic, A., & Capy, P. (2012). How does selfing affect the dynamics of selfish transposable elements? *Mobile DNA*, *3*(1), 1–9. http://dx.doi.org/10.1186/1759-8753-3-5.

Brandt, M. E., Hutwagner, L. C., Kuykendall, R. J., & Pinner, R. W. (1995). Comparison of multilocus enzyme electrophoresis and random amplified polymorphic DNA analysis for molecular subtyping of *Cryptococcus neoformans*. *Journal of Clinical Microbiology*, *33*(7), 1890–1895.

Bui, T., Lin, X., Malik, R., Heitman, J., & Carter, D. A. (2008). Isolates of *Cryptococcus neoformans* from infected animals reveal genetic exchange in unisexual, alpha mating type populations. *Eukaryotic Cell*, *7*(10), 1771–1780. http://dx.doi.org/10.1128/EC.00097-08.

Byrnes, E. J., 3rd, Bartlett, K. H., Perfect, J. R., & Heitman, J. (2011). *Cryptococcus gattii*: An emerging fungal pathogen infecting humans and animals. *Microbes and Infection*, *13*(11), 895–907. http://dx.doi.org/10.1016/j.micinf.2011.05.009.

Byrnes, E. J., 3rd, Bildfell, R. J., Frank, S. A., Mitchell, T. G., Marr, K. A., & Heitman, J. (2009). Molecular evidence that the range of the Vancouver Island outbreak of *Cryptococcus gattii* infection has expanded into the Pacific Northwest in the United States. *Journal of Infectious Diseases*, *199*(7), 1081–1086. http://dx.doi.org/10.1086/597306.

Byrnes, E. J., 3rd, Li, W., Lewit, Y., Ma, H., Voelz, K., Ren, P., et al. (2010). Emergence and pathogenicity of highly virulent *Cryptococcus gattii* genotypes in the Northwest United States. *PLoS Pathogens*, *6*(4), e1000850. http://dx.doi.org/10.1371/journal.ppat.1000850.

Callen, D. F. (1974). Recombination and segregation of mitochondrial genes in *Saccharomyces cerevisiae*. *Molecular and General Genetics*, *134*(1), 49–63. http://dx.doi.org/10.1007/BF00332812.

Cambareri, E. B., Jensen, B. C., Schabtach, E., & Selker, E. U. (1989). Repeat-induced G–C to A–T mutations in *Neurospora*. *Science*, *244*(4912), 1571–1575.

Carriconde, F., Gilgado, F., Arthur, I., Ellis, D., Malik, R., van de Wiele, N., et al. (2011). Clonality and α–a recombination in the Australian *Cryptococcus gattii* VGII population – An emerging outbreak in Australia. *PLoS ONE*, *6*(2), e16936. http://dx.doi.org/10.1371/journal.pone.0016936.

Cavalier-Smith, T. (2004). Only six kingdoms of life. *Proceedings of the Royal Society of London. Series B: Biological Sciences*, *271*(1545), 1251–1262. http://dx.doi.org/10.1098/rspb.2004.2705.

Chan, R. K., & Otte, C. A. (1982). Physiological characterization of *Saccharomyces cerevisiae* mutants supersensitive to G1 arrest by a factor and α factor pheromones. *Molecular and Cellular Biology*, *2*(1), 21–29.

Charlesworth, D., & Wright, S. I. (2001). Breeding systems and genome evolution. *Current Opinion in Genetics & Development*, *11*(6), 685–690. http://dx.doi.org/10.1016/S0959-437X(00)00254-9.

Chen, J., Chen, J., Lane, S., & Liu, H. (2002). A conserved mitogen-activated protein kinase pathway is required for mating in *Candida albicans*. *Molecular Microbiology*, *46*(5), 1335–1344.

Crow, J. F., & Kimura, M. (1965). Evolution in sexual and asexual populations. *The American Naturalist*, *99*(909), 439–450. http://dx.doi.org/10.2307/2459132.

Dacks, J., & Roger, A. J. (1999). The first sexual lineage and the relevance of facultative sex. *Journal of Molecular Evolution*, *48*(6), 779–783. http://dx.doi.org/10.1007/PL00013156.

Dalgaard, J. Z., & Klar, A. J. S. (1999). Orientation of DNA replication establishes mating-type switching pattern in *S. pombe*. *Nature*, *400*(6740), 181–184.

Daniels, K. J., Srikantha, T., Lockhart, S. R., Pujol, C., & Soll, D. R. (2006). Opaque cells signal white cells to form biofilms in *Candida albicans*. *The EMBO Journal*, *25*(10), 2240–2252. http://dx.doi.org/10.1038/sj.emboj.7601099.

Datta, K., Bartlett, K. H., Baer, R., Byrnes, E. J., 3rd, Galanis, E., Heitman, J., et al. (2009). Spread of *Cryptococcus gattii* into Pacific Northwest region of the United States. *Emerging Infectious Diseases*, *15*(8), 1185–1191. http://dx.doi.org/10.3201/Eid1508.081384.

Davidson, R. C., Nichols, C. B., Cox, G. M., Perfect, J. R., & Heitman, J. (2003). A MAP kinase cascade composed of cell type specific and non-specific elements controls mating and differentiation of the fungal pathogen *Cryptococcus neoformans*. *Molecular Microbiology*, *49*(2), 469–485.

Debuchy, R., & Turgeon, B. G. (2006). Mating-type structure, evolution, and function in euascomycetes. In U. Kües, & R. Fischer (Eds.), *Growth, Differentiation and Sexuality* (Vol. 1) (pp. 293–323). Springer Berlin Heidelberg.

Delsing, C. E., Bleeker-Rovers, C. P., Kullberg, B.-J., & Netea, M. G. (2012). Treatment of candidiasis: Insights from host genetics. *Expert Review of Anti-infective Therapy*, *10*(8), 947–956. http://dx.doi.org/10.1586/eri.12.79.

Derelle, E., Ferraz, C., Rombauts, S., Rouzé, P., Worden, A. Z., Robbens, S., et al. (2006). Genome analysis of the smallest free-living eukaryote *Ostreococcus tauri* unveils many unique features. *Proceedings of the National Academy of Sciences of the United States of America*, *103*(31), 11647–11652. http://dx.doi.org/10.1073/pnas.0604795103.

Dietrich, F. S., Voegeli, S., Brachat, S., Lerch, A., Gates, K., Steiner, S., et al. (2004). The *Ashbya gossypii* genome as a tool for mapping the ancient *Saccharomyces cerevisiae* genome. *Science*, *304*(5668), 304–307. http://dx.doi.org/10.1126/science.1095781.

Dietrich, F. S., Voegeli, S., Kuo, S., & Philippsen, P. (2013). Genomes of *Ashbya* fungi isolated from insects reveal four mating-type loci, numerous translocations, lack of transposons, and distinct gene duplications. *G3: Genes, Genomes, Genetics*, *3*(8), 1225–1239. http://dx.doi.org/10.1534/g3.112.002881.

Dobzhansky, T., Levene, H., Spassky, B., & Spassky, N. (1959). Release of genetic variability through recombination. III. *Drosophila prosaltans*. *Genetics*, *44*(1), 75–92.

Dodge, B. O. (1927). Nuclear phenomena associated with heterothallism and homothallism in the Ascomycete *Neurospora*. *Journal of Agricultural Research*, *35*, 289–305.

Dodge, B. O., Singleton, J. R., & Rolnick, A. (1950). Studies on lethal E gene in *Neurospora tetrasperma*, including chromosome counts also in races of *N. sitophila*. *Proceedings of the American Philosophical Society*, *94*(1), 38–52. http://dx.doi.org/10.2307/3143250.

Dolgin, E. S., Charlesworth, B., & Cutter, A. D. (2008). Population frequencies of transposable elements in selfing and outcrossing *Caenorhabditis* nematodes. *Genetics Research*, *90*(04), 317–329. http://dx.doi.org/10.1017/S0016672308009440.

Dumitru, R., Navarathna, D. H., Semighini, C. P., Elowsky, C. G., Dumitru, R. V., Dignard, D., et al. (2007). *In vivo* and *in vitro* anaerobic mating in *Candida albicans*. *Eukaryotic Cell*, *6*(3), 465–472. http://dx.doi.org/10.1128/EC.00316-06.

Dunham, M. J., Badrane, H., Ferea, T., Adams, J., Brown, P. O., Rosenzweig, F., et al. (2002). Characteristic genome rearrangements in experimental evolution of *Saccharomyces cerevisiae*. *Proceedings of the National Academy of Sciences of the United States of America*, *99*(25), 16144–16149. http://dx.doi.org/10.1073/pnas.242624799.

Ellis, D. H., & Pfeiffer, T. J. (1990). Natural habitat of *Cryptococcus neoformans* var. *gattii*. *Journal of Clinical Microbiology*, *28*(7), 1642–1644.

Farrer, R. A., Henk, D. A., Garner, T. W. J., Balloux, F., Woodhams, D. C., & Fisher, M. C. (2013). Chromosomal copy number variation, selection and uneven rates of recombination reveal cryptic genome diversity linked to pathogenicity. *PLoS Genetics*, *9*(8), e1003703. http://dx.doi.org/10.1371/journal.pgen.1003703.

Farrer, R. A., Weinert, L. A., Bielby, J., Garner, T. W. J., Balloux, F., Clare, F., et al. (2011). Multiple emergences of genetically diverse amphibian-infecting chytrids include a globalized hypervirulent recombinant lineage. *Proceedings of the National Academy of Sciences of the United States of America*, *108*(46), 18732–18736. http://dx.doi.org/10.1073/pnas.1111915108.

Faugeron, G., Rhounim, L., & Rossignol, J. L. (1990). How does the cell count the number of ectopic copies of a gene in the premeiotic inactivation process acting in *Ascobolus immersus? Genetics*, *124*(3), 585–591.

Felsenstein, J. (1974). The evolutionary advantage of recombination. *Genetics*, *78*(2), 737–756.

Feretzaki, M., & Heitman, J. (2013). Genetic circuits that govern bisexual and unisexual reproduction in *Cryptococcus neoformans*. *PLoS Genetics*, *9*(8), e1003688. http://dx.doi.org/10.1371/journal.pgen.1003688.

Fields, S., & Herskowitz, I. (1985). The yeast STE12 product is required for expression of two sets of cell-type-specific genes. *Cell*, *42*(3), 923–930. http://dx.doi.org/10.1016/0092-8674(85)90288-0.

Forche, A., Alby, K., Schaefer, D., Johnson, A. D., Berman, J., & Bennett, R. J. (2008). The parasexual cycle in *Candida albicans* provides an alternative pathway to meiosis for the formation of recombinant strains. *PLoS Biology*, *6*(5), e110. http://dx.doi.org/10.1371/journal.pbio.0060110.

Fraser, J. A., Diezmann, S., Subaran, R. L., Allen, A., Lengeler, K. B., Dietrich, F. S., et al. (2004). Convergent evolution of chromosomal sex-determining regions in the animal and fungal kingdoms. *PLoS Biology*, *2*(12), e384. http://dx.doi.org/10.1371/journal.pbio.0020384.

Fraser, J. A., Giles, S. S., Wenink, E. C., Geunes-Boyer, S. G., Wright, J. R., Diezmann, S., et al. (2005). Same-sex mating and the origin of the Vancouver Island *Cryptococcus gattii* outbreak. *Nature*, *437*(7063), 1360–1364. http://dx.doi.org/10.1038/nature04220.

Fraser, J. A., Subaran, R. L., Nichols, C. B., & Heitman, J. (2003). Recapitulation of the sexual cycle of the primary fungal pathogen *Cryptococcus neoformans* variety *gattii*: Implications for an outbreak on Vancouver Island. *Eukaryotic Cell*, *2*, 1036–1045.

Freedman, T., & Pukkila, P. J. (1993). De novo methylation of repeated sequences in *Coprinus cinereus*. *Genetics*, *135*(2), 357–366.

Fritz-Laylin, L. K., Prochnik, S. E., Ginger, M. L., Dacks, J. B., Carpenter, M. L., Field, M. C., et al. (2010). The genome of *Naegleria gruberi* illuminates early eukaryotic versatility. *Cell*, *140*(5), 631–642. http://dx.doi.org/10.1016/j.cell.2010.01.032.

Galagan, J. E., Calvo, S. E., Cuomo, C. A., Ma, L.-J., Wortman, J. R., Batzoglou, S., et al. (2005). Sequencing of *Aspergillus nidulans* and comparative analysis with *A. fumigatus* and *A. oryzae*. *Nature*, *438*(7071), 1105–1115.

Gerstein, A. C., & Otto, S. P. (2009). Ploidy and the causes of genomic evolution. *Journal of Heredity*, *100*(5), 571–581. http://dx.doi.org/10.1093/jhered/esp057.

Ghannoum, M. A., Jurevic, R. J., Mukherjee, P. K., Cui, F., Sikaroodi, M., Naqvi, A., et al. (2010). Characterization of the oral fungal microbiome (mycobiome) in healthy individuals. *PLoS Pathogens*, *6*(1), e1000713. http://dx.doi.org/10.1371/journal.ppat.1000713.

Giles, S. S., Dagenais, T. R. T., Botts, M. R., Keller, N. P., & Hull, C. M. (2009). Elucidating the pathogenesis of spores from the human fungal pathogen *Cryptococcus neoformans*. *Infection and Immunity*, *77*(8), 3491–3500. http://dx.doi.org/10.1128/IAI.00334-09.

Gioti, A., Mushegian, A. A., Strandberg, R., Stajich, J. E., & Johannesson, H. (2012). Unidirectional evolutionary transitions in fungal mating systems and the role of transposable elements. *Molecular Biology and Evolution, 29*(10), 3215–3226. http://dx.doi.org/10.1093/molbev/mss132.

Gioti, A., Stajich, J. E., & Johannesson, H. (2013). *Neurospora* and the dead-end hypothesis: Genomic consequences of selfing in the model genus. *Evolution, 67*(12), 3600–3616. http://dx.doi.org/10.1111/evo.12206.

Glass, N. L., Metzenberg, R. L., & Raju, N. B. (1990). Homothallic *Sordariaceae* from nature: The absence of strains containing only the *a* mating type sequence. *Experimental Mycology, 14*(3), 274–289. http://dx.doi.org/10.1016/0147-5975(90)90025-O.

Glass, N. L., & Smith, M. L. (1994). Structure and function of a mating-type gene from the homothallic species *Neurospora africana. Molecular and Cellular Biology, 244*(4), 401–409.

Gordon, J. L., Armisén, D., Proux-Wéra, E., ÓhÉigeartaigh, S. S., Byrne, K. P., & Wolfe, K. H. (2011). Evolutionary erosion of yeast sex chromosomes by mating-type switching accidents. *Proceedings of the National Academy of Sciences of the United States of America, 108*(50), 20024–20029. http://dx.doi.org/10.1073/pnas.1112808108.

Goyon, C., & Faugeron, G. (1989). Targeted transformation of *Ascobolus immersus* and de novo methylation of the resulting duplicated DNA sequences. *Molecular and Cellular Biology, 9*(7), 2818–2827.

Graia, F., Lespinet, O., Rimbault, B., Dequard-Chablat, M., Coppin, E., & Picard, M. (2001). Genome quality control: RIP (repeat-induced point mutation) comes to *Podospora. Molecular Microbiology, 40*(3), 586–595.

Gräser, Y., Kühnisch, J., & Presber, W. (1999). Molecular markers reveal exclusively clonal reproduction in *Trichophyton rubrum. Journal of Clinical Microbiology, 37*(11), 3713–3717.

Halary, S., Malik, S.-B., Lildhar, L., Slamovits, C. H., Hijri, M., & Corradi, N. (2011). Conserved meiotic machinery in *Glomus* spp., a putatively ancient asexual fungal lineage. *Genome Biology and Evolution, 3*, 950–958. http://dx.doi.org/10.1093/gbe/evr089.

Haldane, J. B. S. (1937). The effect of variation of fitness. *The American Naturalist, 71*(735), 337–349. http://dx.doi.org/10.2307/2457289.

Haldane, J. B. S. (1949). Disease and evolution. *Supplement to La Ricerca Scientifica* (19), 68–76.

Halliday, C. L., Bui, T., Krockenberger, M., Malik, R., Ellis, D. H., & Carter, D. A. (1999). Presence of α and a mating types in environmental and clinical collections of *Cryptococcus neoformans* var. *gattii* strains from Australia. *Journal of Clinical Microbiology, 37*(9), 2920–2926.

Hamann, A., Feller, F., & Osiewacz, H. D. (2000). The degenerate DNA transposon Pat and repeat-induced point mutation (RIP) in *Podospora anserina. Molecular and Cellular Biology, 263*(6), 1061–1069.

Hamilton, W. D. (1980). Sex versus non-sex versus parasite. *Oikos, 35*(2), 282–290. http://dx.doi.org/10.2307/3544435.

Hamilton, W. D., Axelrod, R., & Tanese, R. (1990). Sexual reproduction as an adaptation to resist parasites (a review). *Proceedings of the National Academy of Sciences of the United States of America, 87*(9), 3566–3573.

Hartwell, L. H. (1980). Mutants of *Saccharomyces cerevisiae* unresponsive to cell division control by polypeptide mating hormone. *The Journal of Cell Biology, 85*(3), 811–822. http://dx.doi.org/10.1083/jcb.85.3.811.

Heitman, J. (2006). Sexual reproduction and the evolution of microbial pathogens. *Current Biology, 16*(17), R711–R725. http://dx.doi.org/10.1016/j.cub.2006.07.064.

Heitman, J. (2010). Evolution of eukaryotic microbial pathogens via covert sexual reproduction. *Cell Host & Microbe, 8*(1), 86–99. http://dx.doi.org/10.1016/j.chom.2010.06.011.

Heitman, J., Sun, S., & James, T. Y. (2013). Evolution of fungal sexual reproduction. *Mycologia, 105*(1), 1–27. http://dx.doi.org/10.3852/12-253.

Hickman, M. A., Zeng, G., Forche, A., Hirakawa, M. P., Abbey, D., Harrison, B. D., et al. (2013). The 'obligate diploid' *Candida albicans* forms mating-competent haploids. *Nature*, *494*(7435), 55–59. http://dx.doi.org/10.1038/nature11865.

Hicks, J. B., & Herskowitz, I. (1977). Interconversion of yeast mating types II. Restoration of mating ability to sterile mutants in homothallic and heterothallic strains. *Genetics*, *85*(3), 373–393.

Hicks, J. B., Strathern, J. N., & Herskowitz, I. (1977). Interconversion of yeast mating types III. Action of the homothallism (*HO*) gene in cells homozygous for the mating type locus. *Genetics*, *85*(3), 395–405.

Hill, W. G., & Robertson, A. (1966). The effect of linkage on limits to artificial selection. *Genetical Research*, *8*(3), 269–294.

Hiremath, S. S., Chowdhary, A., Kowshik, T., Randhawa, H. S., Sun, S., & Xu, J. (2008). Long-distance dispersal and recombination in environmental populations of *Cryptococcus neoformans* var. *grubii* from India. *Microbiology*, *154*(Pt 5), 1513–1524. http://dx.doi.org/10.1099/mic.0.2007/015594-0.

Hsueh, Y.-P., Lin, X., Kwon-Chung, K. J., & Heitman, J. (2011). Sexual reproduction of *Cryptococcus*. In J. Heitman, T. R. Kozel, K. J. Kwon-Chung, J. R. Perfect, & A. Casadevall (Eds.), *Cryptococcus: from human pathogen to model yeast* (pp. 81–96). Washington, DC: ASM Press.

Hua-Van, A., Hericourt, F., Capy, P., Daboussi, M. J., & Langin, T. (1998). Three highly divergent subfamilies of the impala transposable element coexist in the genome of the fungus *Fusarium oxysporum*. *Molecular and Cellular Biology*, *259*(4), 354–362.

Huang, G., Srikantha, T., Sahni, N., Yi, S., & Soll, D. R. (2009). CO(2) regulates white-to-opaque switching in *Candida albicans*. *Current Biology*, *19*(4), 330–334. http://dx.doi.org/10.1016/j.cub.2009.01.018.

Huang, G., Wang, H., Chou, S., Nie, X., Chen, J., & Liu, H. (2006). Bistable expression of *WOR1*, a master regulator of white-opaque switching in *Candida albicans*. *Proceedings of the National Academy of Sciences of the United States of America*, *103*(34), 12813–12818. http://dx.doi.org/10.1073/pnas.0605270103.

Huang, G., Yi, S., Sahni, N., Daniels, K. J., Srikantha, T., & Soll, D. R. (2010). N-acetylglucosamine induces white to opaque switching, a mating prerequisite in *Candida albicans*. *PLoS Pathogens*, *6*(3), e1000806. http://dx.doi.org/10.1371/journal.ppat.1000806.

Hull, C. M., Boily, M.-J., & Heitman, J. (2005). Sex-specific homeodomain proteins Sxi1α and Sxi2α coordinately regulate sexual development in *Cryptococcus neoformans*. *Eukaryotic Cell*, *4*(3), 526–535.

Hull, C. M., Davidson, R. C., & Heitman, J. (2002). Cell identity and sexual development in *Cryptococcus neoformans* are controlled by the mating-type-specific homeodomain protein Sxi1α. *Genes & Development*, *16*(23), 3046–3060. http://dx.doi.org/10.1101/gad.1041402.

Hull, C. M., & Heitman, J. (2002). Genetics of *Cryptococcus neoformans*. *Annual Review of Genetics*, *36*, 557–615. http://dx.doi.org/10.1146/annurev.genet.36.052402.152652.

Hull, C. M., & Johnson, A. D. (1999). Identification of a mating type-like locus in the asexual pathogenic yeast *Candida albicans*. *Science*, *285*(5431), 1271–1275.

Hull, C. M., Raisner, R. M., & Johnson, A. D. (2000). Evidence for mating of the "asexual" yeast *Candida albicans* in a mammalian host. *Science*, *289*(5477), 307–310.

Idnurm, A., Bahn, Y.-S., Nielsen, K., Lin, X., Fraser, J. A., & Heitman, J. (2005). Deciphering the model pathogenic fungus *Cryptococcus neoformans*. *Nature Reviews Microbiology*, *3*(10), 753–764. http://dx.doi.org/10.1038/nrmicro1245.

Ikeda, K., Nakayashiki, H., Kataoka, T., Tamba, H., Hashimoto, Y., Tosa, Y., et al. (2002). Repeat-induced point mutation (RIP) in *Magnaporthe grisea*: Implications for its sexual cycle in the natural field context. *Molecular Microbiology*, *45*(5), 1355–1364.

Inderbitzin, P., Harkness, J., Turgeon, B. G., & Berbee, M. L. (2005). Lateral transfer of mating system in *Stemphylium*. *Proceedings of the National Academy of Sciences of the United States of America*, *102*(32), 11390–11395. http://dx.doi.org/10.1073/pnas.0501918102.

Iwasa, Y., & Sasaki, A. (1987). Evolution of the number of sexes. *Evolution*, *41*(1), 49–65. http://dx.doi.org/10.2307/2408972.

Jackson, C. L., & Hartwell, L. H. (1990a). Courtship in *S. cerevisiae*: Both cell types choose mating partners by responding to the strongest pheromone signal. *Cell*, *63*(5), 1039–1051. http://dx.doi.org/10.1016/0092-8674(90)90507-B.

Jackson, C. L., & Hartwell, L. H. (1990b). Courtship in *Saccharomyces cerevisiae*: An early cell–cell interaction during mating. *Molecular and Cellular Biology*, *10*(5), 2202–2213. http://dx.doi.org/10.1128/mcb.10.5.2202.

James, T. Y., Litvintseva, A. P., Vilgalys, R., Morgan, J. A. T., Taylor, J. W., Fisher, M. C., et al. (2009). Rapid global expansion of the fungal disease chytridiomycosis into declining and healthy amphibian populations. *PLoS Pathogens*, *5*(5), e1000458. http://dx.doi.org/10.1371/journal.ppat.1000458.

Janbon, G., Maeng, S., Yang, D.-H., Ko, Y.-J., Jung, K.-W., Moyrand, F., et al. (2010). Characterizing the role of RNA silencing components in *Cryptococcus neoformans*. *Fungal Genetics and Biology*, *47*(12), 1070–1080. http://dx.doi.org/10.1016/j.fgb.2010.10.005.

Jones, S. K., Jr., & Bennett, R. J. (2011). Fungal mating pheromones: Choreographing the dating game. *Fungal Genetics and Biology*, *48*(7), 668–676. http://dx.doi.org/10.1016/j.fgb.2011.04.001.

Kämper, J., Reichmann, M., Romeis, T., Bölker, M., & Kahmann, R. (1995). Multiallelic recognition: Nonself-dependent dimerization of the bE and bW homeodomain proteins in *Ustilago maydis*. *Cell*, *81*(1), 73–83. http://dx.doi.org/10.1016/0092-8674(95)90372-0.

Kano, R., Isizuka, M., Hiruma, M., Mochizuki, T., Kamata, H., & Hasegawa, A. (2013). Mating type gene (*MAT1-1*) in Japanese isolates of *Trichophyton rubrum*. *Mycopathologia*, *175*(1–2), 171–173. http://dx.doi.org/10.1007/s11046-012-9603-2.

Kaykov, A., & Arcangioli, B. (2004). A programmed strand-specific and modified nick in *S. pombe* constitutes a novel type of chromosomal imprint. *Current Biology*, *14*(21), 1924–1928. http://dx.doi.org/10.1016/j.cub.2004.10.026.

Keller, S. M., Viviani, M. A., Esposto, M. C., Cogliati, M., & Wickes, B. L. (2003). Molecular and genetic characterization of a serotype A *MAT*a *Cryptococcus neoformans* isolate. *Microbiology*, *149*(Pt 1), 131–142.

Kibbler, C. C., Seaton, S., Barnes, R. A., Gransden, W. R., Holliman, R. E., Johnson, E. M., et al. (2003). Management and outcome of bloodstream infections due to *Candida* species in England and Wales. *Journal of Hospital Infection*, *54*(1), 18–24. http://dx.doi.org/10.1016/S0195-6701(03)00085-9.

Kidd, S. E., Bach, P. J., Hingston, A. O., Mak, S., Chow, Y., MacDougall, L., et al. (2007a). *Cryptococcus gattii* dispersal mechanisms, British Columbia, Canada. *Emerging Infectious Diseases*, *13*(1), 51–57. http://dx.doi.org/10.3201/eid1301.060823.

Kidd, S. E., Chow, Y., Mak, S., Bach, P. J., Chen, H., Hingston, A. O., et al. (2007b). Characterization of environmental sources of the human and animal pathogen *Cryptococcus gattii* in British Columbia, Canada, and the Pacific Northwest of the United States. *Applied and Environmental Microbiology*, *73*(5), 1433–1443. http://dx.doi.org/10.1128/AEM.01330-06.

Kidd, S. E., Hagen, F., Tscharke, R. L., Huynh, M., Bartlett, K. H., Fyfe, M., et al. (2004). A rare genotype of *Cryptococcus gattii* caused the cryptococcosis outbreak on Vancouver Island (British Columbia, Canada). *Proceedings of the National Academy of Sciences of the United States of America*, *101*(49), 17258–17263. http://dx.doi.org/10.1073/pnas.0402981101.

Klein, R. S., Harris, C. A., Small, C. B., Moll, B., Lesser, M., & Friedland, G. H. (1984). Oral candidiasis in high-risk patients as the initial manifestation of the acquired immunodeficiency syndrome. *New England Journal of Medicine*, *311*(6), 354–358. http://dx.doi.org/10.1056/NEJM198408093110602.

Kronstad, J. W., & Leong, S. A. (1990). The *b* mating-type locus of *Ustilago maydis* contains variable and constant regions. *Genes & Development*, *4*(8), 1384–1395. http://dx.doi.org/10.1101/gad.4.8.1384.

Kruzel, E. K., Giles, S. S., & Hull, C. M. (2012). Analysis of *Cryptococcus neoformans* sexual development reveals rewiring of the pheromone-response network by a change in transcription factor identity. *Genetics*, *191*(2), 435–449. http://dx.doi.org/10.1534/genetics.112.138958.

Kvaal, C., Lachke, S. A., Srikantha, T., Daniels, K. J., McCoy, J., & Soll, D. R. (1999). Misexpression of the opaque-phase-specific gene *PEP1 (SAP1)* in the white phase of *Candida albicans* confers increased virulence in a mouse model of cutaneous infection. *Infection and Immunity*, *67*(12), 6652–6662.

Kwon-Chung, K. J. (1975). A new genus, *Filobasidiella*, the perfect state of *Cryptococcus neoformans*. *Mycologia*, *67*, 1197–1200.

Kwon-Chung, K. J. (1976a). Morphogenesis of *Filobasidiella neoformans*, the sexual state of *Cryptococcus neoformans*. *Mycologia*, *68*(4), 821–833.

Kwon-Chung, K. J. (1976b). A new species of *Filobasidiella*, the sexual state of *Cryptococcus neoformans* B and C serotypes. *Mycologia*, *68*(4), 943–946.

Kwon-Chung, K. J., & Bennett, J. E. (1978). Distribution of α and a mating types of *Cryptococcus neoformans* among natural and clinical isolates. *American Journal of Epidemiology*, *108*(4), 337–340.

Kwon-Chung, K. J., Chang, Y. C., Bauer, R., Swann, E. C., Taylor, J. W., & Goel, R. (1995). The characteristics that differentiate *Filobasidiella depauperata* and *Filobasidiella neoformans*. *Studies of Mycology*, *38*, 67–79.

Lachke, S. A., Srikantha, T., & Soll, D. R. (2003). The regulation of *EFG1* in white-opaque switching in *Candida albicans* involves overlapping promoters. *Molecular Microbiology*, *48*(2), 523–536.

Ladle, R. J., Johnstone, R. A., & Judson, O. P. (1993). Coevolutionary dynamics of sex in a metapopulation: Escaping the Red Queen. *Proceedings: Biological Sciences*, *253*(1337), 155–160. http://dx.doi.org/10.2307/49803.

Lang, G. I., Murray, A. W., & Botstein, D. (2009). The cost of gene expression underlies a fitness trade-off in yeast. *Proceedings of the National Academy of Sciences of the United States of America*, *106*(14), 5755–5760. http://dx.doi.org/10.1073/pnas.0901520106.

Lang, G. I., Rice, D. P., Hickman, M. J., Sodergren, E., Weinstock, G. M., Botstein, D., et al. (2013). Pervasive genetic hitchhiking and clonal interference in forty evolving yeast populations. *Nature*, *500*(7464), 571–574. http://dx.doi.org/10.1038/nature12344.

Legrand, M., Lephart, P., Forche, A., Mueller, F.-M. C., Walsh, T., Magee, P. T., et al. (2004). Homozygosity at the *MTL* locus in clinical strains of *Candida albicans*: Karyotypic rearrangements and tetraploid formation. *Molecular Microbiology*, *52*(5), 1451–1462. http://dx.doi.org/10.1111/j.1365-2958.2004.04068.x.

Lengeler, K. B., Cox, G. M., & Heitman, J. (2001). Serotype AD strains of *Cryptococcus neoformans* are diploid or aneuploid and are heterozygous at the mating-type locus. *Infection and Immunity*, *69*(1), 115–122. http://dx.doi.org/10.1128/IAI.69.1.115-122.2001.

Lengeler, K. B., Wang, P., Cox, G. M., Perfect, J. R., & Heitman, J. (2000). Identification of the *MAT*a mating-type locus of *Cryptococcus neoformans* reveals a serotype A *MAT*a strain thought to have been extinct. *Proceedings of the National Academy of Sciences of the United States of America*, *97*(26). 14555–14460.

Levene, H. (1959). Release of genetic variability through recombination. IV. Statistic theory. *Genetics*, *44*(1), 93–104.

Lewis, W. M., Jr. (1985). Nutrient scarcity as an evolutionary cause of haploidy. *The American Naturalist*, *125*(5), 692–701. http://dx.doi.org/10.2307/2461479.

Li, W., Metin, B., White, T. C., & Heitman, J. (2010). Organization and evolutionary trajectory of the mating type (*MAT*) locus in dermatophyte and dimorphic fungal pathogens. *Eukaryotic Cell*, *9*(1), 46–58. http://dx.doi.org/10.1128/ec.00259-09.

Lin, C. H., Kabrawala, S., Fox, E. P., Nobile, C. J., Johnson, A. D., & Bennett, R. J. (2013). Genetic control of conventional and pheromone-stimulated biofilm formation in *Candida albicans*. *PLoS Pathogens, 9*(4), e1003305. http://dx.doi.org/10.1371/journal. ppat.1003305.

Lin, X., & Heitman, J. (2007). Mechanisms of homothallism in fungi and transitions between heterothallism and homothallism. In J. Heitman, J. W. Kronstad, J. W. Taylor, & L. A. Casselton (Eds.), *Sex in Fungi* (pp. 35–58). Washington, DC: ASM Press.

Lin, X., Huang, J. C., Mitchell, T. G., & Heitman, J. (2006). Virulence attributes and hyphal growth of *C. neoformans* are quantitative traits and the *MATα* allele enhances filamentation. *PLoS Genetics, 2*(11), e187. http://dx.doi.org/10.1371/journal.pgen.0020187.

Lin, X., Hull, C. M., & Heitman, J. (2005). Sexual reproduction between partners of the same mating type in *Cryptococcus neoformans*. *Nature, 434*(7036), 1017–1021. http://dx.doi. org/10.1038/nature03448.

Lin, X., Jackson, J. C., Feretzaki, M., Xue, C., & Heitman, J. (2010). Transcription factors Mat2 and Znf2 operate cellular circuits orchestrating opposite- and same-sex mating in *Cryptococcus neoformans.*. *PLoS Genetics, 6*(5), e1000953. http://dx.doi.org/10.1371/ journal.pgen.1000953.

Lin, X., Litvintseva, A. P., Nielsen, K., Patel, S., Floyd, A., Mitchell, T. G., et al. (2007). αADα hybrids of *Cryptococcus neoformans*: Evidence of same-sex mating in nature and hybrid fitness. *PLoS Genetics, 3*(10), 1975–1990. http://dx.doi.org/10.1371/journal. pgen.0030186.

Lin, X., Patel, S., Litvintseva, A. P., Floyd, A., Mitchell, T. G., & Heitman, J. (2009). Diploids in the *Cryptococcus neoformans* serotype A population homozygous for the α mating type originate via unisexual mating. *PLoS Pathogens, 5*(1), e1000283. http://dx.doi. org/10.1371/journal.ppat.1000283.

Lindegren, C. C., & Lindegren, G. (1943a). Legitimate and illegitimate mating in *Saccharomyces cerevisiae*. *Genetics, 28*(81).

Lindegren, C. C., & Lindegren, G. (1943b). Segregation, mutation, and copulation in *Saccharomyces cerevisiae*. *Annals of the Missouri Botanical Garden, 30*(4), 453–468. http://dx.doi. org/10.2307/2394308.

Lindegren, C. C., & Lindegren, G. (1944). Instability of the mating type alleles in *Saccharomyces*. *Annals of the Missouri Botanical Garden, 31*(2), 203–217. http://dx.doi. org/10.2307/2394338.

Litvintseva, A. P., Kestenbaum, L., Vilgalys, R., & Mitchell, T. G. (2005). Comparative analysis of environmental and clinical populations of *Cryptococcus neoformans*. *Journal of Clinical Microbiology, 43*(2), 556–564. http://dx.doi.org/10.1128/jcm.43.2.556-564.2005.

Litvintseva, A. P., Lin, X., Templeton, I., Heitman, J., & Mitchell, T. G. (2007). Many globally isolated AD hybrid strains of *Cryptococcus neoformans* originated in Africa. *PLoS Pathogens, 3*(8), e114. http://dx.doi.org/10.1371/journal.ppat.0030114.

Litvintseva, A. P., Marra, R. E., Nielsen, K., Heitman, J., Vilgalys, R., & Mitchell, T. G. (2003). Evidence of sexual recombination among *Cryptococcus neoformans* serotype A isolates in sub-Saharan Africa. *Eukaryotic Cell, 2*(6), 1162–1168.

Lloyd, D. G. (1988). Benefits and costs of biparental and uniparental reproduction in plants. In R. E. Michod, & B. R. Levin (Eds.), *The Evolution of Sex* (pp. 233–252). Sunderland, Massachusetts: Sinauer Associates.

Lockhart, S. R., Daniels, K. J., Zhao, R., Wessels, D., & Soll, D. R. (2003). Cell biology of mating in *Candida albicans*. *Eukaryotic Cell, 2*(1), 49–61.

Lockhart, S. R., Pujol, C., Daniels, K. J., Miller, M. G., Johnson, A. D., Pfaller, M. A., et al. (2002). In *Candida albicans*, white-opaque switchers are homozygous for mating type. *Genetics, 162*(2), 737–745.

Loftus, B. J., Fung, E., Roncaglia, P., Rowley, D., Amedeo, P., Bruno, D., et al. (2005). The genome of the basidiomycetous yeast and human pathogen *Cryptococcus neoformans*. *Science, 307*(5713), 1321–1324. http://dx.doi.org/10.1126/science.1103773.

Lynch, M. (2007). The frailty of adaptive hypotheses for the origins of organismal complexity. *Proceedings of the National Academy of Sciences of the United States of America, 104*(Suppl 1), 8597–8604. http://dx.doi.org/10.1073/pnas.0702207104.

Mable, B. K. (2001). Ploidy evolution in the yeast *Saccharomyces cerevisiae*: A test of the nutrient limitation hypothesis. *Journal of Evolutionary Biology, 14*(1), 157–170. http://dx.doi.org/10.1046/j.1420-9101.2001.00245.x.

Mable, B. K., & Otto, S. P. (2001). Masking and purging mutations following EMS treatment in haploid, diploid and tetraploid yeast (*Saccharomyces cerevisiae*). *Genetical Research, 77*(1), 9–26.

Magee, B. B., Legrand, M., Alarco, A. M., Raymond, M., & Magee, P. T. (2002). Many of the genes required for mating in *Saccharomyces cerevisiae* are also required for mating in *Candida albicans*. *Molecular Microbiology, 46*(5), 1345–1351.

Magee, B. B., & Magee, P. T. (2000). Induction of mating in *Candida albicans* by construction of *MTLa* and *MTLα* strains. *Science, 289*(5477), 310–313.

Maor, R., & Shirasu, K. (2005). The arms race continues: Battle strategies between plants and fungal pathogens. *Current Opinion in Microbiology, 8*(4), 399–404. http://dx.doi.org/10.1016/j.mib.2005.06.008.

Martinez, D. A., Oliver, B. G., Gräser, Y., Goldberg, J. M., Li, W., Martinez-Rossi, N. M., et al. (2012). Comparative genome analysis of *Trichophyton rubrum* and related dermatophytes reveals candidate genes involved in infection. *mBio, 3*(5), e00259–00212. http://dx.doi.org/10.1128/mBio.00259-12.

Martins, M., Henriques, M., Ribeiro, A. P., Fernandes, R., Goncalves, V., Seabra, A., et al. (2010). Oral *Candida* carriage of patients attending a dental clinic in Braga, Portugal. *Revista Iberoamericana De Micologia, 27*(3), 119–124. http://dx.doi.org/10.1016/j.riam.2010.03.007.

Maynard Smith, J. (1978). *The Evolution of Sex*. Cambridge: Cambridge University Press.

McCusker, J. H. (2006). *Saccharomyces cerevisiae*: An emerging and model pathogenic fungus. In J. Heitman, S. G. Filler, J. E. Edwards, & A. P. Mitchell (Eds.), *Molecular Principles of Fungal Pathogenesis*. (pp. 245–259). Washington, DC: ASM Press.

Melton, J. J., Redding, S. W., Kirkpatrick, W. R., Reasner, C. A., Ocampo, G. L., Venkatesh, A., et al. (2010). Recovery of *Candida dubliniensis* and other *Candida* species from the oral cavity of subjects with periodontitis who had well-controlled and poorly controlled type 2 diabetes: A pilot study. *Special Care in Dentistry, 30*(6), 230–234. http://dx.doi.org/10.1111/j.1754-4505.2010.00159.x.

Miller, M. G., & Johnson, A. D. (2002). White-opaque switching in *Candida albicans* is controlled by mating-type locus homeodomain proteins and allows efficient mating. *Cell, 110*(3), 293–302.

Morgan, J. A. T., Vredenburg, V. T., Rachowicz, L. J., Knapp, R. A., Stice, M. J., Tunstall, T., et al. (2007). Population genetics of the frog-killing fungus *Batrachochytrium dendrobatidis*. *Proceedings of the National Academy of Sciences of the United States of America, 104*(34), 13845–13850. http://dx.doi.org/10.1073/pnas.0701838104.

Morgan, M. T. (2001). Transposable element number in mixed mating populations. *Genetics Research, 77*(03), 261–275. http://dx.doi.org/10.1017/S0016672301005067.

Muller, H. J. (1932). Some genetic aspects of sex. *The American Naturalist, 66*(703), 118–138. http://dx.doi.org/10.2307/2456922.

Muller, H. J. (1964). The relation of recombination to mutational advance. *Mutation Research, 1*(1), 2–9. http://dx.doi.org/10.1016/0027-5107(64)90047-8.

Murphy, H. A., & Zeyl, C. W. (2010). Yeast sex: Surprisingly high rates of outcrossing between asci. *PLoS ONE, 5*(5), e10461. http://dx.doi.org/10.1371/journal.pone.0010461.

Nakayashiki, H., Nishimoto, N., Ikeda, K., Tosa, Y., & Mayama, S. (1999). Degenerate MAGGY elements in a subgroup of *Pyricularia grisea*: A possible example of successful capture of a genetic invader by a fungal genome. *Molecular and General Genetics, 261*(6), 958–966.

Neuveglise, C., Sarfati, J., Latge, J. P., & Paris, S. (1996). Afut1, a retrotransposon-like element from *Aspergillus fumigatus*. *Nucleic Acids Research*, *24*(8), 1428–1434.

Ni, M., Feretzaki, M., Li, W., Floyd-Averette, A., Mieczkowski, P., Dietrich, F. S., et al. (2013). Unisexual and heterosexual meiotic reproduction generate aneuploidy and phenotypic diversity *de novo* in the yeast *Cryptococcus neoformans*. *PLoS Biology*, *11*(9), e1001653. http://dx.doi.org/10.1371/journal.pbio.1001653.

Nygren, K., Strandberg, R., Wallberg, A., Nabholz, B., Gustafsson, T., García, D., et al. (2011). A comprehensive phylogeny of *Neurospora* reveals a link between reproductive mode and molecular evolution in fungi. *Molecular Phylogenetics and Evolution*, *59*(3), 649–663. http://dx.doi.org/10.1016/j.ympev.2011.03.023.

Odds, F. C., Bougnoux, M.-E., Shaw, D. J., Bain, J. M., Davidson, A. D., Diogo, D., et al. (2007). Molecular phylogenetics of *Candida albicans*. *Eukaryotic Cell*, *6*(6), 1041–1052. http://dx.doi.org/10.1128/EC.00041-07.

Orr, H. A., & Otto, S. P. (1994). Does diploidy increase the rate of adaptation? *Genetics*, *136*(4), 1475–1480.

Otto, S. P., & Lenormand, T. (2002). Resolving the paradox of sex and recombination. *Nature Reviews Genetics*, *3*(4), 252–261. http://dx.doi.org/10.1038/nrg761.

Otto, S. P., & Michalakis, Y. (1998). The evolution of recombination in changing environments. *Trends in Ecology & Evolution*, *13*(4), 145–151. http://dx.doi.org/10.1016/S0169-5347(97)01260-3.

Paoletti, M., Seymour, F. A., Alcocer, M. J. C., Kaur, N., Calvo, A. M., Archer, D. B., et al. (2007). Mating type and the genetic basis of self-fertility in the model fungus *Aspergillus nidulans*. *Current Biology*, *17*(16), 1384–1389. http://dx.doi.org/10.1016/j.cub.2007.07.012.

Patterson, D. J. (1999). The diversity of eukaryotes. *The American Naturalist*, *154*(S4), S96–S124. http://dx.doi.org/10.1086/303287.

Pavelka, N., Rancati, G., Zhu, J., Bradford, W. D., Saraf, A., Florens, L., et al. (2010). Aneuploidy confers quantitative proteome changes and phenotypic variation in budding yeast. *Nature*, *468*(7321), 321–325. http://dx.doi.org/10.1038/nature09529.

Peever, T. L., Canihos, Y., Olsen, L., Ibanez, A., Liu, Y. C., & Timmer, L. W. (1999). Population genetic structure and host specificity of *Alternaria* spp. causing brown spot of *Minneola tangelo* and rough lemon in Florida. *Phytopathology*, *89*(10), 851–860. http://dx.doi.org/10.1094/PHYTO.1999.89.10.851.

Pendrak, M. L., Yan, S. S., & Roberts, D. D. (2004). Hemoglobin regulates expression of an activator of mating-type locus alpha genes in *Candida albicans*. *Eukaryotic Cell*, *3*(3), 764–775. http://dx.doi.org/10.1128/EC.3.3.764-775.2004.

Pfaller, M. A., Jones, R. N., Messer, S. A., Edmond, M. B., & Wenzel, R. P. (1998). National surveillance of nosocomial blood stream infection due to *Candida albicans*: Frequency of occurrence and antifungal susceptibility in the SCOPE program. *Diagnostic Microbiology and Infectious Disease*, *31*(1), 327–332. http://dx.doi.org/10.1016/s0732-8893(97)00240-x.

Phadke, S. S., Feretzaki, M., & Heitman, J. (2013). Unisexual reproduction enhances fungal competitiveness by promoting habitat exploration via hyphal growth and sporulation. *Eukaryotic Cell*, *12*(8), 1155–1159. http://dx.doi.org/10.1128/ec.00147-13.

Pöggeler, S., Risch, S., Kück, U., & Osiewacz, H. D. (1997). Mating-type genes from the homothallic fungus *Sordaria macrospora* are functionally expressed in a heterothallic ascomycete. *Genetics*, *147*(2), 567–580.

Pollak, E. (1987). On the theory of partially inbreeding finite populations. I. Partial selfing. *Genetics*, *117*(2), 353–360.

Ramesh, M. A., Malik, S.-B., & Logsdon, J. M., Jr. (2005). A phylogenomic inventory of meiotic genes: Evidence for sex in *Giardia* and an early eukaryotic origin of meiosis. *Current Biology*, *15*(2), 185–191. http://dx.doi.org/10.1016/j.cub.2005.01.003.

Rancati, G., Pavelka, N., Fleharty, B., Noll, A., Trimble, R., Walton, K., et al. (2008). Aneuploidy underlies rapid adaptive evolution of yeast cells deprived of a conserved cytokinesis motor. *Cell*, *135*(5), 879–893. http://dx.doi.org/10.1016/j.cell.2008.09.039.

Raper, J. R. (1966). *Genetics of Sexuality in Higher Fungi.* New York: Ronald Press Co.

Reedy, J. L., Floyd, A., & Heitman, J. (2009). Mechanistic plasticity of sexual reproduction and meiosis in the *Candida* pathogenic species complex. *Current Biology, 19*(11), 891–899. http://dx.doi.org/10.1016/j.cub.2009.04.058.

Rodriguez-Carres, M., Findley, K., Sun, S., Dietrich, F. S., & Heitman, J. (2010). Morphological and genomic characterization of *Filobasidiella depauperata*: A homothallic sibling species of the pathogenic *Cryptococcus* species complex. *PLoS ONE, 5*(3), e9620.

Rydholm, C., Dyer, P., & Lutzoni, F. (2007). DNA sequence characterization and molecular evolution of *MAT1* and *MAT2* mating-type loci of the self-compatible ascomycete mold *Neosartorya fischeri. Eukaryotic Cell, 6*(5), 868–874. http://dx.doi.org/10.1128/EC.00319-06.

Sahni, N., Yi, S., Daniels, K. J., Huang, G., Srikantha, T., & Soll, D. R. (2010). Tec1 mediates the pheromone response of the white phenotype of *Candida albicans*: Insights into the evolution of new signal transduction pathways. *PLoS Biology, 8*(5), e1000363. http://dx.doi.org/10.1371/journal.pbio.1000363.

Sahni, N., Yi, S., Daniels, K. J., Srikantha, T., Pujol, C., & Soll, D. R. (2009). Genes selectively up-regulated by pheromone in white cells are involved in biofilm formation in *Candida albicans. PLoS Pathogens, 5*(10), e1000601. http://dx.doi.org/10.1371/journal.ppat.1000601.

Saul, N., Krockenberger, M., & Carter, D. A. (2008). Evidence of recombination in mixed-mating-type and α-only populations of *Cryptococcus gattii* sourced from single eucalyptus tree hollows. *Eukaryotic Cell, 7*(4), 727–734.

Schaefer, D., Cote, P., Whiteway, M., & Bennett, R. J. (2007). Barrier activity in *Candida albicans* mediates pheromone degradation and promotes mating. *Eukaryotic Cell, 6*(6), 907–918. http://dx.doi.org/10.1128/EC.00090-07.

Schulz, B., Banuett, F., Dahl, M., Schlesinger, R., Schäfer, W., Martin, T., et al. (1990). The *b* alleles of *U. maydis*, whose combinations program pathogenic development, code for polypeptides containing a homeodomain-related motif. *Cell, 60*(2), 295–306. http://dx.doi.org/10.1016/0092-8674(90)90744-Y.

Seitz-Mayr, G., Wolf, K., & Kaudewitz, F. (1978). Extrachromosomal inheritance in *Schizosaccharomyces pombe. Molecular and General Genetics, 164*(3), 309–320. http://dx.doi.org/10.1007/BF00333162.

Selker, E. U., Cambareri, E. B., Jensen, B. C., & Haack, K. R. (1987). Rearrangement of duplicated DNA in specialized cells of *Neurospora. Cell, 51*(5), 741–752. http://dx.doi.org/10.1016/0092-8674(87)90097-3.

Selker, E. U., & Stevens, J. N. (1985). DNA methylation at asymmetric sites is associated with numerous transition mutations. *Proceedings of the National Academy of Sciences of the United States of America, 82*(23), 8114–8118. http://dx.doi.org/10.1073/Pnas.82.23.8114.

Selmecki, A. M., Dulmage, K., Cowen, L. E., Anderson, J. B., & Berman, J. (2009). Acquisition of aneuploidy provides increased fitness during the evolution of antifungal drug resistance. *PLoS Genetics, 5*(10), e1000705. http://dx.doi.org/10.1371/journal.pgen.1000705.

Selmecki, A. M., Forche, A., & Berman, J. (2006). Aneuploidy and isochromosome formation in drug-resistant *Candida albicans. Science, 313*(5785), 367–370. http://dx.doi.org/10.1126/science.1128242.

Selmecki, A. M., Forche, A., & Berman, J. (2010). Genomic plasticity of the human fungal pathogen *Candida albicans. Eukaryotic Cell, 9*(7), 991–1008. http://dx.doi.org/10.1128/ec.00060-10.

Sharpley, M. S., Marciniak, C., Eckel-Mahan, K., McManus, M., Crimi, M., Waymire, K., et al. (2012). Heteroplasmy of mouse mtDNA is genetically unstable and results in altered behavior and cognition. *Cell, 151*(2), 333–343. http://dx.doi.org/10.1016/j.cell.2012.09.004.

Sherwood, R. K., & Bennett, R. J. (2009). Fungal meiosis and parasexual reproduction–lessons from pathogenic yeast. *Current Opinion in Microbiology, 12*(6), 599–607. http://dx.doi.org/10.1016/j.mib.2009.09.005.

Shiu, P. K. T., & Metzenberg, R. L. (2002). Meiotic silencing by unpaired DNA: Properties, regulation and suppression. *Genetics, 161*(4), 1483–1495.

Shiu, P. K. T., Raju, N. B., Zickler, D., & Metzenberg, R. L. (2001). Meiotic silencing by unpaired DNA. *Cell, 107*(7), 905–916. http://dx.doi.org/10.1016/S0092-8674(01)00609-2.

Simmons, E. G. (1986). *Alternaria* themes and variations (22–26). *Mycotaxon, 25*(1), 287–308.

Simmons, E. G. (1999). *Alternaria* themes and variations (226–235) – Classification of citrus pathogens. *Mycotaxon, 70,* 263–323.

Simpson, A. G. B., & Roger, A. J. (2004). The real 'kingdoms' of eukaryotes. *Current Biology, 14*(17), R693–R696. http://dx.doi.org/10.1016/j.cub.2004.08.038.

Sionov, E., Lee, H., Chang, Y. C., & Kwon-Chung, K. J. (2010). *Cryptococcus neoformans* overcomes stress of azole drugs by formation of disomy in specific multiple chromosomes. *PLoS Pathogens, 6*(4), e1000848. http://dx.doi.org/10.1371/journal.ppat.1000848.

Slutsky, B., Staebell, M., Anderson, J., Risen, L., Pfaller, M. A., & Soll, D. R. (1987). "White-opaque transition": A second high-frequency switching system in *Candida albicans. Journal of Bacteriology, 169*(1), 189–197.

Snetselaar, K. M., Bölker, M., & Kahmann, R. (1996). *Ustilago maydis* mating hyphae orient their growth toward pheromone sources. *Fungal Genetics and Biology, 20*(4), 299–312. http://dx.doi.org/10.1006/Fgbi.1996.0044.

Son, H., Min, K., Lee, J., Raju, N. B., & Lee, Y. W. (2011). Meiotic silencing in the homothallic fungus *Gibberella zeae. Fungal Biology, 115*(12), 1290–1302. http://dx.doi.org/10.1016/j.funbio.2011.09.006.

Spassky, B., Spassky, N., Levene, H., & Dobzhansky, T. (1958). Release of genetic variability through recombination. I. *Drosophila pseudoobscura. Genetics, 43*(5), 844–867.

Speed, B., & Dunt, D. (1995). Clinical and host differences between infections with the two varieties of *Cryptococcus neoformans. Clinical Infectious Diseases, 21*(1), 28–34. http://dx.doi.org/10.1093/clinids/21.1.28.

Spellig, T., Bolker, M., Lottspeich, F., Frank, R. W., & Kahmann, R. (1994). Pheromones trigger filamentous growth in *Ustilago maydis. The EMBO Journal, 13*(7), 1620–1627.

Spiess, E. B. (1959). Release of genetic variability through recombination. II. *Drosophila persimilis. Genetics, 44*(1), 43–58.

Springer, D. J., Saini, D., Byrnes, E. J., 3rd, Heitman, J., & Frothingham, R. (2013). Development of an aerosol model of *Cryptococcus* reveals humidity as an important factor affecting the viability of *Cryptococcus* during aerosolization. *PLoS ONE, 8*(7), e69804. http://dx.doi.org/10.1371/journal.pone.0069804.

Stewart, J. E., Kawabe, M., Abdo, Z., Arie, T., & Peever, T. L. (2011). Contrasting codon usage patterns and purifying selection at the mating locus in putatively asexual *Alternaria* fungal species. *PLoS ONE, 6*(5), e20083. http://dx.doi.org/10.1371/journal.pone.0020083.

Stewart, J. E., Thomas, K. A., Lawrence, C. B., Dang, H., Pryor, B. M., Timmer, L. M. P., et al. (2013). Signatures of recombination in clonal lineages of the citrus brown spot pathogen, *Alternaria alternata* sensu lato. *Phytopathology, 103*(7), 741–749. http://dx.doi.org/10.1094/phyto-08-12-0211-r.

Strausberg, R. L., & Perlman, P. S. (1978). The effect of zygotic bud position on the transmission of mitochondrial genes in *Saccharomyces cerevisiae. Molecular and General Genetics, 163*(2), 131–144. http://dx.doi.org/10.1007/BF00267404.

Sturm, N. R., Vargas, N. S., Westenberger, S. J., Zingales, B., & Campbell, D. A. (2003). Evidence for multiple hybrid groups in *Trypanosoma cruzi. International Journal for Parasitology, 33*(3), 269–279. http://dx.doi.org/10.1016/S0020-7519(02)00264-3.

Sukroongreung, S., Kitiniyom, K., Nilakul, C., & Tantimavanich, S. (1998). Pathogenicity of basidiospores of *Filobasidiella neoformans* var. *neoformans. Medical Mycology, 36*(6), 419–424.

Tavanti, A., Davidson, A. D., Fordyce, M. J., Gow, N. A. R., Maiden, M. C. J., & Odds, F. C. (2005). Population structure and properties of *Candida albicans*, as determined by multi-locus sequence typing. *Journal of Clinical Microbiology*, *43*(11), 5601–5613. http://dx.doi. org/10.1128/jcm.43.11.5601-5613.2005.

Torres, E. M., Dephoure, N., Panneerselvam, A., Tucker, C. M., Whittaker, C. A., Gygi, S. P., et al. (2010). Identification of aneuploidy-tolerating mutations. *Cell*, *143*(1), 71–83. http://dx.doi.org/10.1016/j.cell.2010.08.038.

Torres, E. M., Sokolsky, T., Tucker, C. M., Chan, L. Y., Boselli, M., Dunham, M. J., et al. (2007). Effects of aneuploidy on cellular physiology and cell division in haploid yeast. *Science*, *317*(5840), 916–924. http://dx.doi.org/10.1126/science.1142210.

Trapero-Casas, A., & Kaiser, W. J. (2007). Differences between ascospores and conidia of *Didymella rabiei* in spore germination and infection of chickpea. *Phytopathology*, *97*(12), 1600–1607. http://dx.doi.org/10.1094/PHYTO-97-12-1600.

Tscharke, R. L., Lazera, M., Chang, Y. C., Wickes, B. L., & Kwon-Chung, K. J. (2003). Haploid fruiting in *Cryptococcus neoformans* is not mating type α-specific. *Fungal Genetics and Biology*, *39*(3), 230–237. http://dx.doi.org/10.1016/S1087-1845(03)00046-X.

Tzung, K.-W., Williams, R. M., Scherer, S., Federspiel, N., Jones, T., Hansen, N., et al. (2001). Genomic evidence for a complete sexual cycle in *Candida albicans*. *Proceedings of the National Academy of Sciences of the United States of America*, *98*(6), 3249–3253. http:// dx.doi.org/10.1073/pnas.061628798.

Ünal, E., Kinde, B., & Amon, A. (2011). Gametogenesis eliminates age-induced cellular damage and resets life span in yeast. *Science*, *332*(6037), 1554–1557. http://dx.doi. org/10.1126/science.1204349.

Van Valen, L. (1974a). Molecular evolution as predicted by natural selection. *Journal of Molecular Evolution*, *3*(2), 89–101.

Van Valen, L. (1974b). Two modes of evolution. *Nature*, *252*(5481), 298–300.

Velagapudi, R., Hsueh, Y.-P., Geunes-Boyer, S. G., Wright, J. R., & Heitman, J. (2009). Spores as infectious propagules of *Cryptococcus neoformans*. *Infection and Immunity*, *77*(10), 4345–4355. http://dx.doi.org/10.1128/iai.00542-09.

Vengrova, S., & Dalgaard, J. Z. (2004). RNase-sensitive DNA modification(s) initiates *S. pombe* mating-type switching. *Genes & Development*, *18*(7), 794–804. http://dx.doi. org/10.1101/gad.289404.

Vengrova, S., & Dalgaard, J. Z. (2006). The wild-type *Schizosaccharomyces pombe mat1* imprint consists of two ribonucleotides. *EMBO Reports*, 7(1), 59–65. http://dx.doi.org/10.1038/ sj.embor.7400576.

Vinces, M. D., Haas, C., & Kumamoto, C. A. (2006). Expression of the *Candida albicans* morphogenesis regulator gene *CZF1* and its regulation by Efg1p and Czf1p. *Eukaryotic Cell*, *5*(5), 825–835. http://dx.doi.org/10.1128/EC.5.5.825-835.2006.

Viviani, M. A., Nikolova, R., Esposto, M. C., Prinz, G., & Cogliati, M. (2003). First European case of serotype A *MATa Cryptococcus neoformans* infection. *Emerging Infectious Diseases*, *9*(9), 1179–1180. http://dx.doi.org/10.3201/eid0909.020770.

Voelz, K., Ma, H., Phadke, S., Byrnes, E. J., 3rd, Zhu, P., Mueller, O., et al. (2013). Transmission of hypervirulence traits via sexual reproduction within and between lineages of the human fungal pathogen *Cryptococcus gattii*. *PLoS Genetics*, *9*(9), e1003771. http://dx.doi. org/10.1371/journal.pgen.1003771.

Vrijenhoek, R. C. (1998). Animal clones and diversity. *Bioscience*, *48*(8), 617–628. http:// dx.doi.org/10.2307/1313421.

Wang, X., Darwiche, S., & Heitman, J. (2013). Sex-induced silencing operates during opposite-sex and unisexual reproduction in *Cryptococcus neoformans*. *Genetics*, *193*(4), 1163–1174. http://dx.doi.org/10.1534/genetics.113.149443.

Wang, X., Hsueh, Y.-P., Li, W., Floyd, A., Skalsky, R., & Heitman, J. (2010). Sex-induced silencing defends the genome of *Cryptococcus neoformans* via RNAi. *Genes & Development*, *24*(22), 2566–2582. http://dx.doi.org/10.1101/gad.1970910.

Wang, X., Wang, P., Sun, S., Darwiche, S., Idnurm, A., & Heitman, J. (2012). Transgene induced co-suppression during vegetative growth in *Cryptococcus neoformans*. *PLoS Genetics*, *8*(8), e1002885. http://dx.doi.org/10.1371/journal.pgen.1002885.

Weiss, R. L., Kukora, J. R., & Adams, J. (1975). The relationship between enzyme activity, cell geometry, and fitness in *Saccharomyces cerevisiae*. *Proceedings of the National Academy of Sciences of the United States of America*, *72*(3), 794–798.

Wendland, J., Dünkler, A., & Walther, A. (2011). Characterization of α-factor pheromone and pheromone receptor genes of *Ashbya gossypii*. *FEMS Yeast Research*, *11*(5), 418–429. http://dx.doi.org/10.1111/j.1567-1364.2011.00732.x.

Wendland, J., & Walther, A. (2011). Genome evolution in the *Eremothecium* clade of the *Saccharomyces* complex revealed by comparative genomics. *G3: Genes, Genomes, Genetics*, *1*(7), 539–548. http://dx.doi.org/10.1534/g3.111.001032.

Wendte, J. M., Miller, M. A., Lambourn, D. M., Magargal, S. L., Jessup, D. A., & Grigg, M. E. (2010). Self-mating in the definitive host potentiates clonal outbreaks of the apicomplexan parasites *Sarcocystis neurona* and *Toxoplasma gondii*. *PLoS Genetics*, *6*(12), e1001261. http://dx.doi.org/10.1371/journal.pgen.1001261.

West, S. A., Lively, C. M., & Read, A. F. (1999). A pluralist approach to sex and recombination. *Journal of Evolutionary Biology*, *12*(6), 1003–1012. http://dx.doi.org/10.1046/j.1420-9101.1999.00119.x.

White, M. J. D. (1978). *Modes of Speciation*. San Francisco, CA: WH Freeman.

Whitehouse, H. L. K. (1949). Heterothallism and sex in the fungi. *Biological Reviews of the Cambridge Philosophical Society*, *24*(4), 411–447. http://dx.doi.org/10.1111/J.1469-185x.1949.Tb00582.X.

Whitton, J., Sears, C. J., Baack, E. J., & Otto, S. P. (2008). The dynamic nature of apomixis in the angiosperms. *International Journal of Plant Sciences*, *169*(1), 169–182. http://dx.doi.org/10.1086/523369.

Wickes, B. L., Mayorga, M. E., Edman, U., & Edman, J. C. (1996). Dimorphism and haploid fruiting in *Cryptococcus neoformans*: Association with the α-mating type. *Proceedings of the National Academy of Sciences of the United States of America*, *93*(14), 7327–7331.

Wik, L., Karlsson, M., & Johannesson, H. (2008). The evolutionary trajectory of the mating-type (*mat*) genes in *Neurospora* relates to reproductive behavior of taxa. *BMC Evolutionary Biology*, *8*, 109. http://dx.doi.org/10.1186/1471-2148-8-109.

Williams, G. C. (1975). *Sex and Evolution*. Princeton, NJ: Princeton University Press.

Winge, Ö. (1935). On haplophase and diplophase in some *Saccharomycetes*. *Comptes-Rendus des travaux du Laboratoire Carlsberg. Série Physiologique*, *21*, 34.

Winge, Ö., & Laustsen, O. (1937). On two types of spore germination, and on genetic segregations in *Saccharomyces*: Demonstrated through single-spore cultures. *Comptes-Rendus des travaux du Laboratoire Carlsberg. Série Physiologique*, *22*, 99–117.

Winge, Ö., & Roberts, C. (1949). A gene for diploidization in yeasts. *Comptes-Rendus des travaux du Laboratoire Carlsberg. Série Physiologique*, *24*, 5.

Wright, S. I., Ness, R. W., Foxe, J. P., & Barrett, S. C. H. (2008). Genomic consequences of outcrossing and selfing in plants. *International Journal of Plant Sciences*, *169*(1), 105–118. http://dx.doi.org/10.1086/523366.

Wright, S. I., & Schoen, D. J. (1999). Transposon dynamics and the breeding system. *Genetica*, *107*(1–3), 139–148. http://dx.doi.org/10.1023/A:1003953126700.

Xie, J., Tao, L., Nobile, C. J., Tong, Y., Guan, G., Sun, Y., et al. (2013). White-opaque switching in natural *MTLa/α* Isolates of *Candida albicans*: Evolutionary implications for roles in host adaptation, pathogenesis, and sex. *PLoS Biology*, *11*(3), e1001525. http://dx.doi.org/10.1371/journal.pbio.1001525.

Xu, J. (2002). Estimating the spontaneous mutation rate of loss of sex in the human pathogenic fungus *Cryptococcus neoformans*. *Genetics*, *162*(3), 1157–1167.

Xu, J. (2005). Cost of interacting with sexual partners in a facultative sexual microbe. *Genetics*, *171*(4), 1597–1604. http://dx.doi.org/10.1534/genetics.105.045302.

Xu, J., Ali, R. Y., Gregory, D. A., Amick, D., Lambert, S. E., Yoell, H. J., et al. (2000a). Uniparental mitochondrial transmission in sexual crosses in *Cryptococcus neoformans*. *Current Microbiology*, *40*(4), 269–273. http://dx.doi.org/10.1007/s002849910053.

Xu, J., Vilgalys, R., & Mitchell, T. G. (2000b). Multiple gene genealogies reveal recent dispersion and hybridization in the human pathogenic fungus *Cryptococcus neoformans*. *Molecular Ecology*, *9*(10), 1471–1481. http://dx.doi.org/10.1046/j.1365-294x.2000.01021.x.

Yan, Z., Hull, C., Sun, S., Heitman, J., & Xu, J. (2007). The mating type-specific homeodomain genes *SXI1α* and *SXI2a* coordinately control uniparental mitochondrial inheritance in *Cryptococcus neoformans*. *Current Genetics*, *51*(3), 187–195. http://dx.doi.org/10.1007/s00294-006-0115-9.

Yan, Z., Li, X., & Xu, J. (2002). Geographic distribution of mating type alleles of *Cryptococcus neoformans* in four areas of the United States. *Journal of Clinical Microbiology*, *40*(3), 965–972. http://dx.doi.org/10.1128/jcm.40.3.965-972.2002.

Yan, Z., & Xu, J. (2003). Mitochondria are inherited from the *MATa* parent in crosses of the basidiomycete fungus *Cryptococcus neoformans*. *Genetics*, *163*(4), 1315–1325.

Yi, S., Sahni, N., Daniels, K. J., Pujol, C., Srikantha, T., & Soll, D. R. (2008). The same receptor, G protein, and mitogen-activated protein kinase pathway activate different downstream regulators in the alternative white and opaque pheromone responses of *Candida albicans*. *Molecular Biology of the Cell*, *19*(3), 957–970. http://dx.doi.org/10.1091/mbc.E07-07-0688.

Yona, A. H., Manor, Y. S., Herbst, R. H., Romano, G. H., Mitchell, A., Kupiec, M., et al. (2012). Chromosomal duplication is a transient evolutionary solution to stress. *Proceedings of the National Academy of Sciences of the United States of America*, *109*(51), 21010–21015. http://dx.doi.org/10.1073/pnas.1211150109.

Yue, C., Cavallo, L. M., Alspaugh, J. A., Wang, P., Cox, G. M., Perfect, J. R., et al. (1999). The STE12α homolog is required for haploid filamentation but largely dispensable for mating and virulence in *Cryptococcus neoformans*. *Genetics*, *153*(4), 1601–1615.

Yun, S. H., Berbee, M. L., Yoder, O. C., & Turgeon, B. G. (1999). Evolution of the fungal self-fertile reproductive life style from self-sterile ancestors. *Proceedings of the National Academy of Sciences of the United States of America*, *96*(10), 5592–5597. http://dx.doi.org/10.1073/pnas.96.10.5592.

Zeyl, C., Vanderford, T., & Carter, M. (2003). An evolutionary advantage of haploidy in large yeast populations. *Science*, *299*(5606), 555–558. http://dx.doi.org/10.1126/science.1078417.

Zordan, R. E., Galgoczy, D. J., & Johnson, A. D. (2006). Epigenetic properties of white-opaque switching in *Candida albicans* are based on a self-sustaining transcriptional feedback loop. *Proceedings of the National Academy of Sciences of the United States of America*, *103*(34), 12807–12812. http://dx.doi.org/10.1073/pnas.0605138103.

Zordan, R. E., Miller, M. G., Galgoczy, D. J., Tuch, B. B., & Johnson, A. D. (2007). Interlocking transcriptional feedback loops control white-opaque switching in *Candida albicans*. *PLoS Biology*, *5*(10), e256. http://dx.doi.org/10.1371/journal.pbio.0050256.

Zörgö, E., Chwialkowska, K., Gjuvsland, A. B., Garré, E., Sunnerhagen, P., Liti, G., et al. (2013). Ancient evolutionary trade-offs between yeast ploidy states. *PLoS Genetics*, *9*(3), e1003388. http://dx.doi.org/10.1371/journal.pgen.1003388.

# INDEX

*Note:* Page numbers followed by "*f*" indicate figures; "*t*" tables.

Edwards Brothers Malloy
Ann Arbor MI. USA
June 3, 2014